HOW THINGS ARE

PHILOSOPHICAL STUDIES SERIES IN PHILOSOPHY

VOLUME 29

HOW THINGS ARE

Studies in Predication and the History of Philosophy and Science

Edited by

JAMES BOGEN

Pitzer College, Claremont, U.S.A.

and

JAMES E. McGUIRE

University of Pittsburgh, Pittsburgh, U.S.A.

D. REIDEL PUBLISHING COMPANY

A MEMBER OF THE KLUWER ACADEMIC PUBLISHERS GROUP

DORDRECHT / BOSTON / LANCASTER

Library of Congress Cataloging in Publication Data
Main entry under title:

How things are.

(Philosophical studies series in philosophy ; v. 29)
Papers from a conference sponsored by Pitzer College in 1981.
Bibliography: p.
Includes index.
1. Predicate (Logic)–History–Congresses. 2. Ontology–History–Congresses. I. Bogen, James, 1935– . II. McGuire, J. E. III. Pitzer College. IV. Series.
BC181.H58 1984 111 84-8331
ISBN 90-277-1583-1

Published by D. Reidel Publishing Company,
P.O. Box 17, 3300 AA Dordrecht, Holland.

Sold and distributed in the U.S.A. and Canada
by Kluwer Academic Publishers
190 Old Derby Street, Hingham, MA 02043, U.S.A.

In all other countries, sold and distributed
by Kluwer Academic Publishers Group,
P.O. Box 322, 3300 AH Dordrecht, Holland.

Printed in The Netherlands

For Thelonius Sphere Monk
– σέ λέγω

CONTENTS

ACKNOWLEDGEMENTS ix

JAMES BOGEN / Introduction 1

ROBERT G. TURNBULL / Zeno's Stricture and Predication in Plato, Aristotle, and Plotinus 21

FRANK A. LEWIS / Form and Predication in Aristotle's *Metaphysics* 59

D. K. MODRAK / Forms and Compounds 85

ALAN CODE / On the Origins of Some Aristotelian Theses About Predication 101

FRANK A. LEWIS / Plato's Third Man Argument and the 'Platonism' of Aristotle 133

MARILYN McCORD ADAMS / Things versus 'Hows', or Ockham on Predication and Ontology 175

CALVIN NORMORE / Buridan's Ontology 189

JAMES E. McGUIRE / Phenomenalism, Relations, and Monadic Representation: Leibniz on Predicate Levels 205

ROBERT M. ADAMS / Predication, Truth, and Transworld Identity in Leibniz 235

WILFRID SELLARS / Towards a Theory of Predication 285

ALAN CODE / On the Origins of Some Aristotelian Theses About Predication: Appendix on 'The Third Man Argument' 323

NOTES ON THE CONTRIBUTORS 327

BIBLIOGRAPHY 329

INDEX OF LABELED EXPRESSIONS 337

NAME INDEX 339

SUBJECT INDEX 343

ACKNOWLEDGEMENTS

The papers in this volume arose in various ways from a working conference on predication and the history of philosophy and science sponsored by Pitzer College in 1981. The conference, in which all of the authors participated, was funded by grants from the Haines Foundation and Pitzer College, and the preparation of the manuscripts of this volume, by a grant from the Pitzer College Research and Development Committee.

We are greatly indebted to the following people for their hospitality, encouragement, and help: Frank Ellsworth, Karin Ellsworth, Ronald and Janet Macaulay, Donald Brenneis and Wynne Furth, Amy Bogen, Morton O. Beckner, Jay Atlas, Charles Young, Richard McKirihan, Beverly Scales, Theresa Hidalgo, Patsy Sampson, Nicholas J. Sandro, Enzo Biscardi, Ellen Chase, Charles Zetterberg, and a number of Pitzer College students including Eileen Fagan, Kiamara Ludwig, Barbara Kohlenberg, Valery Krieger, Peter Forrester, Neepa Chowdury, and the staff of the Grove House. And, Kathy Dahlgren and Jeane Pease, the typing maching.

JAMES BOGEN

INTRODUCTION

I

One of the earliest and most influential treatises on the subject of this volume is Aristotle's *Categories*. Aristotle's title is a form of the Greek verb for speaking against or submitting an accusation in a legal proceeding. By the time of Aristotle, it also meant: to signify or to predicate. Surprisingly, the "predicates" Aristotle talks about include not only bits of language, but also such nonlinguistic items as the color white in a body and the knowledge of grammar in a man's soul. (*Categories* I/ii) Equally surprising are such details as Aristotle's use of the terms 'homonymy' and 'synonymy' in connection with things talked about rather than words used to talk about them. Judging from the evidence in the *Organon*, the *Metaphysics*, and elsewhere, Aristotle was both aware of and able to mark the distinction between using and mentioning words; and so we must conclude that in the *Categories*, he was not greatly concerned with it. For our purposes, however, it is best to treat the term 'predication' as if it were ambiguous and introduce some jargon to disambiguate it. Code, Modrak, and other authors of the essays which follow use the terms 'linguistic predication' and 'metaphysical predication' for this.[1]

The sentence 'Socrates is a man', used to say something about Socrates, is an example of *linguistic predication*; and – depending upon how one chooses to parse it – the word 'man', the phrase 'a man', or the phrase 'is a man' is a linguistic predicate. Someone who is interested in linguistic predication would be concerned with the syntax and semantics of the sentence and the bits of language which go to make it up. In contrast, Socrates (the person, not the name, 'Socrates') being a man is an example of *metaphysical* predication, as are such items as Whitey Herzog spitting on the dugout steps; Aubrey Bogen being a redhead (and her hair being red); Carlota Bogen being bigger than a breadbox and younger than her uncle; Thelonius Monk dying; the inner city of Claremont, California, undergoing a mild economic upswing; etc., etc. Someone who is interested in metaphysical predication would be concerned with what kinds of nonlinguistic things there are and how they stand when various sorts of linguistic predications are true or false. That

James Bogen and James E. McGuire (eds.), How Things Are, 1–19.

linguistic predication is different from metaphysical predications should be clear from the fact that the things we talk about are different from the words we use to talk about them.

It will be obvious to readers of this volume that Plato, Aristotle, and other predication theorists it discusses were primarily concerned with metaphysical predication. They paid close attention to linguistic predication when they believed it provided data requiring an account, and clues an ontologist should take seriously. But they did not believe that metaphysical predication is uniquely determined by the syntax or semantics of linguistic predication, let alone that a theory of linguistic predication could take the place of an ontology. That is why this book is called '*How Things Are*' instead of 'How We Talk about Things'.

II

While it should not be difficult to see the difference between linguistic and metaphysical predication, the connections between linguistic and metaphysical or ontological investigations are complex and hard to make out. The problem is seeing just how ontological or metaphysical conclusions are warranted by facts about linguistic predication.

An extreme view, suggested by such logical positivist writings as Carnap's "Empiricism, Semantics, and Ontology"[2] is that sentences which purport to make claims about metaphysics are meaningless unless understood as disguised or misleadingly phrased claims about language. Thus the apparently ontological thesis 'there are numbers' is, Carnap thought, to be understood as recommending us to speak a language which includes numerals, variables for which they can be substituted, etc. And, if meaningful, the sentence 'there is a prime number greater than 100' should be understood as saying of some specified language that at least one value of the function '*x* is a prime number greater than 100' is provable in that language. This view arose from constraints (e.g. that no sentence should be considered meaningful unless it is either empirically verifiable (or confirmable) or analytic) which are, for the most part, no longer accepted.[3] None of the authors represented in this volume would subscibe to it.

Another extreme view is that an adequate account of the syntax and semantics of a natural language would show us straightway what items should be included in an ontological inventory and what kinds of connections between them constitute metaphysical predications. The idea is to take a sentence, say,

(1) Alice's nose is red,

list the items which would be contained in its intended model if it were true, and classify these extra-linguistic items into groups which parallel the syntactical categories to which the parts of the sentence belong. Thus, we might say that if (1) is true, the world contains a nose belonging to Alice, something which corresponds to the predicate 'red', and a relation (e.g. inclusion) between the two.

But it's not that easy. Ordinary English provides sentences whose truth conditions are the same as those of (1) but whose syntactical forms seem to be quite different from one another. Among these are

(2) Red is the color of Alice's nose,

(3) The color of Alice's nose is red,

(4) Alice's nose is a red nose, and

(5) Alice's nose has the color red (has a red color).

on the view just sketched, we would have to say that a world in which these sentences are true contains not only a relation corresponding to the word 'is' in (1), (2), and (3), but also another relation corresponding to 'has' in (5). Furthermore, the 'is' relation would have to connect different items (red and the nose) for (1) than it connects for (3) (red, and the color of the nose).

The resulting ontological bloat could be avoided by finding a canonical form for the linguistic predication. We would look for a sentence in English, or an artificial language (e.g. predicate calculus notation) whose syntax mirrors the real structure of the predication. This suggestion is by no means anachronistic. Aristotle selected sentences from Greek and sometimes invented technical locutions for this purpose. Ockham and Buridan believed that the structures of metaphysical predications are mirrored by a language of thought whose syntax differs from that of spoken languages. But the grammatical sentences of a natural language can be rendered into different formulae of different syntactical forms. On the program under construction, different ontologies should result from the selection of different notations as canonical. To pick the right notation, the ontologist would need guidance which syntactical theory cannot provide all by itself.

III

A natural place to look for guidance is semantics. Familiar accounts of truth characterize 'is true' as a predicate to be applied to sentences of a language, L, just in case they stand in some specified relation ("satisfaction", for example) to a set of objects, O, whose members constitute the intended model for L. If L is a natural language, the program calls for translating the sentences of L into a canonical language, $L1$, whose syntax can be characterized precisely. $L1$ must be logically consistent and it must preclude the construction of the Liar and the other semantic paradoxes. $L1$ must provide translations for all of the sentences of L which are considered to be true or false by speakers of L. And the syntax of $L1$ should allow us to construct a simple and elegant semantical theory in which the truth conditions of all of its sentences can be specified. We wondered earlier if any of the sentences (1)... (5) available in English for saying that Alice's nose is red should be taken as determining what to include in an ontological inventory. If $L1$ were the only, or by far the best canonical notation available for translating L sentences and defining their truth conditions, our worries would be over. Sentences (1)–(5) would be given one and the same translation in $L1$, and the ontologist would posit different kinds of extralinguistic items for syntactically different verbal elements in $L1$. If two different notations, $L1$ and and $L2$, allowed for equally good definitions of truth in L, we could at least rule out notations which differed significantly from them.

Suppose the canonical translation of some natural language sentence is '*Fa*' and that '*Fa*' belongs to language $L1$. Suppose the semantical theory for $L1$ treats the intended model, O, as including objects $o_1, o_2, \ldots, o_n$, and subsets of objects, $S_1, S_2, \ldots, S_n$. Suppose it assigns object o_i to the name '*a*', and subset S_k to the predicate '*F*'.[4] Then '*Fa*' (and the natural language sentence it translates) will be true just in case o_i belongs to S_k. The ontology should then include the objects and subsets of O and should characterize them in such a way as to allow for the possibility that o_i belongs to S_k.

If we agree that '*Fa*' is true just in case object o_i belongs to subset S_k, this tells us something about what an adequate theory of metaphysical predication should be like. But it does not tell us terribly much. Unfortunately, it is compatible with drastically different theories of what it is for an object like o_i to belong to a subset like S_k. Thus, it does not tell us how to adjudicate the following dispute between a Platonist, an Aristotelian, and a highly austere nominalist. When he wrote the *Phaedo*, Plato would have said (for some properties at least) that unless o_i is a Form (The Beautiful Itself, The Large Itself, etc., etc.), o_i belongs to S_k just in case (1) there is a Form, the *F* Itself, which is not identical to o_i and which exists whether or not o_i exists, and (2) o_i stands in some relation, *R* ("participates in," "has a share in," etc.), to The *F* Itself. This is one account of what it is for o_i to belong to S_k, and hence, of what makes '*Fa*' true. But Aristotle held that there can be no relation of the kind "participation" was supposed to be and excluded Forms (like The *F* Itself) from his ontology. For some predicates, he holds that if '*F*' signifies anything, it signifies something in the individual to which '*F*' applies and which can *not* exist without the individual. In such cases, o_i belongs to S_k because of the item signified by '*F*' and belonging to o_i. In other cases (for example, predications of 'is a man' and 'two-footed animal' to Socrates, discussed by Lewis, Code, and Modrak in this volume), different accounts of o_i's membership in S_k are offered, but they do not require either Platonic Forms or participation. An austere nominalist would deny that the linguistic predicate '*F*' signifies *anything* in terms of which the inclusion of o_i in S_k is to be explained.

A theory of metaphysical predication may be considered to be a theory of the intended model for a canonical language like L_1. The Aristotelian, the Platonist, and the austere nominalist may cheerfully agree to accept L_1 as the canonical notation of choice and *O* as its intended model. They may also accept the semantic assignment of '*a*' to o_i and '*F*' to S_k. And, they may agree that '*Fa*' is true just in case o_i belongs to S_k. But, because they can still disagree about what it is for o_i to belong to S_k, the requirements of an adequate semantical definition of truth in L_1 are far too weak to tell us which theory of metaphysical predication – which account of the intended model – to accept.

In order to see how features of linguistic predication can be used to support or oppose specific accounts of metaphyiscal predication, we must consider further constraints on ontological theories. The papers in this volume can be understood in part as articulating constraints on theories of metaphysical predication and strategies by which various theories attempt to abide by them.[5]

IV

An historically early constraint on metaphysical theories of predication can be appreciated by considering that where '*a*', '*b*', and '*c*', etc., name numerically different individuals and '*F*' is a linguistic predicate, it should be possible that

(1) even if '*Fa*', '*Fb*', etc., are true, *a* is different from *b* and all of the other individuals of which '*F*' is truly predicated, and

(2) what '*F*' predicates of *a* is the same as what it predicates of the other individuals.

A natural way of accounting for (2) is to say that, when '*Fa*', '*Fb*', etc., are true, what the individuals have in common is that they are all *F*, e.g. that red individuals are all red, individuals who are men are all men, etc., etc. But if *a* is *F*, and *b* is *F*, how can *a* and *b* differ from *F* (from whatever the linguistic predicate '*F*' signifies)? And, if they are not different from *F*, how can *a* differ from *b*, as required by (1)? This problem gives rise to a constraint which Turnbull calls 'Zeno's Stricture':

(C1) Where *F* is any metaphysical predicate, no individual *is F* if it is different from *F*.

(C1) allows true *linguistic* predications of the form '*x* if *F*' only because the structures of linguistic and metaphyiscal predications can differ. What (C1) rules out are accounts of (2) under which *a* and *F* are different extra-linguistic items even though *a* is *F*. And, if canonical notations are supposed to reveal the structures of metaphysical predications, (C1) makes expressions of the form '*x* is *F*' noncanonical where '*F*' and values of *x* signify different items. Turnbull argues that characteristic features of theories of Plato, Aristotle, and Plotinus, as well as fundamental disagreements between them can be understood as stemming from differences in their understandings of and attempts to abide by (C1).

As noted, Plato maintained (in *Phaedo*, for example) that where *a* is different from *F*, the metaphysical predication which makes '*a* is *F*' true is the obtaining of a two-term relation, "participation", between *a* (a lump of burning coal, for example) and The *F* Itself (e.g. The Hot Itself). This meets Zeno's stricture. The Hot Itself *is* hot because hot is just what it is. And Plato does not have to say the coal *is* hot; the linguistic predication is true because the coal *has* heat, i.e. it stands in the participation relation to the Form. A different version of the same basic maneuver is attempted by Aristotle for accidental (κατὰ συμβεβγκὸς) predication. According to the *Categories*, there is a bit of heat in each hot thing. The *heat* in the coal *is* hot; but the linguistic predication 'the coal is hot' is true, not because the coal *is* hot, but because a bit of heat is *in* it. Instead of a single form, there is heat in each hot thing. Instead of participation, there is the "in" relation between each hot thing and the heat it has. But, as in Plato, the truth of '*a* is hot' is explained by a relation between *a* and what *is* hot. In contrast, the formula 'two-legged animal', predicated of the species man, tells us what the species *is*; the linguistic predicate does not signify something which is in the species in the way heat can be in a piece of coal. This leaves open the question of how to understand the metaphysical predications which make it true to say of Socrates both that he is a man and that he is a two-legged animal. The difficulties involved in this area are discussed by Code and Lewis and will be mentioned briefly in the next section of this Introduction.

V

The early sections of *Parmenides* show that Plato had terrible trouble and little success in trying to say what sort of relation participation is. The difficulties provided Aristotle with an arsenal of arguments against the theory of Forms. The most severe lead to the following dilemma: on some accounts of participation, the Form which participates cannot be a single individual, whereas accounts which preserve Form's individuality do not seem to allow participation to be the kind of relation Plato needs.

Why did Plato have so much trouble with this? It certainly isn't a general problem for two term relations. After all, Bach retained his unity while standing in the "father of" relation to many different children; and a few years ago, Deborah Harry maintained her unity while standing in the "admired and imitated by" relation to hordes of young women. The difficulty is peculiar to "participation", and it comes from an additional constraint on theories of metaphysical predication:

(C2) Extra-linguistic entities, relations, etc., posited to account for the truth values of linguistic predications must to able to figure informatively in adequate scientific explanations of features of the things of which the linguistic predications are true.

A good deal of Plato's troubles with participation come from this, together with the intuition, shared by ancient thinkers, that the characteristics (e.g. heat, justice, size, etc.) to which a science appeals to explain features (including behaviors) of individuals had better be constituents of those individuals.[6] Explaining why a cake is sweet by saying it contains sugar, and why a wine is thin by saying it contains more water than it should are instances of the pattern of explanation which fits this intuition. If the sugar were not contained in the cake, or the water in the wine, they could not provide satisfactory explanations. This, I think, is an important reason why Plato considered treating participation as a relation of containment or inclusion. While this meets (C2), it runs afoul of Plato's account of (2), which requires that true linguistic predications of the form '*x* is *F*' should predicate the same thing. If the *F* Itself is a single, countable individual, it cannot be wholly contained in each of the different individuals of which the relevant linguistic predications are true. If it is thought to consist of discrete parts or bits, each one of which is contained in a different individual, it is hard to see how the Form can be one thing and thus how (2) is to be accounted for. On the other hand, if participation is not some sort of inclusion or containment, and The *F* Itself stands apart from the individuals which have it, it is hard to see how it could explain features or behaviors of those individuals. How, for example, could Socrates's wisdom explain the clever things he does if Wisdom (The Wise Itself, or part of it) isn't somehow contained within Socrates?

Aristotle's account of the truth of such linguistic predications as 'Socrates is musical' in terms of knowledge in his soul, 'Socrates is pale' in terms of a bit of color in his skin, etc., treats the metaphysical predicates (e.g. pallor) as constituents of the individual whose features they are used to explain. The importance he attached to this is shown by his repeated objection against Plato that because Platonic Forms are entirely removed from the items a science should explain, nothing we could know about the Forms (even if there were such things) could be of any help to science. But the other horn of Plato's dilemma remains. If the *F* in one individual is not the same as the *F* in another, what prevents the explanations '*a* has (does) such and such because *Fa*' '*b* has (does) such and such because *Fb*' from mentioning *different* explanatory factors? This would be unacceptable to Aristotle and anyone else who thinks the explanatory principles of a science should hold universally. And, how can linguistic predications of '*F*' say the same thing of different individuals if each one has its own *F* in it and there is no numerically one thing they are share? Aristotle's answer seems to have been that what the instances of *F* to be found in different individuals have to do with one another is that they cause similar perceptions in normal observers. The perceptions they cause give rise to a single concept of *F* in the soul of a person who perceives and comes to know something about various objects which have *F*. The instances of *F* are called the same thing ('*F*') because they fall under the same concept.[7] But this is not worked out in enough detail by Aristotle to allow us to say exactly what it amounts to or how satisfactory a response it is to the question whether all explanations of the form ' . . . because *x* has *F*' are real and not just trivial instances of the same explanatory principle.

A parallel problem arises for nonaccidental predication. The essence of a living thing is supposed to explain what it does by nature (e.g. its reproduction, development, and growth) in addition to grounding its classification in a biological taxonomy. The essential features of such inanimate natural things as earth, air, fire, and water are supposed to explain their unforced motions. And some of the features of manufactured articles may be explained by their essences. This means that the essence of each thing must be intimately connected to it; a single Platonic form separated from all of the individuals whose essence it purports to be won't do. Aristotle's *Metaphysics* provides a very intimate connection between individuals and their essences. Each thing is a compound of form (essence) and matter, and the form (essence) is the substance of the compound. But each member of a kind should have the same essence so that the same explanation can be given, e.g. for the normal growth patterns of each one. For example, the essence of two parents is supposed to explain why their offspring should belong to their species.[8] Accordingly, sentences of the form '*x* and *y* are armadillos', together with an account of the essence of armadillos, should explain the species of normal baby armadillos. But, then the essence should be the same for every member of the species. That makes it look as though (a) essences should be universals metaphysically predicated of a number of different individuals. But (b) Aristotle also calls them substances, and substances are not supposed to be metaphysical predicates.

The apparent incompatibility of (a) and (b) is part of the motivation for Lewis' discussion of essential predication. He suggests that essential features are predicated universally, not of the individuals whose essences they constitute, but of the matters of these individuals. Each individual is a composite of matter and the essence or form predicated of its matter. The formula or definition of a thing's form can then be thought of as signifying a universal which is predicated of as many parcels of matter as there are individuals which have the same essence. But this is consistent with the theses that the form of an individual is that individual's substance and that an individual's substance is not predicated of it. It is consistent because the form is not predicated of the compound, but only of its material component. In her reply to Lewis Deborah Modrak grants that this would resolve the apparent contradiction, but questions whether Lewis's resolution was available to Aristotle. She argues that batches or parcels of matter cannot be considered fit subjects for the predication of form and, therefore, that Aristotle could not have said that the form which is an individual's essence is a universal predicated of its matter.

Related issues are discussed by Code. The term 'man' and its definition, 'two-legged animal', signify the essence of the species to which Socrates belongs. According to *Metaphysics Zeta*, a thing's essence is identical to that thing. Thus, 'man' and its definition signify what is identical to the species, man. But, Code argues, since Socrates is not identical to the species to which he belong, what 'man' and its definition signify is not identical to Socrates. Nor do 'man' or the definition signify what is in Socrates, like such accidental features as his color, his knowledge of grammar, etc. This seems an unhappy result. If Socrates is neither identical to the essence (he is not connected to the essence as is the species) nor *has* the essence (it is not *in* him as are his accidental metaphysical predicates), what can the essence have to do with the natural features, behaviors, etc., of Socrates which it should explain? To thicken the plot further, Aristotle seems to claim that 'man' and the definition 'two-legged animal' are true of both Socrates *and* the species in the same way. If that were correct, (C2) would be satisfied; but it is obviously difficult to see how 'man' and 'two-legged animal' can tell us what Socrates is if they signify or characterize something (the species or the essence to which the species is identical) which is not identical to Socrates. If so many difficulties can be traced to the thesis of Zeta vi (that an essence is identical to the thing whose essence it is), one wants to ask why Aristotle committed himself to it. Code's paper attempts to explain this, and Lewis's second paper comments on his account.

VI

The nominalist theories of Ockham, Buridan, and Sellars – discussed respectively by Marilyn Adams, Calvin Normore, and Wilfrid Sellars are natural responses to the problems of participation and the question of how to account for the fact that '*Fa*' says the same thing of *a* that '*Fb*' says of *b*. Sellars argues that what is said in a natural language by subject-predicate sentences can be said in an artificial language by sentences which lack predicates altogether. The syntax and semantics of the artificial language sentence do not warrant or require the positing of individual things (like Platonic Forms) to serve as referents for such words as 'white' and 'whiteness', Nor do they require us to posit individual whites of the peculiar kind supposed by Aristotle to inhere somehow in white things. Sellars's paper is not, it should be emphasized, anything like a full sketch of the complex metaphysics he advocates. It is intended to show how a number of crucial semantic notions can be understood, and how certain ontological issues can be resolved by a theory which avoids the snares of Platonic and Aristotelian accounts of universal predication.

Early forms of nominalism are discussed by Normore and Adams. Buridan and Ockham held that such things as individual whites do somehow inhere in white objects. According to both of them, forms of a term like 'white' should be understood as standing in some sentences for things which have the colors, and sometimes, for the colors they have. Where Plato thinks that '*a* is *F*' is true just in case the individual designated by '*a*' stands in the "participation" relation to the individual (the Form) designated by '*F*', Buridan and Ockham think it is true just in case '*a*' stands for one of the things '*F*' stands for. And *a* is one of the things named by '*F*' just in case one of the *F*s inheres in it. For example, '*a* is white' is true just in case 'white signifies *a* (along with other white things), as it does just in case one of the whites is in *a*. In a sentence like 'white is a color', on the other hand, 'white' is used to designate whites and not the things they inhere in. The sentence is true because 'white' stands for some of the items designated by 'color'. Like Sellars, Ockham and Buridan have disposed of the question how can a "universal" (The *F* Itself, *F*ness, etc.) be one thing and still belong to all of the different things of which the predicate '*F*' is true by expelling "universals" from their ontologies. If there is no "universal" for individuals to participate in, there are no problems of participation to be solved. But, if there are no "universals", how do we explain why all of the things of which the linguistic predicate 'white' is true resemble one another insofar as they are white? According to Ockham and Buridan, all the whites in white things just are similar and their similarity does not need an explanation, let alone an explanation in terms of Platonic Forms, "whiteness", or any other "universal" entities of their ilk.

According to the clichés and stereotypes of which surveys of the history of philosophy are made, the nominalists were supposed to have an overriding concern for parsimony. It hardly seems parismonious to include individual qualities like the white in this paper, the white in the next page, the white in the next, and so on, in an ontological inventory. Why not just say that 'whiteness' or 'white' signifies all and only the members of a certain set of objects (the set which contains sheets, pieces of paper, snowflakes, etc.) and that the linguistic predication '*a* is white' is true just in case and only because *a* is a member of that set? If Ockham and Buridan didn't need to posit additional entities to explain the similarity between individual whites, why should they posit the whites to explain the similarity between my white sheets and towels? In this respect, Buridan is even less parsimonious than Ockhan, for his qualities (e.g. the whites in white things) are capable of existing without inhering in anything. Accordingly, they seem to be very much like full-fledged individuals. Something more than parsimony is clearly at work here. According to Normore, Buridan believed that nothing short of free-floating qualities could account for some of the dogmas of the Eucharist. The requirement of conformity with theological doctrine was as strong a constraint on the theories of Ockham, Buridan, and Leibniz as the requirement of conformity with scientific explanations was for other philosophers.

A further departure from parsimony appears in connection with "modes". Marilyn Adams argues that Ockham's treatment of linguistic terms for quantity, relations, actions, and passions threatens to require the inclusion of "hows", "ways", or "modes" in his ontology. These are strange items to find in any ontology, let alone a nominalist metaphysics. Adams and Normore point out that Buridan not only includes modes, but does so cheerfully, offering arguments for their inclusion. And, Normore suggests that there may be considerable theoretical gains for a nominalist who does so.

VII

The final group of papers in this volume includes discussions of Leibniz by Robert Adams and James E. McGuire. They deal with two areas in which Leibniz departs from assumptions shared by Plato, Aristotle, and the nominalists (with extensive qualifications in the case of Sellars).

The first is the assumption that truth should be explained as some sort of correspondence between sentences, thoughts, beliefs, or propositions, and the extra-linguistic items we talk about. For Leibniz, true sentences do correspond to the actual world. For example, he would not deny that Arnauld is celebate if the sentence 'Arnauld is celibate' is true, and therefore that certain extra-linguistic items stand to one another in different ways than they would have if the sentence had been false. But truth is not to be *defined* or *characterized* in terms of correspondence. Instead, Leibniz thinks, the truth value of a sentence depends upon whether the concept of what is signified by its linguistic predicate is "contained" in the concept of what is signified by its linguistic subject. If there is containment, then there is correspondence: things are as the sentence says they are. But truth is defined in terms of conceptual containment and not in terms of correspondence.

The conceptual containment theory of truth raises two problems discussed by Adams. The first concerns the distinction between necessary and contingent truths. Intuitively, this seems to be a real distinction for which a theory of predication should account. It was accounted for by Plato and

Aristotle in terms of the differences between the ways in which a thing may be connected with its metaphysical predicates. For example, Plato can say it is necessarily true that The Large Itself is large because the predicate 'large' signifies just what that Form is. On the other hand, it is contingently true that Wilt is large because 'large' signifies a Form in which Wilt need not have participated. This marks the distinction if truth is correspondence which depends upon how a given thing stands with regard to particular Form. If there are different sorts of connections between metaphysical subjects and predicates, it should be informative to characterize necessity and contingency by appeal to these differences. But this sort of explanation is unavailable to Leibniz for whom truth depends upon relations between concepts instead of the things of which they are concepts. Furthermore, Leibniz cannot very well explain the distinction by saying that the concepts which figure in contingent truths are connected differently than those which figure in necessary truths. If some predicate, '*F*', is contained in the concept of something *a*, then no matter what kind of containment it is, nothing can be *a* unless it is *F*, and so '*a* is *F*' cannot be false. Leibniz's way out is to recast the distinction as an epistemological one. In fact, all truths are necessary; and so anyone who possessed a complete and clear concept, e.g. of Arnauld, would see the sentence 'Arnauld is celibate' as expressing a necessary truth. But anyone (including all of us) whose concept of Arnauld is incomplete or unclear would classify this as a contingent truth, for celibacy is not contained in an incomplete or confused concept of Arnauld. Contingency is thus relativized to incomplete concepts, and the general agreement of our intuitions about which truths are and which are not contingent is to be explained in terms of similarities between the incomplete concepts we possess.

If contingent truths are contingent *only* from the standpoint of incomplete concepts, Leibniz faces a number of theological difficulties. God is supposed by Leibniz to have considered all possible worlds and created the best of them. We should expect this to mean, for example, that because Arnauld is celibate, at least one world in which he is celibate is better than those in which he is not. But if celibacy is part of God's complete concept of Arnauld, the idea of a world in which Arnauld is not celibate should be as inconsistent as the idea of a world in which triangles lack three sides. Adams argues that Leibniz accepted this. How, then, are we to understand what God had to choose from in deciding to create this world? And how are we to understand it as a free choice (which is what it should be, if God is ominipotent)? If it is true of God that He chose this world, why doesn't it follow that it is necessarily true and, therefore, that He had no freedom to choose?

Finally, if all truths are necessary from the standpoint of the complete concepts God possesses, what room is left for the possibility of miracles? These, together with Leibniz's reason for granting the troublesome consequences of the conceptual containment theory of truth are discussed in Robert Adams's paper.

McGuire's paper takes up issues raised by Leibniz's rejection of a second assumption. With the exception of Plotinus (and extensive qualifications in the case of Sellars), physical objects are generally considered full-fledged inhabitants of the world in good ontological standing. Of course, Plato thought physical objects were less deserving of a philosopher's attention than the nonphysical Forms in which they participate. But he still considered physical things (including plants and animals) to be genuine individuals, and he appeared to have no doubt that they are the subjects of metaphysical predication. For example, if the linguistic predication 'this statue is beautiful' is true, that is because a physical thing – the statue – participates in The Beautiful Itself. Plato does not suggest any reduction of the statue to more basic nonphysical items whose dispositions constitutes the "real" nature of the participation of the statue in the Form. Similarly, even though Aristotle treats Socrates as a compound with one nonphysical component (a form), Socrates's pallor is said to belong to Socrates, the compound, not its formal or material component.

Against this sort of background, Leibniz's treatment of physical things is startling: physical objects are not genuine individuals in Leibniz's ontology. Instead, they are "phenomena" to be accounted for in terms of monads which are not spatially extended, let alone physical things. Two of Leibniz's reasons for thinking this are of particular interest. First, he thought that to treat physical objects as genuine and unqualifiedly real individuals would violate constraint (C2). Leibniz held that, unless they are considered phenomena to be reduced or explained in terms of monads and their features, physical objects are not the right kinds of things to appeal to in order to explain completely and correctly the facts with which physics, astronomy, and other sciences must deal. Secondly, Leibniz thought that physical objects lack the sort of unity genuine individuals must have.

The explanatory shortcomings of a science whose ontology is limited to physical objects would appear most clearly, he thought, in connection with teleological explanations. Besides that, Leibniz was keenly aware that, as characterized by Descartes, the occasionalisits, and others, physical ("material") objects must, by nature, be inert or inactive, and are therefore incapable of acting as *efficient* causes. A symptom of the difficulties involved in thinking of physical objects as efficient causes is the extraordinary role assigned to God by the Occasionalists (and, most likely, by Descartes) in accounting for what look to be efficient causal interactions among physical things. A further symptom is the apparent failure of the Newtonians (later insisted on vehemently by Berkeley and Hume) to assign any satisfactory ontological status to forces. Leibniz thought that explanations in physics must be founded upon genuine agency. If, as Descartes and the others believed, mere spatial extension constitutes the essence of physical things, there is nothing in the nature of a physical object to account for full-blooded agency.

Even if they are efficient causes, why shouldn't physical things be genuine individuals? Because physical objects (including Cartesian corpuscles and Newtonian atoms) are spatially extended, they are divisible. But what is divisible cannot have the kind of unity required by Leibniz of individuals in his ontology. What is spatially divisible should be considered, Leibniz thought, as composed of extended parts. Why, then, should it be considered one thing, rather than an aggregate of discrete things? More imporantly, on pain of what Liebniz considered a vicious regress, a collection of parts must eventually divide up into simple constituents. But if everything extended is divisible and extension is essential to physical things, no physical constituent – no matter how small – could be simple (indivisible). If there are simple individuals, as Leibniz thinks there have to be, they must therefore be unextended and hence nonphysical. This line of thought, together with the idea that physical things cannot be efficient causes, led Leibniz to posit nonextended, simple individuals – the monads – whose theoretical role is to provide the ultimate basis for metaphysical predication and scientific explanation.

But what becomes of tables, chairs, the bodies of living things, sticks and stones, and other extended things? They can't be wholes whose parts are monads, for no accumulation of extensionless items can be extended. Leibniz holds that physical objects are "phenomena" which result from the extensionless monads' perceptions or representations of each other. At the same time, he attempts to block the inference that physical things are *mere* appearances. Leibniz considered science a serious enterprise and believed that physical things merit the careful attention given to them by physics, biology, etc. Unlike *mere* appearances, physical things are "well-founded phenomena", connected so directly and closely to monads and their properties, that they should in some way be considered to be aggregates of monads. As noted, since monads are extensionless, this cannot mean that extended things are monads standing in spatial relations to one another. According to McGuire, the required aggregation is constituted by one to many correspondences between monadic representation of the various states that pertain to different monads. In order to understand just what this amounts to, and how it allows Leibniz to distinguish the well-founded phenomena he considers physical objects to be from mere appearances to be dismissed (i.e. as unindicative of the facts with which a science should deal, and in terms of which it should offer explanations), at least two themes must be developed. The first is Leibniz' treatment of relations. Accordingly, McGuire asks whether Leibniz's account of monads allows for the kinds of relations which must obtain among monads, monadic states, and their representations in order for physical things to be well-founded phenomena. Monads are supposed to be intrinsically non-relational, i.e. lacking in what McGuire calls "first order relational properties". How, then, can physical things which are thought of as having relational properties be well founded phenomena?

The second theme taken up by McGuire is the notion of levels of predication. Because monads are not spatial, they cannot have extension, shape, size, and other features which presuppose extension and which we ascribe to physical things. At the same time, activity, thought, consciousness, and other features Leibniz ascribed to monads do not apply to physical things. Furthermore, if physical objects and their features are phenomena which depend upon the monads and their representations, the connections between physical things and their features must be explained in terms of connections between monads and the metaphysical predicates which belong to them. Predicates may thus be assigned to levels on the basis of what sorts of things they apply to and what sorts of things have the features which explain them. A full explanation of the way in which physical things are well founded would require an exhaustive assignment of predicates to their proper levels, and a full account of how the predicates of each level stand with regard to those of the other levels.

VIII

The attempt to account systematically for different levels of predication is by no means peculiar to Leibniz. For example, Plato's response to Zeno's stricture and to (C2) involved distinguishing between the ways in which heat is predicated of The Hot Itself, of fire, and of the object (e.g. a lump of burning coal) which is hot because it contains fire and therefore participates in The Hot Itself. Corresponding to the different kinds of predication, there are differences in explanations given for heat: the Form is hot by its very nature;

the heat of bits of fire is explained in terms of their relation to the Form; and the heat of a burning coal is explained (in one way) by appeal to the fire it contains. A much more elaborate version of the distinction between predicate levels was developed by Aristotle. The matter of Socrates is the aggregate of his limbs and organs, each one of which is analyzed as a composit of flesh or bone (e.g. the flesh of his heart) and form (e.g. what is given by a definition of the human heart). Flesh and bone are, in turn, matter (i.e. two or more of the four elements – earth, air, water, and fire) with forms (the ratios which characterize the combinations of the elements). And some of Aristotle's writings (e.g. his treaties on coming to be and passing away) suggest that each element may itself be considered a form-matter compound. Among the things that are true of Socrates (e.g. that he is a man, that he is two legged, that he is pale) are some which are not predicated and others which are not predicated in the same way of his matter (his organs and limbs). Some of the predicates of his matter (e.g. the property of having five fingers, which belongs to Socrates's hands) are not predicated of flesh, (the matters of the hands). And so on. Furthermore, predications on one level are explained in terms of predications on another. Thus, features of an animal's growth and reproductive patterns are explained partly in terms of the animal's matter – his organs and limbs – and partly in terms of his form or essence.

In these cases and in the case of Leibniz, the assignment of predicates to levels is as much a metaphysical classification of predicates required for an explanatorily adequate science as a syntactical distinction between kinds of grammatical and ungrammatical, deviant and nondeviant linguistic predications.

The requirement of distinctions between levels of predication is yet another constraint under which some ontologists developed their theories of metaphysical predication. The constraint operates in several ways which a contemporary philosopher should consider congenial. For example, consider the linguistic oddity of the sentences 'the flesh in Socrates's eye sockets is focussed' (in contrast to 'Socrates's eyes are focussed') and 'these organs and limbs are thinking about Xantippe' (as opposed to 'Socrates is thinking about Xantippe'). Such linguistic data must, of course, be considered in assessing the adequacy of notations. But the requirement of levels of predication allows us to use such data as indicative of differences between the subjects of, and kinds of connections which constitute different sorts of metaphysical predication. At the same time, it allows arguments from programs of scientific explanation to claims about what should be included in the syntax and semantics of a canonical notation. Like Zeno's stricture, and the other constraints we have considered, the requirement of predicate levels provides a partial answer to the question of how linguistic data should be deployed in helping to determine the adequacy of a theory of metaphysical predication.

NOTES

1 I hasten to add that the term 'metaphysical' is by no means used here as a pejorative. The term is meant only to mark the distinction between features of a language and features of the extra-linguistic items we use language to talk about. Thus, the terms 'metaphysics' and 'ontology' are interchangeable throughout this volume.

2 Included in Rudolf Carnap, *Meaning and Necessity* (Chicago: 1956), p. 205ff.

3 Such claims as those made by Plato and Aristotle concerning the existence and ontological status of physical objects and various kinds of abstract entities are clearly neither analytic nor contradictory. Nor are they subject to anything like the kind of empirical verification or confirmation required by the logical positivists for the meaningfulness of synthetic propositions or sentences. Accordingly, the positivists had no alternatives besides throwing them out as if they were meaningless, or drastically reinterpreting them in some version of the way Carnap treated existence claims. It goes without saying that the demise of the logical positivists' criteria for meaningfulness and of the idea that confirmation must require inference from sense data, sensations, and their ilk opened up alternatives unavailable to the logical positivists for understanding metaphysical claims.

4 Note that the assignment of a linguistic predicate to a set leaves open the question what, if anything, the predicate refers to or designates. In particular, we can assign 'F' to S_k and say that 'Fa' is true if $O_i \in S_k$ *without* commitment to the view that 'F' designates or refers to the set.

5 Plotinus and Leibniz are exceptions to the policy of explaining truth in terms of satisfaction or any other sort of correspondence between sentences and extralinguistic (and extra-mental) models or worlds. But in Liebniz's case at least, something like what I have just been saying can still be made to apply. For Leibniz, the linguistic predication 'Fa' is true just in case F, or its concept belongs to the concept of a. But if it does, then whatever is signified by 'F' is, in some way, metaphysically predicated of whatever is signified by 'a'. Thus, even though metaphysical predication does not *explain* the truth of linguistic predications, Leibniz can still ask what sorts of things there are and how they stand when various linguistic predications are true. Nothing this clear can be said about Plotinus, who believed – as Turnbull argues in chapter 1 – that all of what I am calling metaphysical predication is a matter of Being's exfoliation and its articulation of itself into a hierarchy of species, subspecies, and further sub-classes. But even so, we must suppose he held some version of the idea that the truth of at least some kinds of linguistic predications allows us to draw conclusions of some sorts about metaphysical predication.

6 See Julius Moravcsik, 'Heracleitian Explanation' forthcoming in *The Monist*'s issue on the Presocratics.

7 See Robert G. Turnbull, '*Physics I*: Sense Universals, Principles, Multiplicity, and Motion'.

[8] See James Lennox, 'Teleology, Chance, and Aristotle's Theory of Spontaneous Generation', *Journal of the History of Philosophy* (July 1982).

[9] Though anachronistic, it is worthwhile to note that Sellers has developed a strategy for accounting for the truth of such sentences as 'white is a color' without positing universals. His program is developed in a series of articles beginning with 'Grammar and Existence: A Preface to Ontology', in Wilfrid Sellars, *Science, Perception and Reality* (London: 1971), and including the essays in *Naturalism and Ontology* (Reseda, California: 1981).

ROBERT G. TURNBULL

ZENO'S STRICTURE AND PREDICATION IN PLATO, ARISTOTLE, AND PLOTINUS

Much philosophical insight (to say nothing of scientific insight) has been the result of taking very literally and even simple-mindedly the things which we say and think. Having so taken them, one may, with minimum risk of verbal bewitchment, hold them up to critical scrutiny. What seemed obvious in its standard rhetorical garb may then suddenly appear trivial, highly questionable, or even impossible. Socrates' practice in the early and middle Platonic dialogues is, of course, a standard example of this common philosophical procedure. Sometimes the insight garnered from this procedure comes only after it has been used aporetically or even to promote apparent absurdity. I think, for example, that a number of arguments for philosophical scepticism are of this latter sort. Though an argument designed to enforce or clarify a distinction between knowledge and opinion does not as such promote absurdity, surely one designed to show that knowledge as such is impossible does. I readily acknowledge, however, that the promotion of apparent absurdity has actually led to insight, as the responses of, for example, Augustine, Descartes, or Berkeley to apparently absurd scepticism show. But my concern in his paper is not with scepticism or any responses to it. It is rather with an argument which Plato attributes to Zeno and which promotes quite a different absurdity. What I wish to do is, first, to state the argument and, second, to look rather carefully at what I take to the responses of Plato, Aristotle, and Plotinus to it.

Zeno's argument, as Plato states it, is designed to show that what we ordinarily take to be *predication* is impossible. This is, I submit, an apparent, if not a real, absurdity. I hope to show, however, that the argument imposes a provocative constraint on the efforts of Plato, Aristotle, and Plotinus to show how predication is possible. That constraint I shall be calling 'Zeno's Stricture'.

The philosophical content of Plato's *Parmenides* commences with Socrates' summing up Zeno's treatise (which, as the dialogue opens, Zeno has just finished reading) as follows:

Socrates Zeno, how does your argument go? If the beings are many, then they must be likes and unlikes. But this is impossible, for un-

James Bogen and James E. McGuire (eds.), How Things Are, 21–58.

likes cannot be likes, and likes cannot be unlikes. Is this not the thrust of your argument?

Zeno It is.

Socrates And so, if it is impossible for unlikes to be likes and for likes to be unlikes, it is also impossible for many to be. For, if the beings were many, they would be impossibly characterized. And so, is the burden of your arguments nothing other than to contend – against everything commonly said – that the beings are not many?

Zeno Yes.[1]

The purport, in context, of this rather obscure argument is, I believe, as follows:

> Suppose that *A*, *B*, *C*, etc. are "beings". Their being many is simply a matter of their being more than one. If '*A*', '*B*', '*C*', etc., are not simply to be names for the same thing, the beings must be different from one another and, as different, unlike each other. Suppose further that we allow predication among these beings, i.e., that one or more of them can truly be said to *be* one of the others. Thus, e.g., *A* may truly be said to be *B*. And so *A* is *B*. But, if *A* really is *B*, it would seem to have to be the same as *B* and, as same, like it. And so, if we allow it to be said that *A* is *B*, we shall have to say that *A* is both like (as being *B*) *B* and unlike (since '*A*' and '*B*' are names for differents) *B*. *A* and *B* would then be both "likes" and "unlikes". But this is impossible, so it cannot be the case that *A* is *B*.[2]

Plato makes it quite clear in context that the argument is used to defend there being *no many* and thus to support Parmenides' insistence upon *One*. Even so, the argument does imply the principle that, if *A* is *B*, then *A* and *B* cannot be different. And it is this principle which I call 'Zeno's Stricture'.

Several comments are in order.

First, every Greek, like the rest of us, spoke the language and simply took a variety of forms of predication for granted. Their language, like ours, obviously "worked" for a large number of purposes. With it one could tell what is happening or has happened, describe people and things, explain changes and procedures, give instructions, praise, blame, issue orders, lay plans, express joy, regret, and sorrow. And many other things. In doing most of these commonplace things predication of some sort is necessary. For example, to describe a person without predicating something of him or her is virtually impossible. In predication, a good share of the time at least, Greeks used expressions of the form, '*A* is *B*', even as we do. Into this comfortable situation – "against everything commonly said" – enters Zeno. In effect he says,

> When you stop to think a minute about what you are saying when you say, '*A* is *B*', it is obvious that something peculiar is going on. If *A* is *B*, strictly speaking, must it not *really be B*? But, if it is really *B*, then *A* and *B* must be the same; and, if they are the same, is anything being said? If they are different, how can they possibly be the same?

Second, if one should quite naturally respond that the above conflates the 'is' of identity and the 'is' of predication, he/she should be asked to explain just what predication might be. For Zeno's sophisticated and contrived naivete is precisely the sort of philosophical procedure I briefly described at the beginning of this paper. He is being deliberately literal and simple-minded in the interest of demanding an intelligible account of predication. Indeed, he thinks that none can be given and is quite prepared to embrace the apparently absurd denial of the intelligibility of predication. As I noted earlier, Plato sees him as arguing for *no many* as a kind of negative defense of Parmenides' *One*, responding to some unknown opponents who find Parmenides' doctrine absurd. He wishes to show *their* doctrine even more absurd.[3] For the time being at least I forbear comment upon the so-called 'is' of identity except to note that, if Zeno had considered it at all, he would surely have thought it simply a matter of there being two conventional names for a single thing.

Third, there are passages in Plato, notably in *Sophist* and *Philebus*, in which persons not named seem to be contending for the peculiarity, if not the impossibility, of intelligibly stringing "names" together in a sentence.[4] In yet others, most ludicrously in the antics of the comedy team of Euthydemus and Dionysodorus recounted in *Euthydemus*, there is perverse eristic exploitation of apparent sameness and difference in predication.[5] Indeed, Plato treats Zeno, in *Philebus* and *Sophist*, as an eristic (the "Eleatic Palamedes") engaged in the same sort of exploitation.[6] Though I think that Plato, in *Parmenides* at least, is directly responsive to Zeno's Stricture, I am not in this paper contending that Plato, Aristotle, and Plotinus are, in every phase of their theories of predication, directly responding to the Zeno of *Parmenides*. The problem of intelligibly accounting for predication was obviously in the air, and my contention is that all three attempt their accounts in the light of the peculiarity of saying that *A* is *B tout court*.[7]

Fourth, the distinction between "particulars" and "universals" which *we* learned at our philosophical mother's knee, with whatever sophisticated rationale, is simply not part of the inherited philosophical baggage of Plato or Aristotle. Plato did not know of or use the distinction (as I shall shortly show), and Aristotle (so far as I can tell) *invented* a form of it, a form which Plotinus explicitly rejected. That form construed the "universal" (*τὸ καθόλου*) as holding for many and the "particular" as one of those held for. As such, it contains a slope towards nominalism or conceptualism which we shall duly note. None of our three philosophers had the slightest inkling of the notion of the propositional function or the theory of types, in which terms the distinction between particulars and universals has been and continues to be commonly made. "Exemplification" or "instancing" as somehow involving a value for a zero-level variable and as also involving a value for a first-level function-variable – or some animadversion on that theme – is a foreign notion in Greek philosophy. And it is no simple matter to import it into Greek philosophy.[8]

I. PLATO

(1) Plato was struck by the fact that the language contains names (*ὀνόματα*)[9] not merely for the commonplace things around us but also for things which are not commonplace and of which we cannot plausibly be said to have sense experience. These latter are such names as 'beauty', 'justice', 'triangularity', 'equality', and other like substantives. Terms of this kind are, in English, commonly formed from adjectives by adding the suffixes, '-ity', '-ness', or '-hood', and Greek has similar means of forming them. Plato also used auxiliary terms for forming such substantives, as 'The *F* Itself' ('The Beautiful Itself', 'The Equal Itself'). He took these terms to be names of intelligible (and non-sensible) things which he called '*forms*' ('*εἴδη*' '*ἰδέαι*'). In his middle period at least, Plato thought that there was no predication problem with respect to them, for he sternly limited such predications to those having the form, 'The *F* Itself is *F*' ('The Beautiful Itself is beautiful'), a limitation imposed by the requirement that a form – to be a form – could, only be what it is. Though there is considerable controversy about how such "self-predication" is to be understood,[10] it does not (or does not obviously) run afoul of Zeno's Stricture, i.e. the requirement that, if *A* is *B*, then *A* and *B* cannot be different.

(2) It is not germane to the present enterprise to offer serious explanation or defense of Plato's theory of forms (at whatever stage), and I shall, for the moment at least, forbear. It is relevant, however to note that middle Plato

takes the *F* Itself, the *G* Itself, etc. to constitute reality with a capital '*R*' and the commonplace things around us to be derivative and humdrum by comparison. Minimally, a form *is* what it is and in no way is not. A commonplace thing *is* and is *not* (beautiful now, not beautiful later; beautiful for a woman, not beautiful for a goddess; etc.). And for the latter to *be* at all is for it to "have a share" of a form. This doctrine gives or accompanies Plato's means of explaining predications made of humdrum things without running afoul of Zeno's Stricuture. As suggested just now, Plato had in Attic Greek a handy metaphor, the employment of which helped considerably in explaining such predications, namely, 'μετέχειν' whose literal meaning was 'to have a share or shares of' and whose extended employment allowed one to have a share of a public occasion or even of courage or timidity. Using that term and his theory of forms, Plato argued that such sentences as 'Phaedo is tall' are to be understood as having the sense of 'Phaedo has a share of The Tall Itself.' This is commonly spoken of as "Plato's doctrine of participation", but the use of 'participation' obscures an important feature of the explanation, namely, that, if Phaedo has a share of The Tall Itself, then he has *a tall* in him – so to speak, his "share" of The Tall Itself.

Ignoring troublesome features of the explanation,[11] it is clear enough that, if one understands 'Phaedo is tall' as 'Phaedo has a share of The Tall Itself', there is no violation of Zeno's Stricture. Though the surface grammar of the sentence has the form, '*A* is *B*', the logical grammar on Plato's theory has the form, '*A* has *B* in it'. Thus properly understood the sentence is not used to claim that Phaedo *is* anything; it claims rather that he *has* something.

(3) Still fairly early in *Parmenides* Socrates uses this doctrine of *having a share or shares* to respond to the Zeno argument we have quoted, that is, Socrates points out that commonplace things may *have* any number of shares of different forms in them without being thereby required to *be* different and thus violate Zeno's Stricture. And he challenges Zeno to show how the theory of forms, given the predicational restriction noted in (1) above, constitutes any violation of the Stricture.[12]

Immediately following Socrates' response to Zeno, Parmenides commences a provocative series of questions. After establishing that Socrates is uncomfortable about the universal applicability of the doctrine of *having shares*, in particular, its applicability to shares of sortal or "substance" forms and to "undignified and trifling things",[13] Parmenides tries to elicit from Socrates a literal sense for the metaphorical 'have a share of' (μετέχειν). If several commonplace things are such that each has a whole form as its share, how can the form be one or single? If each has a part as its share, then the form must already, as divided, be multiple. If the literal grounding for *having shares* is located in the possibility that several things may have one and the same appearance or look (ἰδέα), then the rationale which permits the distinction between things and their appearance or look will allow a similar distinction between the things and their appearance on the one hand and *their* appearance or look on the other. So again the form will be multiple. If the form is taken to be a thought which can be one while being *of* many, then, given the required connection between form and share, it will follow that everything, *qua* having a share of any form, thinks. And this will not do. If the share a thing has is taken as a copy and the form as pattern or paradigm, then, given the requirement that copy and pattern must be like and the explication of 'like' as 'having a share of a common form', then once again the form will be many. Parmenides ends his critique with an illustration of the consequences of being unable to supply an intelligible sense to 'having shares', the distressing ones for Socrates being that the gods could neither know us nor be our masters.[13] This highly compressed and, lacking detailed argument, biased summary of a highly controversial text, namely, the so-called "first part" of *Parmenides*, is given here only (a) to hint at what I take to be a proper interpretation of the text and (b) to make it quite clear that Plato himself challenged his own rather neat response to Zeno, i.e. the *has* as opposed to *is* interpretation of predications made of commonplace things.

Interestingly enough, though Plato continues in later dialogues to speak of commonplace things having shares of forms, I can find no indication in later dialogues of "shares" like "the tall in Phaedo", "the beautiful in Helen", or "the equal in these sticks" – the stock shares of *Phaedo*.[14] Nor can I find in those later dialogues any usage of the *has* interpretation of predications made of commonplace things. I think that (and will assume in the next sections of this paper that) Plato modified his account of *having shares*, thus obviating the criticisms recounted in the paragraph above. He also allows, as has often been noted, more interesting predications to be made of his forms – in effect and at a very sophisticated level, answering his own challenge to Zeno. I turn to the matter of the forms first, then to that concerning commonplace things.

(4) In later dialogues, most notably *Sophist*, we get the rationale for the more interesting predications to be made of the forms. The key to that rationale is the recognition of *Being, Same,* and *Different* as themselves forms, forms which "run through" and can be predicated of every form, including themselves. Thus for any form, the *F* Itself, it *is*, is the *same* (as itself), and is *different* (from any other). Full-scale interpretation of the much-discussed metaphysical heartland of *Sophist* is obviously both inappropriate and

impossible here.[15] For what I am about, however, we must note that predicating *Being, Same,* and *Different* of a form does not demand that the form be different from itself or the same as some different, and thus does not produce a violation of Zeno's Stricture. And we must, to accommodate the insights of *Sophist* and Plotinus' use of them,[16] be prepared, as we shall see in (6) below, to shift the focus of our standard way of looking at genera and species. For the moment, however, it will help to stay within the normal focus.

I think that, for Plato, when one says that The *F* Itself *is*, one is not predicating *existence* of it. (Indeed, any attempt to import the "is/exists" distinction in Plato's text needs justification.)[17] One is rather recognizing *Being* as a super-genus, as the principle of everything which in any way *is*. In this respect, it is predicable of every form in the same way that The Plane Figure Itself is predicable of all of the two-dimensional shape forms (as species of that genus). *Same* and *Different* are not in themselves genera. They are rather the principles which, while predicable of everything, provide for differentiation of a genus (form) into species and sub-species. Thus Plane Figure is the *same* in all its species (as they are all Plane Figures), but each of the species is *different* from every other.

(5) Socrates, in *Philebus*, astonishes his interlocutor, Protarchus, by claiming that pleasure can be and is different from pleasure – and knowledge different from knowledge.[18] (He might even have said that being is different from being!) What he means is, of course, that pleasures can be different species while all are generically pleasures (and thus the same). Illustrating this with the non-controversial example of The Plane Figure Itself, it is a genus of which The *Straight* Plane Figure Itself and the *Curved* Plane Figure Itself are proximate species. The Straight Plane Figure Itself divides into species by sidedness, one being The *Three-Sided* Straight Plane Figure Itself. This, in turn, divides, one species being The *Equal-Sided* Three-Sided Straight Plane Figure Itself. Thus The Equilateral Triangle Itself (our last species) is both *different* (from its fellow species) and the *same* (as itself and its higher-level species – as being Plane Figurate). But it is not in any paradoxical way both different from and the same as anything.

The above is, as will be readily recognized, an illustration of the division procedure outlined in *Philebus* (the discovery of "a certain Prometheus").[19] It is brought into *Philebus* not merely to show how pleasure can be different from pleasure but also to show how something can be *one* (as Equilateral Triangle is Plane Figure) and also a certain definite number (as Plane Figure, Straight Plane Figure, Triangle, and Equal-Sided Triangle) – thus to illustrate the procedure for arriving at definitions by genus and differentia. Such definitions, showing just "how many" something is, are, I believe, sophistications upon the "safe and stupid" form-explanations of *Phaedo.*[20] In the light of the *Philebus* account, the *Phaedo* account asks one to go too quickly from The *F* Itself to the *F* in *x* (e.g. from The Beautiful Itself to the beautiful in Helen) before making it clear what The *F* itself *is*. From the point of view of *Philebus*, the *Phaedo* account fails to exhibit the genera/species structuring of the forms. So failing, it deprives us of definitional explanation.

Our focus is, however, on predication, and *Philebus*, besides allowing for increased predications for forms (beyond the austere self-predication of "The Beautiful Itself is beautiful"), suggests an interesting expansion of possibilities for commonplace things. This is found in the idea that *one and the same thing may be named by more or less specific names*. Thus the same thing may be named by 'equilateral triangle', 'triangle', and 'plane figure' without violation of Zeno's Stricture – though it is most perspicuously named by 'equilateral triangle'.

(6) Strictly speaking, matters have to be a bit more complicated than (5) suggests. After all, Plato's forms are natures and *not* natured things (a fact which puzzled Aristotle and exhilarated Plotinus – as we shall see). The *F* Itself is not *an F*. The Triangle Itself is not *a* triangle, nor is it triangular. It would therefore be a mistake to think of Plato's forms as a mere convenience for classifying commonplace things. Plato's forms are not concepts. They are, if you please, reified natures, and *their* being or being what they are is in no way dependent upon there being any individuals which have shares of them. Predication for the forms considered in themselves is, therefore, independent of predications made of natur*ed* things.

It would thus seem that predication for the forms considered in themselves is either a definitional identity or a marking out of relative place(s) in a genus/species ordering. And so Triangularity *is* Plane-Figure-Bounded-by-Three-Straight-Lines-hood, or The Triangle Itself *is* The Plane Figure Bounded by Three Straight Lines Itself. Conceived realistically (i.e. not merely nominally) such identities unpack forms and, in so doing, indicate their places in the genera/species structure of forms. Other predications made of forms considered in themselves also exhibit (unless false) place in that structure. If relative independence is taken as a criterion for subjecthood, proper subjects would be genera or higher species (i.e. higher than the predicate). In our continuing example, The Plane Figure Itself is subject relative to its species and sub-species. Such predication reflects the division of a genus into species. Such division, "becoming" different while remaining the same in all, can be thought of as a timeless exfoliation. Nor is the division arbitrary; it simply articulates what is "already" in the genus or in it subject to division by Different and Same. Since, as noted earlier, *Being* is for Plato a (indeed, *the*) genus, the whole intelligible world of Platonic forms may be thought of as the articulate exfoliation of *Being*.

Except in the so-called "second part" of *Parmenides*[20] (and there especially in Hypothesis 2), Plato does not give us much detail for the articulation, though examples of division abound in *Sophist* and *Statesman.*[21] In Hypothesis 2 of *Parmenides* the articulate exfoliation is from or of the One Being with the partitioning of the One Being coming first, then the "generation" of numbers, then shapes, and so on.[22] This invites a highly mathematical conception of the forms, one which, I think, gives a correct picture of the later Plato and which is attested by Aristotle.[23] As we shall see later in discussing Plotinus, it is, with slight revision, the major Platonic picture in the *Enneads.*

(7) In the later dialogues, Plato maintains a dualistic stance. There is, on the one hand, the unsullied world of forms, the fully articulated *natures*, whose articulation at once provides our standard of intelligibility and delineates what, strictly speaking, *is*. There is, on the other hand, the sensible world, that of commonplace things, the unique embodiment of the world of forms, thus natur*ed* things.

It is a major contention of this paper that, while Plato maintains this dualistic stance, Aristotle adopts and adapts Plato's scheme for the sensible world *sans* the world of forms, while Plotinus adopts and adapts Plato's scheme for the world of forms *sans* any sort of independence for the sensible world. For predication matters, both schemes can conform to Zeno's Stricture. The sensible world scheme makes use of the possibility of the predicate term's naming with equal or less specificity what the subject term names, as in 'This (some, every) man is an animal'. On the world of forms scheme, predication makes use of the need for every species' being a partial articulation of a genus and thus containing it or the need for genus's exfoliation into species which it, in a different way, contains. We may speak of the species, The *F* Itself as *G*-ate (The Triangle Itself as Plane Figur*ate*), we may speak of the genus, The *G* Itself as *F*-ated (The Plane Figure Itself as Triangul*ated*). Both ways of speaking may be allowed, since each expresses one direction, as it were, of the same eternal order – though genera on the scheme are principles or causes relative to species.

Three comments may be helpful. *First*, we are looking at a scheme for which some of Aristotle's comments about "genus as matter" may be apropos.[24] Thus one may speak of the genus as common to all of its species in a manner analogous to gold's being common to variously shaped golden things.[25] Being would then, as the ultimate genus, be a sort of "prime matter". The analogy fails, however, in that gold does not determine the shapes it will take, whereas Platonic genera, split up by *Different*, do determine their species. *Second*, one must not confuse *our* means of investigating the world of forms with the world of forms itself. Sense-ridden beings that we are, we need to make regular reference to natured things in testing definitions for counter-examples, in grouping things of an apparent kind in the search for a common species, and the like. So doing, one may mistake the aids to *our* investigation for the *objects* one is looking for, as one might mistake the diagrams of geometers for geometrical objects.[26] *Third*, the by-now time-honoured formula of definition by genus *and differentia* may lead one to looking for some sort of independent status for differentiae. This would, I believe, be a mistake. Platonic genera are articulated into species *qua* the genera they are by *Difference*. Differentiae are not, on this scheme, "properties" in any modern sense of that term; they are rather modes of differentiating or 'speciating' genera. The Plane Figure Itself, as just that nature and as susceptible to "internal" differentiation, speciates, though this is not a process but an eternal order. The habit of looking at differentiae as "properties" is largely due to attention to what an individual "has" and the investigation of forms or "universals" by way of individuals, a habit encouraged both by classical Aristotelianism and recent analytic philosophy. But, of course, the "property/individual" dichotomy is hardly appropriate for talking about Plato's world of forms *as such.*

(8) But what shall we say for the later Plato concerning what I have been calling 'commonplace things'? I think that, though clearly recognizing the difference between forms as *natures* and the commonplace as *natured things*, he allows the latter, when exhibiting appropriate configurations, to be quite legitimately called by the form-names. Thus, given an appropriate configuration of matter (ultimately, the elementary particles of *Timaeus),*[27] reference can be made to *this* or *a certain* equilateral triangle as a natured thing. And appropriate predications can be made of it, as *equilateral* or (a) *plane figure.* Such predications do not run afoul of Zeno's Stricture, for they make use of possibility that the same natured thing may be named by more or less specific names. Thus, assuming that the subject term succeeds in naming, 'This equilateral triangle is a plane figure' names one and the same thing more specifically in the subject and less so in the predicate.

As this last hints and as noted earlier, the same thing can be named by more or less specific *subject terms*, as 'This equilateral triangle' and 'This plane figure.' The latter term is, of course, less perspicuous in that it narrows the range of necessary or *καθ' αὐτό* (of which more shortly) predications and unduly limits definitory explanation of the sort invoked by the *Philebus* account. But there is nothing to block the intelligibility of its being appropriately used as a subject expression.

It is not uncommon for Plato interpreters to understand Plato as holding a doctrine of "particulars" like that of the Russell of "The Lectures in Logical Atomism",[28] with the model furnished by values for zero-level variables in propositional functions. Such particulars are very "bare" indeed and hardly describable as 'natured things', and I can find no evidence that Plato held any doctrine remotely like it. Though he does use proper names in crucial illustrations, as 'Phaedo' in *Phaedo* and 'Theaetetus' in *Sophist*, it is quite clear that 'This man' would have served his purposes as well.[29] There is simply no evidence of an appeal to a Russell-type distinction between proper names and descriptive expression.

The chief reason for emphasizing this point is that it aids in understanding predications concerning commonplace things in later dialogues (to say nothing of Aristotle's logical works!). If 'This triangle', 'This ox', and the like are appropriate terms for commonplace things, then, whenever they are appropriately used, some predicate expressions like 'plane figure' and 'animal' will hold simply in virtue of those things' being what they are. And this is, of course, the very meaning of '*καθ' αὐτὸ*'. There could be no such necessary or *καθ' αὐτὸ* predication for Russellian bare particulars.

(9) In an earlier paper,[30] I argued that Platonic forms (in the later dialogues at least) are profitably thought of as principles of structure and commonplace things as structur*ed* things. On that view *having a share* of a form is a matter of being structured in a certain way and *not* a matter of literally having a share, where shares are taken (as in *Phaedo* and the first part of *Parmenides)*[31] as *separated* from the forms. Things can quite literally be said to have the *same structure* without requiring that the structure be divided up, indefinitely replicated, or in any intelligible sense separated from that or those of which it is the structure. And the concept of a certain structure just by itself and quite apart from anything's having that structure is intelligible as, if nothing else, the examples of pure mathematics and logic make quite clear. There is, particularly in *Parmenides, Timaeus, Philebus*, and Aristotle's *Metaphysics,*[32] considerable evidence for attributing this sort of doctrine of the forms and having shares to the later Plato. But this is hardly the place to present it, especially since the sense of difficult passages does not leap out at one.

I mention the structure/structured things interpretation here to flesh out the account of καθ' αὐτό predications made of (things in) the sensible world. The demiurgos of *Timaeus* sees to the maximum organization of "matter" in the receptacle, including, of course, the mathematical construction of the Platonic "atoms" (tetrahedra, cubes, octahedra, and icosahedra).[33] That organized cosmos is a "copy" of the organization of the forms and as much like the latter as possible. With the ensouling of bodies, including the celestial ones and noting that souls are "made of" Being, Same, and Different, and thus the introduction of rational motion, the cosmos becomes a "moving image of eternity".[34] And so, insofar as one can grasp either the articulate exfoliation of form-genera or the structure(s) inherent in the cosmos, he/she is well-positioned to grasp the other. There is, with all this, the genuine possibility of attaching the right form-term to a thing and being forthwith able to predicate something necessarily or καθ' αὐτό of it.

(10) There are, of course, a number of predications which we regularly make of sensible things which are not necessary or καθ' αὐτό. Though I know no explicit supporting texts of Plato, I think that we can safely speculate what his general account seems to require for one class of such predications. These are predications in which the subject term is a genus or a higher species term and the predicate is a lower species term or a differentia term, for example, 'This plane figure is a triangle' or 'This triangle is equiangular'. Plane figures need not be triangles nor need triangles be equiangular, but the former could be a triangle and the latter equiangular in a way in which neither could be, say, grammatical. The pertinent point for our issue of predication is that the subject and predicate terms are form-terms for forms which fall within the same form articulation or genus/species structuring. Thus those terms fall within a set which can be used to express καθ' αὐτό predications. Given that a subject term used to refer to a structured thing can so refer whether it is more or less specific or generic, there is no problem with Zeno's Stricture. In the cases at hand, when the sentence is true, the subject and predicate terms refer more or less generically to the same thing.

Another sort of non-necessary predication involves terms which are not either less general or differentia terms, but rather terms which, as we might say, are *compatible* with the subject terms in question. (Aristotle would have said that the subject terms "admit" those predicates – δέχομαι.)[35] Thus triangles can be large or small, but not grammatical. Men can be older or younger, geometers or salesmen, but not divisible by three or equiangular. In an earlier paper which did not emphasize Zeno's Stricture,[36] I argued, in effect, that one and the same structured thing could have or embody a subordinate structure which is *compatible* with the structure invoked to identify it (with whatever purpose in mind) as a thing of a particular kind. 'Compatibility' could thus be construed as 'falling within a range specifiable by the nature indicated by the subject term'. Though Plato addresses this issue indirectly, in deciding what to include in a definition or which differences are appropriate in making divisions,[37] I know no place in which it is systematically addressed. With respect to conformity to Zeno's Stricture, it is, I think, clear enough that a predicate term which names a subordinate structure (i.e. to that picked out by the subject term) does not name something *different* from the structured thing picked out by the subject term.

(11) My final comments on Plato concern failed predications, that is, false sentences. To make those comments it is necessary to say something about negative sentences. The source of the comments is once again *Sophist*.[38] In it, Plato accounts for negative predications by invoking *Different* and its "parts". The parts of *Different* have the form, *Different from F*. Thus sentences of the form, '*G* is not *F*', are taken as having the sense of '*G* is different from *F*', and, supposing the sentence true, 'Different from *F*' holds for *G*. And Different from *F* is itself a form, as it were, one of the many species of The Different Itself, where such species are unmediated by higher species or genera except for The Different Itself. (This reflects what many have noted, namely, Plato's not distinguishing in *Sophist* between what Aristotle would have called 'contraries' and 'contradictories'.)[39] It is, of course, incompatible with *F* or The *F* Itself. And the same rationale applies to sentences about sensible (or structured) things, as in 'This *G* is not *F*'.

Plato obviously needs some means of accounting both for the truth and falsity of sentences and for the sense of negations. These are related, but not identical, issues. As Plato puts it, a sentence is true if it states (or can be used to state) what is or what is not concerning its subject.[40] Thus '*G* is *F*' is true only if *G* is *F* and false if *G* is not *F*. And the latter has the sense of '*G* is different from *F*', i.e. 'Different from *F*' really holds for *G*. In the same way, '*G* is not *F*' is true only if *G* is not *F* (i.e. is different from *F*) and false if *G* is *F*. Unless a sense could be attached to 'is not *F*', the truth conditions could not be stated. And this would not be possible unless, as Plato puts it earlier in *Sophist*, it can be shown that "not-being *is*".[41]

On Plato's view, there is therefore some fact of the matter for every well-formed sentence, though many such facts are "difference" facts, and they constitute both falsifying and truthifying conditions. And there appears again no violation of Zeno's Stricture. Even though it is not in a standard genus/species name series, 'Different from *F*' can (unperspicuously) name what is named by '*G*' or 'This *G*'.

II. ARISTOTLE

(1) As might be expected from a colleague of some twenty years, Aristotle is remarkably close to Plato, and the interpretation of either is helped by reading the other (though the thesis stated in I (7). obtains). The *locus classicus* of Aristotle's theory of predication is obviously *Categories*. In standard Attic Greek a *κατηγορία* was a charge, such as might be brought in a court of law, and the verb, '*κατηγορέω*' meant 'to bring a charge'. Aristotle uses the noun as we should use the noun, 'predicate', and the verb as we should use 'to predicate'. And he attempts a listing in *Categories*[42] of the different sort of predicates, namely: *thing* (οὐσία), e.g. man; *how large* (ποσὸν), e.g. two feet long; *how qualified* (ποὶον), e.g. grammatical; *to* or *of what* (πρός τι), e.g. double, larger; *where* (ποῦ), e.g. in the Lyceum; *when* (ποτὲ), e.g. yesterday: *in what posture* (κεῖσθαι), e.g. is lying, is sitting; *in what condition* (ἔχειν), e.g. is armed; *doing what* (ποιεῖν), e.g. cuts, hits; *undergoing what* (πάσχειν), e.g. is cut, is hit. The list as such is not critical (and Aristotle does not in other works always give the same list) except as suggesting the completeness he obviously has in mind. Some such list is, however, important; for, as we shall see, Aristotle, unlike Plato, does not take *Being* to be a genus or principle. He takes it, as a term, to be in itself ambiguous and "said" in as many ways as there are categories.[43] But we shall give careful attention to this matter a bit later.

(2) Early in *Categories* Aristotle makes an important distinction between *said of* and *in*[44], a distinction between two kinds of predication. Fortunately for ease of understanding, the distinction corresponds to an English language distinction which uses the indefinite article (Greek lacks one), namely, that between *is a* (*n*) and *is* predication. In any true sentence of the form, '*F* is *a G*', the subject term and the predicate term name the same thing as two conventional names for the same thing, as simple repetition of the subject term in the predicate, as individual/species terms, or as species/genus terms. (Thus directly conforming to Zeno's Stricture or conforming by making use of more or less specific names in the manner discussed in the Plato section.) Thus, e.g. 'Dobbin is a horse' or 'A certain horse is an animal'. Both of these predications are, of course, in the category of *thing* (οὐσία). But *said of* or *is a* (*n*) predications occur also in the other categories, as in 'This white is a color'. So in *said of* or *is a* (*n*) predication, the subject term must be the name of something of a kind and the predicate term an appropriate kind-term, or the subject must be a less general kind-term of which the predicate is a more general kind-term. Or, obviously, they may be identical terms.

In *is* or *in* predication, the subject term and the predicate term must be from different categories, though in a true *is* or *in* predication, the two terms must in some manner name the same thing. They can hardly do so, however, as more or less general terms in the same category or sub-category. The paradigm case of *is* or *in* predication has an *οὐσία* or thing term as subject and a term from another category as predicate, as in 'A certain man is white' or 'Socrates is brave'. Both of these have *how qualified* terms as predicates. One of the interesting features of Attic Greek and Aristotle's usage is the sanctioning of *how qualified* (as well as other non-*thing* terms) in the subject place. Thus subject reference by means of 'The white' or 'The brave' is common. And both could be references to the man of our first sentence or to Socrates of our second – though not, of course, *qua* man or *qua* Socrates. And one could go on to say of the white or the brave that he is a father, that he is in the room, that he is lying down, and so on through the categories. Granting this predicational usage and delaying for the moment the matter of the primacy of *thing* or *οὐσία* terms, it would be correct to say that *is* or *in* predication sentences of the form, '*F* is *G*', are true only if '*G*' could be used in the subject place to name the same thing (though not, of course, *qua F*). In this way, though the subject and predicate terms are not related as less or more general kind-terms, the subject and predicate terms in a true sentence must name the same thing. And in this way Zeno's Stricture is not violated even in *is* or *in*.

(3) In various places, though not, oddly enough, in *Categories*, Aristotle says that 'being' or 'is' is "said in many ways", that is, in as many ways as there are categories.[45] But he goes on to insist that such "being saying" is all with regard to one (*πρὸς ἕν*), namely, *thing* or *οὐσία*. Though this doctrine of what is commonly called *πρὸς ἕν* ambiguity' is not expressly stated in *Categories*, it is certainly in the spirit of that work. For *Categories* equally emphasizes the role of *οὐσία* or *thing* as basic.

As I noted in (2), terms from various categories may be used as subject terms, e.g. 'The white', 'The brave'. And I must note in this context that Artistotle is like Plato in this matter of subject terms. Subject terms drawn from whatever category are in themselves characterizing. Though – again like Plato – Aristotle commonly uses proper names ('Socrates', 'Callias') in illustrations, it is clear enough in context that he is not drawing upon a Russellian distinction between proper names and descriptions. He would have been perfectly satisfied with 'The man' instead of 'Socrates', provided that which man is being picked out is clear. And I must note as well that, when non-*thing* or *οὐσία* terms are used as subject terms, though a *thing* may be picked out (not, of course, *qua* a *thing*), there is the possibility of *said of* or *is a* (*n*) predication within the category of the subject term.[46]

When, however, a non-*thing* or *οὐσία* term is used as subject term, as 'The white' or 'The brave', it is always appropriate to ask the question, 'The white what?' or 'The brave what?', expecting some such response as 'Wall' or 'Man'. When, on the other hand, a *thing* or *οὐσία* term is used as a subject term, as 'The man' or 'The horse', that sort of question is unintelligible. This is the linguistic intuition lying behind the *πρὸς ἕν* in *προς ἕν* ambiguity. Though non-*thing* terms may be used either in subject or predicate places in well-formed sentences, their usage presupposes the applicability of some *thing* or *οὐσία* term.

It is this sort of consideration which – leaving some caveats unspoken – leads Aristotle to think that reality with a capital 'R' consists of individual *things* or *οὐσίαι*. Reference to them or the possibility of reference to them is presupposed in all other reference and predication. As we shall soon note, crucial for Aristotle's account of explanation as well as (simple) reference and predication is the idea of reference to things by means of *nature* terms, so that linkages between and among *things*, whether as such or as otherwise characterizable, be assured.

Even so, non-*thing* category words used to name *things* (though not, of course, *as things*) are extremely important, and they provide the basic terms for whole sciences, e.g. mathematics. For, as the applicability of *said of* predication to non-*thing* categories suggests, there are genera/species hierarchies for (most of) the other categories and thus the whole apparatus of definition. But we shall return to this matter after a brief look at Aristotle's logic.

(4) In *On Interpretation* and *Prior Analytics*, Aristotle opts for formal means of expressing and perspicuously exhibiting the interrelationships of predications. He uses capital letters as dummy subject and predicate terms and expresses their linkage by means of 'holds for' (*ὑπάρχειν*).[47] Thus '*A* is *B*' is expressed by '*B* holds for *A*'. Quantification is expressed by the use of 'whole' (*ὅλον*) and 'part' (*μέρος*) added to the subject term. So 'Every *A* is *B*' is expressed by '*B* holds for A_{whole}'; and 'Some *A* is *B*' is expressed by '*B* holds for A_{part}'. The corresponding negative predications are achieved simply by adding 'not' to 'holds for'. With the apparatus of quantification and negation, Artistotle can articulate truth conditions for the so-called "square of opposition".[48] and for "immediate inference". Thus, e.g., if '*B* holds for A_{whole}' is true, '*B* holds for A_{part}' must be true as must '*A* holds for B_{part}'; but '*B* not-holds for A_{whole}' and '*B* not-holds for A_{part}' must be false.

It is noteworthy that, despite the requirement that the ultimate subjects of predication must be individual *things*, Aristotle is perfectly comfortable about convertibility, i.e. the exchange of subject for predicate terms and vice versa, quite regardless of category match. And this should make no difference, if non-*thing* category terms can be used for picking out things (though not *qua things*), i.e. make no difference for purposes of formal logic. He can even develop his account of syllogism without attending to category match. Thus, in his paradigm syllogism, if C holds for B_{whole} and B holds for A_{whole}, then C must hold for A_{whole}, i.e. if the first two are true, the third must be true. It is obviously not my intention here to go at length about syllogism, and I shall confine myself to *Barbara*, i.e. first figure universal affirmative syllogism. Relevant to the general purpose of this paper is the fact that in that syllogism (as in the others as well, but it would be tedious to go through the lot) there is no violation of Zeno's Stricture – given what was said in (3) above.

(5) Equally relevant is Aristotle's use of syllogism in his doctrine of *demonstration* in *Posterior Analytics*. Demonstration requires premisses in which the predications are *καθ' αὑτό* and it will be remembered that I introduced that Aristotelian term in discussing Plato (I(8) and (9)). Those are predications such that, if the subject term holds for or applies to anything, then the predicate term *must* apply. The obvious case is that in which the subject term is a species term and the predicate either that same term (trivial case) or a higher-level species or genus term in the same genus/species hierarchy. Thus *animal* holds for this man and, indeed, *every* man *καθ' αὑτό*, and *plane figure* for this and every triangle. Given a genus/species hierarchy, it is obvious that several *καθ' αὑτό* predications can be made. And it is equally obvious that, properly sequenced, syllogism will result from stringing some of them together. But Aristotle is interested in a somewhat richer form of demonstration than would result from stringing such premisses together.

In the paradigm syllogism form of *Barbara* sketched in (4), '*B*' is the "middle term". If '*C*', '*B*', and '*A*' of that paradigm form were 'Animal', 'Mammal', and 'Man', it is reasonably clear that, were the question, 'Why is this or every man an animal?', the response, 'because he is a mammal', would be explanatory if not completely satisfying. Aristotle, in *Posterior Analytics* and elsewhere,[49] places great emphasis on middles or middle terms as explanatory. And he links middle terms – indeed, the whole matter of determining genera species hierarchies – with *definition*. Much of *Posterior Analytics II* is devoted to discussing means for defining by genus and differentia, for insuring that no intermediate species are left out, for insuring congruence of differentiae,[50] and for contrasting Aristotle's procedures with Plato's procedures for collection and division.[51] Fortunately, since our focus is predication, details of all this are neither appropriate nor necessary. What is of moment for my purpose is the idea of a definitory formula as a middle term in a demonstration and of its being used to pick out *things* by specification of their natures. (Since non-*thing* categories allow as well for such definitory formulae and thus for demonstration, though – as noted earlier – they all link with *thing* or *οὐσία*, at least some non-*things* must be thought of as having "as it were" natures. Geometric demonstration is the conspicuous example. Despite this complication, I shall in the interest of simplicity stay pretty strictly with the *thing* category in what follows.)[52]

(6) Aristotle worries in several different works about the problem of the unity of a formula (λόγος), where the formula consists of several words.[53] The crucial feature of his solution is, I think, definition or formula by genus and differentia. But it is not simply that, for, as the continuing illustration of the division of plane figure suggests, definition by genus and differentia exists (in standard Aristotelian sciences) at several levels. If one starts with a relatively remote genus, one may then divide into species by differentia, then divide each of these in turn into sub-species by differentia, then . . . and so on. It is clear from the plane figure example, looked at from the point of view of wider and narrower names of things, 'plane figure' is wide, 'straight plane figure' narrower, 'three-sided straight plane figure' narrower yet, and 'equal-sided three-sided straight plane figure' even narrower. Each of these can be used (with appropriate auxiliary terms) to refer to an individual thing (indeed, the very same individual thing!), though the more specific is the more perspicuous. In a clear sense, there is only one genus in the series, namely, *plane figure*, with sequentially added differentiae. Nor are these formulaic names or strings arbitrarily chosen series of terms. Their unity lies not simply (if at all) in a single definitional "intuition". It lies rather in their *fit* in a genus/species hierarchy. And, if any individual thing can be referred to by a lower-level species term, then it must (*καθ' αὐτό*) have the differentia for that species *as well as* those for higher-level species and genera. The explanatory power of middle terms in demonstrations is a function of the interesting middle terms' being definitional formulae which are linked with a single genus and a sequence of differentiae.

With such middle terms, says Aristotle in *Posterior Analytics*, the need for Platonic forms is obviated.[54] If definitory formulae can be used to refer to various things, and if such formulae fit into a series of the genera/species kind, then explanations of the connections between extreme terms (i.e. the terms of demonstrated conclusions) require no appeal to "separated" form-structures of the sort delineated in *Philebus*. When one contemplates the possibility of dispositional terms as differentiae terms and the whole Aristotelian apparatus of potentiality and actuality, the scope of middle term explanations is greatly enlarged. And Aristotle enlarges it.

(7) Unlike Plato, Aristotle has a doctrine of *universals* (τὰ καθόλου), and he characterizes the universal as *holding for many*. It would be a mistake, despite the appropriateness of the term, to assimilate Aristotle's universals to those of various realists (e.g. Moore and Russell) of the present century. Indeed, Aristotle writes of his universals as being "in the soul"[55] (a characterization which would hardly do for realists like Moore and Russell). So understood, Aristotelian universals play a role usually played by *concepts* in later philosophy. The last chapter of *Posterior Analytics* speaks of the coming to be of the "first universal in the soul" as a result of a sequence including sense perception, memory and "repeated memory of the same".[56] This sort of text and what I take to be the general drift of Aristotle's thought on kindred matters incline me to understand texts (as in *On the Soul* III)[57] in which Aristotle appears to speak of knowledge as *of the universal* in the sense of being "about" a universal (and thus suggesting Platonism) as appearance only. I think that those texts should be understood as, in effect, using 'of the universal' as a phrase characterizing the knowledge state or condition, with the genitive *not* carrying the weight of intentionality.[58] But I cannot make the case for this contention here.

In *On Interpretation* Aristotle is clear but quite brief in stating the relation between universals in the soul and the conventional vocabulary of language.[59] The latter are expressions of or signs (συμβολά) of non-conventional universals in the soul. If 'in the soul' is unpacked as one naturally would for a philosopher who thinks of soul as "the first actuality of the body",[60] i.e. as a disposition appropriately connected with sense perception and conventional vocabulary, then it is relatively easy to understand how different conventional terms may be expressive of or vehicles for the same disposition and how the latter may not be conventional at all. And the manner of acquisition of the universals in the soul was roughly described (or at least hinted at) in the last paragraph. It would be a mistake, however, to think that the first universals, so acquired, will survive without change in the process of acquiring "scientific knowledge" (ἐπιστήμη). In the process of moving from what is "more intelligible to us" to what is "more intelligible in nature", i.e. in the process of determining genera/species hierarchies in defining, the universals in the soul are considerably changed and systematized.[61] If one adds to the account I have been giving the idea, laid out in most detail in *Metaphysics*, of natures in *things*, then the accounting for predication without any recourse to Platonic, separated, forms acquires considerable plausibility. And, of course, one has an Aristotle working out a complex philosophy which requires no violation of Zeno's Stricture.

III. PLOTINUS

(1) Porphyry's arrangement of the fifty-four treatises of Plotinus into six groups of nine in the *Enneads* offers formidable obstacles to understanding. It mixes treatises from different periods of composition. It mixes technical treatises with exhortations and homely advice. It divides and sometimes separates longer writings of Plotinus.[62] The writings themselves offer equally formidable obstacles. Some attack written and oral traditions of Gnostics and astrologers about whom our knowledge is imperfect or lacking. Some attack Stoic doctrines about which our knowledge is imperfect. Some exhibit excessively crude scientific understanding of natural processes. And most of the treatises, though indebted to Plato and the Platonic tradition, mix together the vocabularies of Plato, Aristotle, Stoics and others in a manner which makes detailed interpretation difficult. Some of the treatises, obviously responding to questions of a small group of disciples and other auditors, hint at or imply a more comprehensive philosophical framework than gets discussed.[63]

Even so and despite some additional problems with Plotinus' Greek, much of Plotinus' philosophy is written with reasonable clarity and precision. Given background in the later Platonic dialogues, the texts of Aristotle (in particular, the *Organon*), the Stoic literature, and (parts of) the great Commentary of Alexander Aphrodisias, it is reasonably clear what Plotinus is arguing for, what he is accepting, and what he is rejecting. Particularly relevant for the present purpose are the three treatises which begin *Ennead* VI and which were probably originally one essay, though divided into three by Porphyry. The first of these is a critical discussion of Aristotle's *Categories*, both the book and the list of ten (mentioned and discussed in the Aristotle section above). Though the argument(s) can hardly be detailed here, the purport of VI(1) is that Aristotle's location of reality with a capital 'R' in individual things or *οὐσίαι* (II(3) above) fails to take proper account of the priority of natures over natured things. But the criticism as such is not my concern. The real interest for our topic begins with the second treatise of *Ennead* VI in which Plotinus systematically lays out his account of the genera/species structuring of natures. I proceed to that account.

(2) Early in the treatise Plotinus insists upon taking *Being* (though others will be added) not merely as a genus but also as a *principle* (ἄρχη). His point in so doing to emphasize *Being*'s role as a source or, if you please, as a cause – and not merely as a classificatory entity. His next move is to introduce *Motion* as a genus and principle coordinate with *Being*, meaning by 'coordinate' a principle which, while neither a species of *Being* nor an independent genus, produces along with *Being* (and other coordinate genera) the species and sub-species of *Being*. *Rest* is introduced as also coordinate on the ground that the produced species and sub-species, as well as the original genera, stay fixed. *Difference* is yet a third coordinate genus-principle, introduced in recognition of a need for a differentiation of *Being* (the latter not being as such or in itself differentiated). Finally, *Same* is introduced as the fourth coordinate genus in recognition of a need for *Being*'s remaining the same through its speciation, as well as the various species and sub-species remaining the same through their respective sub-speciations.

The arguments given for this dividing of *Being* by means of coordinate principles make sense if one assumes with Plotinus that *Being* must be both *one* and *many*. It must be one both in itself, i.e. as *Being*, and also as the generic ingredient in all of its parts or species. That is, all of the species, sub-species, etc. are *beings*. Species of *Motion* would have to be *beings*, but species of *Being* may or may not be *motions*. And the same sort of argument holds for the other coordinate genera *vis a vis Being*. And this would seem to be compatible with requiring all four coordinate genera for the explicit series of species and sub-species of *Being*. So only *Being* divides. One must therefore assume with Plotinus that there can be equally ultimate and coordinate genera which (timelessly) produce a single descending series of species, sub-species, etc.

Given these assumptions, Plotinus' arguments in *Ennead* VI(2) invite comparison with those of Plato in *Sophist*, where Plato attempts to prove that *Being, Motion, Rest, Same*, and *Different* are all different natures, none being "reducible" to any or some of the others. They are, indeed, Plato's "greatest kinds" (*μεγίστα γένη*).[64] And they are Plotinus' greatest kinds. At two or three points in the treatise Plotinus says that, though the greatest kinds are separable for thought, they are not really separate – a point which is implied by my characterization of the need for all five in the first paragraph of this section.[65]

(3) It may be appropriate to mention three matters which, though not strictly germane to the topic of predication as such, may help the exposition and join it to some themes commonly invoked in discussions of Plotinus.

These are: the role of the One, the role of Soul, and the curious matter of Plotinus' characterization of the world of forms or natures as Intellect (νοῦς).

In our treatise (VI2) Plotinus claims that, though *Being* divides into species, *One* cannot. The insight which seems to lie behind this claim is that there are many kinds of *Being*, but there are not many kinds of *One*. An army, a horse, a thought, and an action are not ones of different kinds, though each of them is one. They are, however, beings of different kinds. *One*, so to speak, puts its imprint on everything which in any way *is*, but it does not itself divide into kinds. And *Being* itself (as well as each of the other coordinate genera) is one. They are not, however, different kinds of *One* or, rather, ones of different kinds. Since even the coordinate genera are each one, though not ones of different kinds, the *One* whose mark they bear must be "beyond *Being*"[66] or, what is more to the point, beyond any intelligible understanding by means of a genera/species hierarchy, i.e. by means of predication. Thus the logical or ontological point of the Plotinian insistence upon the "super-essential *One*".

In our treatise (VI(2) Plotinus starts talking about Soul at the same time as he talks about *Motion*. He does so, I believe, from the Platonic association of soul and motion,[67] more particularly, from the passage in *Sophist* in which the Eleatic Stranger demands the recognition of soul and life in *Being* and consequently of *Motion* as a form.[68] Plotinus evidently understands this as requiring a timeless motion in the form of speciation of Being (or exfoliation). In this way Plotinus is able to think of the world of forms as *alive*. But this leads us to the third matter, which is related.

Speciation or exfoliation of *Being* is, of course, a *timeless* articulation of *Being*. Plotinus evidently takes this living articulation as the very paradigm of *knowing*. It is, as it were, intellection itself. There is a sort of dualism of knower and known in the distinction between genus (coordinate genera?) and articulated species, though it is surely not the "standard" dualism between representation and represented or subject and object. Even so, I think Plotinus takes the latter to be a blurred reflection of the former and the former to be what the latter strives to be. The individual soul, i.e. the soul of a human body, while linked to a body and its sensations and while imperfectly aware of the "world", is implicated in the standard dualism; but, when fully come to itself, realizing its identity with Intellect-Soul, then knows in the exfoliation sense.[69] I shall say more on this matter in (7) below, but it is worth noting here that there is some intuitive ground for this rather exotic conception of knowledge. First, when two persons both believe or know an abstract truth like 'Two plus two is four', there is nothing in their separate knowings to separate them from one another, except, perhaps the number of the knowings, namely, two. Second, Plotinus is heir to a tradition (and it has a long subsequent history) which has it that genuine knowledge of principles is not arbitrarily thrust upon us, but rather the welcome and liberating condition of self-realization. There is in this the flavour of Spinoza's *amor intellectualis Dei.*[70]

Given *Being* and its four coordinate genera and the requirement of speciation or exfoliation, it is reasonably clear that Plotinus works rather in the manner of an ontologizing logician who thinks of the intelligible frame of things as produced from a small number of very simple logical principles *and* of the details of the universe as combinatorial outworkings of the intelligible frame. On any such scheme the outworkings must, of course, follow necessarily from the intelligible frame, leaving no place at all for contingency. And on any such scheme, if it purports to be complete in any significant way, there will be problems with space, time, and sensibility. Plotinus is such a necessitarian, and he relegates space, time, and sensibility to mere appearance. And, obviously, he thinks of what *we* might take to be empirical laws that they are, at best, rules of thumb which both reflect and obscure the real causal order. But we shall return a bit later to this last matter.

(4) Returning to predication matters as such, I think it should be obvious from what has been said concerning *Ennead* VI(2) that the predicational scheme for Plotinus' world of natures or the Intellect is the same as that found for Plato's unchanging world of forms. In I(7) above, it will be recalled, I talked of the species, *F*, as "*G*-ate" (i.e. as "containing" the genus, *G*) and of the genus, *G*, as "*F*-ated" (i.e. as exfoliated or exfoliating into the species, *F*), finding these simply two different "directional" ways of expressing a feature of the eternal genera/species structuring. Plotinus is fairly explicit about the two different directional ways. We have, of course, looked at the "downwards" exfoliation way. Plotinus also commonly speaks of an "upwards" way by means of the term, 'contemplation' (θεωρία).[71] Contemplation seems at once an expression of causal aetiology in higher genera and principles and a yearning or desire for that or those causal origins. Like the timeless "activity" of exfoliation, contemplation seems a timeless desire. But this is all part of Plotinus' thinking of the Intellect or world of natures as itself a living thing – though its life is totally free of time and its determinants.

The assignation of life to the world of natures should not occasion the ignoring of the strict and logical character of the genera/species series. If one thinks simply of predication within that structure, one may either think of predicating the species of the genus (and thus as contained in it) or of the genus as predicated of the species (and thus as containing it). And one may think of false predications as mixing genus/species sub-hierarchies and thus as necessarily false. And they would be necessarily false because the subject and predicate would name things which were different in a way which violates Zeno's Stricture. After all, a genus and a species in the same genus/species hierarchy are not relevantly different, and, though remaining strictly within the framework of natures (and not natur*ed* things), one may speak of their names as having wider or narrower scope.

(5) Plotinus' story of the sensible world is somewhat more complicated, has less evident textual foundation, and is thus open to greater interpretative controversy. Let me start with some relatively uncontroversial comments. First, Plotinus' account is, by his own insistence, modelled on Plato's *Timaeus*.[72] Second, he takes very seriously the claim that *Being* and its coordinate genera are principles and causes of everything which in any way *is* – including the sensible world. The latter, as some sort of continuation of the exfoliation of *Being*, must therefore be a completely articulate and causally determined ordering, subject to what was later called 'continuous creation'.[73] Third, the sensible world must contain what, to avoid linguistic over-commitment, I shall call the 'bottoming out' of the series of causes and effects descending from *Being*, and that requires the reaching of entities which are *effects only*. So the Plotinian system has a causal beginning in *Being* and a causal ending – though "creation" is continuous and causation unceasing. Fourth, Plotinus has use for the notion of pure or prime matter, but it is not, as such, the end of the causal chain. It is tempting, however, and, I think, true to say that Plotinus needs matter for theoretical closure to his causal system. Matter is, as it were, non-entity which is, nevertheless, affected (*not e*ffected!) by the last entities in the causal series. Fifth, Soul is given the job of artificer of the sensible world. But, as I noted in (3). above, Soul is not really a separate principle for Plotinus. Soul is the ultimate genera *in act*,[74] not merely for the production of other lower genera and species but also for the production of the sensible world.

It follows from all this, of course, that all *real* causation in in Plotinus' universe is *vertical*, i.e. timeless exfoliation of *Being*. Any regularity which one may observe to occur *in time*, even the regularity of the motions of the heavenly bodies, though it may well obtain, is *causally derivative*. If one thinks of the timeless exfoliation as "vertical" and causality requiring a temporal before and after as "horizontal", one can say quite generally that all horizontal causation is derivative. There is, of course, no problem about horizontal causation in Plotinus' Intellect for the good reason that causation "there" is strictly vertical. With respect to the sensible world, however, where there obviously is horizontal causation, Plotinus owes us an account which explains how horizontal causation can be understood as derivative.

(6) Perhaps the only way in which Plotinus can provide the required account has the exfoliation or speciation of the Intellect as not directly producing the sensible world but rather as productive of what I shall be calling a 'plan' for the sensible world. Intellect could then be thought of as containing the general principles whose realization in the sensible world is mediated by a total plan. I think that Plotinus goes this way and that he assigns the mediation task, as it were, to World Soul.[75] I think, however, that 'World Soul', like the earlier 'Soul', is not used to refer to anything different from the working out of *Being* (and the coordinate genera). By means of it, Plotinus is inviting our attention to the articulation of the timeless world process beyond what he calls 'Intellect.'[76]

Plotinus is, of course, familiar with Plato's mathematical bent – if you wish, Plato's Pythagoreanism. And he, like Plato, associates numbers and number-combinations with points, lines, planes, and solids. One may therefore think of "unrealized" spatial configurational patterns in the plan. These would obviously not be spatial any more than the Platonic Triangle Itself is triangular or a triangle. A "realization" (my term) of them in *matter* (of which more shortly) would, however, be spatial in the standard sense. A dense set of such unrealized configurational patterns in a plan would be such that, given realization, there would be a completely "filled" sensible world. This is, I believe, the picture Plotinus invites us to consider.[77]

One may, as well, think of a mathematical configuration which is itself a configuration of dense sets of configurational patterns – perhaps on the model of a reified algebraic formula for the series or odd or prime numbers. Here, however, one must think of the formula, so to speak, as completely spelled out, with every number "present". What is needed, of course, is the idea of a non-temporal but directional series such that a realization would yield not merely a spatially "filled" but a temporally "filled" sensible world. So we are being invited to consider what I have called a 'plan', which is in itself an incredibly complex combinatorial workout of the more general genera and species of the Intellect and which, on what I have called 'realization', is the sensible world in space and time.

In the plan, of course, are sufficient detail and complex inner structure to yield, on realization, earth, air, fire, and water as basic material constituents, the simpler and more complez organizations of these basic materials, the various bodies from the imperishable ones of the heavens to the perishable ones of earth, the sea below, and the air above, and so on. Plotinus actually thinks of the realization on the model of Plato's *Timaeus*. He therefore thinks of it as involving the vast sphere of the fixed stars, the several circular orbits, and as geocentric – a model which was given elegant mathematical detail by Claudius Ptolemy in the *Syntaxis* – the *Almagest*.[78] And Plotinus thinks of the realized plan as a "repeatable", as it were, running through a complete cycle with a Platonic "great year".[79] thus giving him a doctrine commonly thought of as "the eternal return".[80] Though there is some fascinating historical interest in these details, I think it is reasonably clear that they are not essential to the general scheme of taking vertical causation as real and horizontal causation as derivative.

(7) Obviously something more must be said concerning Soul and (prime) matter.

Soul appears to do double duty for Plotinus in what I have been calling the 'realization' of the plan. First, it is the active realization itself, producing the fully-fledged sensible world, indeed *continually* producing it. Second, it is, as it were divided up as the active principle in any number of organized bodies produced in the active realization (a dividing up which is, of course, prefigured in the plan). Some of these ensoul in highly rational fashion the heavenly bodies; many more ensoul the bodies of animals, most importantly though not exclusively, the bodies of human beings. The realization noted under 'First' really does produce the entire physical or material world with its entire "filling" of space and time. The "division" of Soul noted under 'Second' links Soul more or less immediately with a vast number of organized bodies – including yours and mine. But, of course, Soul is not really divided up, though the linkage with particular bodies and, especially, with sense presentations tends to produce that illusion.[81] And I shall turn to that matter in the next section.

But let us now turn to the matter of matter. Matter, on the scheme, is nothing at all, though it is necessary for the realization of the plan. Without "it" there would be no "bottom" to the exfoliation or, better, what I have been calling the 'plan' would be the end of the exfoliation. And Plotinus would have no means of accounting for what I have been calling 'realization'. And there would be no connection in the system to the experienced sensible world – the colors, sounds, odors, etc. of ordinary sense experience. Put rather bluntly, matter is required for the realization of that ordinary world out of the plan and the Intellect. Plotinus is ambivalent about both the sensible world and matter. Sometimes he writes as though the sensible world were an appropriate completion of the aboriginal exfoliation from *Being*. Sometimes he writes as though there were some sort of perverse necessity which got the great order of things beyond Intellect or Intellect and the plan.[82] At any rate, matter is invoked as the non-being whose receptivity to the being of the plan makes possible the sensible world.

(8) Some final comments concerning perception or sensation before turning directly to the matter of predications concerning the sensible world. Neither Plotinus nor, for that matter, any early Greek philosopher betrays any sort of Cartesian problem concerning the status and place of sensed colors, sounds, hards, softs, flavors, etc. Plato and Aristotle, e.g., address issues concerning their relativity to perceivers, but they seem to have no difficulty in treating "sensa" (i.e. the sensed colors, etc.) as belonging to the physical order (rather than the intelligible order). Roughly, they are treated as features of or states of physical things and sense organs of human bodies which those things and sense organs have only in an appropriate sort of causal interaction with one another.[83] They are, as it were, *emergent* physical features or states.

Plotinus insists in many places that Soul as such and individual souls are in no way passive and are thus in themselves strictly unaffected by physical changes in the bodies of which they are the souls. With respect to Soul generally, the point being made is obvious from all which has been said above; after all, Soul simply *is* the exfoliative production *in act*. Individual souls, however, since they are, as it were, "assigned" to individual bodies which have sense organs, may misconceive their situation. I shall try to explain how Plotinus accounts for this.

As we noted in (6) and (7) above, Soul produces the entire physical or sensible world both from the Intellect by producing the plan and from the plan by realizing the plan in matter, thus engendering the spatial and temporal world of sense. That world, as we have noted, is a *plenum*, filled with elements, bodies of various kinds, etc. Though it is the product of vertical causation, it is so organized as to make an Aristotelian – or, generally, a naturalistic – account of it at least *prima facie* plausible. (And, when larger issues are not are stake, Plotinus himself lapses into naturalistic vocabulary in describing a number of things, some of them rather technically, e.g. vision.)[84]

An individual soul, permeating a physical body, is, as it were, presented with vast numbers of sensa (including pleasures and pains) and, having no other obvious distractions, commonly does what soul does generally for Plotinus: it produces. It uses sensa to "make" an ordered, essentially Aristotelian, physical world, a world to which the categories are applicable. Or one could think of the matter, perhaps a bit more aptly for Plotinus, on the model of Plato's analogy of the line. The soul at first takes the presented sensa natively – as at the stage of *εἰκασία*. Confronted with sense variations which demand more sophisticated taking account of perspective, color variation, and the like, the soul moves to "making" three-dimensional things – at at the stage of *πίστις*.[85] With this sort of construction occurring, basic Aristotelianism seems natural enough.

(9) What this general picture points up is the possibility of construing Plotinus as having, for the sensible world, an account of predication which is essentially that of Aristotle. Indeed, in VI (3), in discussing the sensible world, Plotinus simply uses – with criticisms Aristotle would have understood – the Aristotelian predicational apparatus. Even so, despite Plotinus' aplomb in the use of Aristotelian styles of predication and naturalistic vocabulary, that usage is justified only by assumptions of the sort I have tried to explain in earlier sections. I cannot resist suggesting that Plotinus' use of that vocabulary is justified for much the same reason as Leibniz' use of standard physics vocabulary is justified, viz., that it is adequate enough for "well-founded phenomena".[86] And Plotinus' account of the causal aetiology of the sensible world gives it a status which is, without distortion, akin to that of Leibniz' well-founded phenomena.

But the soul which makes Plotinian progress gets beyond *πίστις* and recognizes the need for natures lying behind natur*ed* things, indeed the need for the Intellect as the timeless causal agency for the sensible world. And it discovers the resources for thinking that causal agency within itself. One may think of this as, to continue the reference to Plato's analogy of the line, reaching the stages of *διάνοια* and *νόησις*.[87] For Plotinus', if not for Plato's, purposes there is some intuitional appeal in the idea that, when we understand or know a general and timeless truth, our personal histories are irrelevant, and, in the understanding, those histories constitute only background clutter. Having gone through the discursive effort of making clear to oneself at least some of the genera/species hierarchies and thus grasped the general idea of such interrelationships, one is presumably in a position to have, as it were, an intuitive grasp of how the general principles work in the exfoliation. Once again, the personal history is irrelevant, and, as well, the subject/object distinction has no application. If there is some sort of representational scheme at work in the soul's use of sensa to represent external objects in the sensible world, there is no such need in the realm of Intellect. There is no gap between intellect and "individual" soul; in what I have called 'an intuitive grasp', one simply *is* the Intellect. There is a flavor here of the idea that, once having grasped a few physical principles, one thinks about and performs tasks in the world as though he/she were, as it were, *willing* those principles and their operation. After all, they are our means of understanding and coping with the world around us. Plotinus puts this sort of insight on a grand scale and thus, as noted earlier, anticipates (and prepares the way for) such remarkable ideas as Spinoza's *amor intellectualis Dei*. To understand Plotinus' world is to join in its constant creation. But, even in such grand-scale vision, Plotinus, in fear of incoherence, stays within the spirit at least of Zeno's Stricture.

Fortunately there is no need here to extend the grand-scale vision beyond *Being* and its coordinates.

IV. CODA

In discussing predication in three seminal figures in Greek philosophy, even trying to stay within the topic of conformity to Zeno's Stricture, I have found it expedient, if not necessary, to sketch and provide some detail of the first philosophy of all three. I think this says something about the centrality of *predication* as a philosophical issue or problem. One can hardly have a doctrine of predication without being prepared to show how that doctrine fits into a general rationale in terms of which we can, without incoherence, make up a world story and give structural sense to distinctions we think important. The contrast between Aristotle and Plotinus is, I think, especially instructive. The later history of Aristotelianism is reasonably well known. But the later history of Plotinian Neo-Platonism, though attended to for Medieval figures, in early modern continental rationalism (with its emphasis on vertical causation) is not much attended to. Though I have in the text alluded only to Spinoza, in a longer account I would include a number of figures and attempt to locate the *substance–attribute–mode–modification* pattern in the Plotinian tradition.

NOTES

1 *Parmenides*, 127E1–10 (my translation).

2 Parmenides's poem stays uniformly with the singular usage of "being" (*on*), thus contrasting remarkably with Zeno's supposition of *beings* (*ta onta*). Obviously, I am understanding Zeno's use of "*onta*" as allowing for the supposition of more than one entity, thus my use of "*A*", "*B*", etc. For the rest, the reconstruction of Zeno's argument depends upon the assumption that Socrates's reply to it, 128E5–130A2, is to the point. *That* reply consists, simply put, of pointing out that an individual may have shares of many different forms without requiring it (or anything else) to *be* both like and unlike itself. Indeed, it may have shares of The Like Itself and also The Unlike Itself. If the reply is pertinent, then Zeno's argument must require the assumption which I have put "*A* is *B*". But see below I. 2.

[3] 128C6–D7: "The truth is that these writings are simply support for the argument of Parmenides, coming to his aid against those who set upon him with caricatures to the the effect that, if one is, many laughable conclusions follow, as well as conclusions contradictory to his argument. This treatise of mine argues against those speaking in behalf of the many, and returns to them the same and more. For it undertakes to show that their supposition, it many are, must allow even more laughable conclusions than that of the one being, if one pursues it sufficiently" (my translation).
[4] *Sophist* 259B7–E5; *Philebus* 15D4–16A4.
[5] Strictly, Euthydemus and Dionysodorus are a comedy team because of their transparent playing with words and meanings. Their "expertise" is a standard eristic one: they will undertake to make any proposition plausible – including the contradictory of one made plausible a moment before. And Socrates gets them to admit the belief which is backdrop to the question of *Sophist*, namely, that no one believes falsely (the question being: "How is false belief possible?").
[6] *Phaedrus* 261D5–8: "Then can we fail to see that the Palamedes of Elea has an art of speaking, such that he can make the same things appear to his audience like and unlike, or one and many, or again at rest or in motion?" (Hackforth translation).
[7] By "peculiarity" I mean, of course, what was meant in the first comment above, namely, taking 'literally" the claim that *A* is *B* in contexts where identity cannot be meant but where the usage is standard.
[8] I shall try to show in the Plato section to follow that Plato has no place for and would be puzzled by the "bare" particulars of Russell and would be equally puzzled by the claim that "basic" sentences are one and all contingent, i.e. those of the form, "$F(x)$". And I can think of no Greek philosophical text which is illuminated by taking Platonic forms or natures as propositional functions and by taking Platonic individuals as non-connotative completions of such functions. I readily admit, however, that if what is at stake is the "structural" use of occasional Platonic or Aristotelian "insights", a philosopher using *Principia Mathematica* as a logical or ontological frame may profit from reading by Plato and Aristotle.
[9] "Name" or "*onoma*", as here used, is very much in the spirit of Aristotle's use of the term in *On Interpretation* where it has very much the sense of the modern "term".
[10] As, for example, the astonishing controversy which developed from Gregory Vlastos Third Man article. I think that Alexander Nehamas, in his 'Self-Predication and Plato's Theory of Forms' (*American Philosophical Quarterly* **16** (1979), 93–103), gives an essentially correct account of the matter.
[11] But only for the moment; the rehearsal of the arguments from the first part of *Parmenides* in (3). (immediately following in the text) makes clear what I take to be the "troublesome features".
[12] *Parmenides* 129C1–3: "If one could show the kinds and forms themselves as admitting opposites in and of themselves, it would be proper to be amazed." *Parmenides* 129E–130A2: "As I say, however, I would be much more impressed if one were to have this same difficulty – complete intercourse in and among themselves – as you encountered going through the visibles also exhibited in those which are apprehended by argument or reasoning".
[13] *Parmenides* 130C1–D9: "*Parmenides*. What of this? A form of man quite separated from us and form all others like us, a form of man, of fire, or of water? *Socrates*. I have often been in perplexity as to whether one must speak of them in this manner or in some

other way. *Parmenides*. And what about these, Socrates, ones which seem even laughable, such as hair, mud, dirt, and any other such unworthy and vile thing – are you perplexed as to whether one must say that there is a separated form for each of these also, one which is other than those we hold in our hands? *Socrates*. Not at all. What we see of them is just what they are. There is little incentive to believe that there is some form for each of these in that it would be strange. Even so, it has occasionally disturbed me that the same should not obtain for everything. But then, when I have taken this stance, I take flight, lest, falling into some abyss of prattling, I perish. So then, coming back to those which we just now said to have forms, I work away at various matters concerning them".

14 *Phaedo* 74A–B; 103B.

15 Though we shall return in the Plotinus section (III below) to the "greatest kinds" of *Sophist*. Plotinus takes them to be the "coordinate genera" of all of the kinds of Being.

16 As we shall see, III. below, Plotinus takes *Being*, *Motion*, *Rest*, *Same*, and *Different* not merely as genera but as *principles* (i.e. *constitutive* principles) of his *Intellect* or world of forms.

17 Despite the frequency of "exists" (and cognates of it) in translations of Plato (and Aristotle), there simply is no linguistic distinction in Attic Greek paralleling that between "is" and "exists" in modern English – to say nothing of philosophical use and interpretations of the modern English distinction. Needless to say, there is nothing in the ancient Greek texts which does the job of the backwards "E" of *Principia Mathematica* logic, whatever philosophical benefits there might be from construing *existence* as an operator.

18 *Philebus* 12C ff.

19 *Philebus* 16C5 ff.

20 *Phaedo* 100E; 105B–C. See also Gregory Vlastos, 'Reasons and Causes in the *Phaedo*', in *Plato I*, ed. by G. Vlastos, Garden City, NY, 1971, pp. 132–66 (originally in *The Philosophical Review* 78 (1969), 291–325).

21 I have in mind, of course, the elaborate attempts to define the sophist at the beginning of *Sophist* (218B–235D or thereabouts). *Statesman* is almost entirely devoted to a very complicated set of divisions.

22 *Parmenides* 142B1–145B5.

23 *Metaphysics M* and *N*; also the "forms as numbers" section of *A*, viz. 991b10–992b19.

24 *Metaphysics* VIII, 1045a 34ff. See Richard Rorty, 'Genus as Matter: A Reading of *Metaphysics* Z–H', in *Exegesis and Argument*, ed. by E. N. Lee, A. P. D. Mourelatos, and R. M. Rorty, Assen, The Netherlands, 1973, pp. 393–420.

25 I am alluding to *Timaeus* 50B–C, Plato's use of gold, which can be molded into a vast variety of shapes, as an image to help make clear his account of the *receptacle*.

26 *Republic* 510D–E, where Plato makes it quite clear that he thinks of geometers as needing diagrams and figures only as aids in fixing attention on the forms.

27 *Timaeus* 53A6–C3: "At first, however, they were all without reason and measure. But when the world began to get into order, fire and water and earth and air did indeed show faint traces of themselves, but were altogether in such a condition as one may expect to find wherever God is absent. Such, I say, being their nature, God now fashioned them by form and number. Let it be consistently maintained by us in all that we say that God made them as far as possible the fairest and best, out of things which were not

fair and good. And now I will endeavor to show you the disposition and generation of them by an unaccustomed argument which I am compelled to use. But I believe that you will be able to follow me, for your education has made you familiar with the methods of science." (Jowett translation). There follows in *Timaeus* the detailed description of the elementary geometrical solids (tetrahedra, cubes, octahedra, and icosahedra), constructed from half-equilateral and half isosceles right-angled triangles which give a mathematically intelligible structure to the elements of Plato's physical universe. That there are such structured things, themselves capable of being structured at the "bottom" of Plato's universe makes that universe capable of apprehension by way of the (mathematical) principles of structure or natures and gives them, *qua* structured, the "borrowed" names of the forms of natures. A case for this interpretation of *Timaeus* is made in my 'Knowledge and the Forms in the Later Platonic Dialogues', *Proceedings and Addresses of The American Philosophical Association* **51** (1978), 735–758; also my review of Gregory Vlastos, *Plato's Universe* (Seattle, 1975), *Journal of the History of Philosophy* **16** (1978), 99–103.

[28] In *Bertand Russell, Logic and Knowledge*, ed. R. D. Marsh (New York: Macmillan 1956), pp. 117–281.

[29] The claim in *Phaedo* 102C that it is not *qua* Socrates that Socrates is short (relative to Simmias) and that it is not *qua* Phaedo that Phaedo is tall (relative to Simmias) could have been made as well with "man" replacing "Socrates" and "Phaedo" respectively. At *Sophist* 262 where the distinction between *onomata* (nouns) and *remata* (verbs) is made to indicate that the simplest sentence needs one of each the Eleatic Stranger first uses "(a) man" and then "Theaetetus" to illustrate the former (*onomata*). Neither in the *Phaedo*, the *Sophist*, nor elsewhere does Plato betray any sign of giving proper names a special status not accorded to definite descriptions.

[30] "Knowledge and the Forms in the Later Platonic Dialogues", cit. in Note 27 above.

[31] *Parmenides* 130A8–B5: "How admirably carried away you are by your zest for arguments, Socrates. Tell me, have you yourself made this division you describe, separating, on the one side, the forms themselves, on the other, those having shares of them? And does it seem to you that there is Likeness itself quite separated from the likeness we have in us, and also one, many, and all the others of which you just now heard Zeno speak?"

[32] For *Parmenides*, see Note 22 above; for *Timaeus*, Note 27 above; for *Philebus*, 24B–27D; for *Metaphysics*, especially Books *M* and *N*. There is also considerable evidence in connection with Plato's famous lecture on the Good, evidence which, to my knowledge, is best marshalled and explained by Kenneth Sayre in a book ms. on the later Platonic ontology. I hope that his ms. is soon published.

[33] I think that it is a mistake to treat the mathematical bodies of *Timaeus* as *atoms* and to assume as Vlastos does (in *Plato's Universe*, *loc. cit.*), that either the triangles or the solids are modelled on the atoms of Leucippus or Democritus. What Plato needs for there to be an intelligible material world are elements which admit of mathematical characterization and understanding and whose combinations admit of the same. Though the latter part of *Timaeus* does, indeed (in reference to *aisthesis*), speak of the tearing produced by tetrahedra and the smooth surface of icosahedra, the work of the demiurge at 53 is to "fashion" the "traces" of orderless earth, air, fire, and water "by form and number". The point of such introduction is obviously to produce a *copy* of the world of forms which can be understood by means of the forms and their structure. There

simply is no such motivation in the fragments of the Greek atomists. Indeed, it is not even evident in Lucretius' poem. And Plato's account of *aistheta* is that of sensibles arising from the interaction of sense organ and material things, where the determinant on the side of the things is surely the mathematical structure of the things or their surfaces or the reflected fire. This sort of sophistication is entirely lacking in the atomist or Epicurean account in terms of "films" constantly flying off things and being squeezed by the medium in order to fit into the eye. Plato's account is compatible with and probably provides the incentive for geometric optics of which the first serious account is highly Platonistic and is in Euclid.

[34] *Timaeus* 37D.

[35] See especially the discussion of *opposites*, in *Categories*, chapter 10. "Socrates is well" and "Socrates is ill" may both be false (if, e.g. Socrates is a stone); but, if Socrates *is dektikos* (i.e. admits) of health or illness, then one must be true and the other false.

[36] 'Knowledge and the Forms, etc.', *loc. cit.* pp. 738 ff.

[37] He comes close in *Statesman* in sections where the Eleatic Stranger lectures Young Socrates on skipping steps in a division or on choosing an inappropriate difference. But, of course, some of the criticisms of attempts at definition in earlier dialogues address the matter of correct or appropriate definitional form.

[38] *Sophist*, Chapter 42.

[39] See Note 35. Roughly, contraries are species under a common genus; contradictories consist of a term and its negation (without for this purpose worrying about negations outside of sentences). From the truth of "*A* is not *B*" nothing follows concerning what *A* is. But from "Socrates is not ill", given that Socrates is appropriately *dektikos*, it does follow that he is well.

[40] *Sophist*, 263A–B..

[41] *Sophist*, 258B–C.

[42] *Categories* 1b25–2a4.

[43] *Metaphysics* IV, ch. 2; V, 1017a, 23 ff.; *Nichomachean Ethics* I, 1096a, 23–33.

[44] 1a20 ff.

[45] See Note 43.

[46] In the same manner as noted above. It will have been noticed that, in effect, I have been offering a rationale for *not* attributing to Aristotle the doctrine of "dependent particulars". By this I mean the rather confusing doctrine that there are "independent particulars" ("substances") like trees, men, and rocks, and also "dependent particulars" like the green in this tree, the white in Socrates, and the hardness of this stone. Stated in just this manner, perhaps, with the addition of the claim that dependent particulars could not exist independently, I do not understand the doctrine at all and am reluctant to attribute it to Aristotle. I think that Aristotle's doctrine *is*, roughly, that, if an *ousia* or *thing* term applies or holds for something, some (at least one) non-*ousia* term will also apply to that same thing. And, generally, no non-*ousia* term will hold for anything unless an *ousia* term will also hold for that same thing. Lying behind this claim is, I believe, something like the doctrine I attributed to Plato in Part I, namely, that there are subordinate structures which structured things have which other structured things of the same kind may have and which structured things even of different kinds may have. As animals of a certain kind may all be the same color, and many other things of quite different kinds may have that very same color. But such subordinate structures could hardly be thought of as "*dependent particulars*". Indeed, I think that this is Aristotle's use of 'in'

for "in" predication. He writes: "By in a subject I mean what is in something *not* as a part and what cannot obtain apart from that in which it is" (*Categories* la 24–46).
47 The term has several meanings. A standard one, used with the dative, is "belong to", "fall to", "accrue to" (Liddell and Scott). My guess is that it is this sense which Aristotle exploits in his technical usage – thus a term *holds for*, i.e. belongs to, falls to, another.
48 *On Interpretation*, Chapter 10.
49 In *Posterior Analytics* II. 19, in particular, he speaks of the "apprehension of middleds" and characterizes *episteme* (usually translated as "scientific knowledge") as "middled knowledge".
50 See especially *Posterior Analytics* II, 96b 30ff, where the prime concern is getting differentiai in the proper order.
51 *Posterior Analytics*, II, Chs, 5, 13, 14.
52 See Note 46. Throughout, I have been talking about Aristotle's doctrine of "abstraction" as conjuring no "abstract entities", but rather as allowing for different ways of referring to (and making ascriptions for) things. Thus, the ability to refer to a certain thing as "this cube" (or "this cubical") is not an ability to refer to an abstraction; it is rather an ability to refer to a thing *qua* shaped. And, of course, the shape of, say, a die is not a part of the die in the standard sense of "part". Even so, given the legitimacy of reference by shape, terms, or concepts and the employment of *said of* predication for such terms, the possibility of geometry without *separated* geometrical objects seems at least *prima facie* assured.
53 See especially chapters 6 and 7 of *Posterior Analytics* II. There (at 92b 32–34), as elsewhere, Aristotle worries that, taking definition merely as showing "what something is or what the name signifies", even the *Iliad* will turn out to be a definition.
54 *Posterior Analytics* 83a31–35. Also, 77a5–10. Ross thinks the latter passage belongs with the former as it seems out of place in the first part of the *Posterior Analytics*.
55 *Posterior Analytics* II. 19. 100a6–7.
56 *Ibid*., 100a4. See my '*Physics* I: Sense Universals, Principles, Multiplicity, and Motion", in *Motion and Time, Space and Matter*, ed. by Peter Machamer and Robert Turnbull (Ohio: Ohio State University Press, 1976), pp. 28–56, for a detailed exposition of *Posterior Analytics* II. 19.
57 *Metaphysics* I, 981a16–17; also *Nichomachean Ethics*, V., Ch. 3.
58 The most interesting text is *Posterior Analytics* II. 19. 100a15–b2: "When one of many undiscriminateds halts, the first universal is in the soul; for, though one perceives the individual, the perception is of-the-universal, for example, of-man, but not of-Callias-man." In this rendering, the perception is *classified* by the genitive, it is an *of-man perception*, using the universal in the soul to apprehend Callias as a man. My contention is, of course, that knowledge states generally for Aristotle are "of-the universal" states – not Russellian acquaintances with abstract entities.
59 *On Interpretation* 16a4–9: "Those in the voice (i.e. spoken words) are signs of states (*pathematon*) in the soul, and those written of those in the voice. And as writing is not the same for all [human beings], neither are vocables. But those of which there are primarily signs. i.e. the states of the soul, are the same for all, as are also those of which they are likenesses, i.e. the things themselves."
60 *On the Soul* II. 2. and 3.
61 See *Physics* I. 1. Especially 184a17–b10: 'The natural way to proceed is from distinctions which are more intelligible and more lucid to us to those which are naturally

more lucid and intelligible. For those which are intelligible to us and those which are absolutely intelligible are not the same. We must, therefore, advance along the route from distinctions which are naturally less lucid but more lucid to us to those which are naturally both more lucid and more intelligible. At first more or less compounded and confused things are clear and lucid to us. Later – from these – the elements come to be intelligible, and the principles separate them out. It is therefore necessary to advance from universals to their constituents. For, as far as sense perception is concerned, the whole is more intelligible, and the universal is a certain kind of whole. For the universal encompasses many – so to speak – parts. The same sort of move holds also for terms in relation to the formula. For here also a whole signifies indiscriminately ("circle" for example); but its definition separates out the constituents. And children at first call all men fathers and all women mothers, but later distinguish each of them."

62 See Porphyry's 'On the Life of Plotinus and the Order of his Books', translated by A. H. Armstrong in *Plotinus I*, the first volume of the *Works of Plotinus* in the Loeb Classical Library (Cambridge, 1966), pp. 3–85. See also Armstrong's Preface in the same volume and also his introduction to each of the treatises. Unfortunately, though Armstrong has completed the translation and introductions, only three of the six volumes have appeared.

63 See especially H.-R. Schwyzer, 'Plotin', in Paulys, *Realencyclopadie der klassischen Alterumswissenschaft*, Band XXI, 1951, col. 471–592.

64 *Sophist*, Ch. 40.

65 *Ennead* VI. 2. 2. 10–14: "If this last is the truth, we have here not merely genera, but genera which are at the same time principles of Being. They are genera because they have subordinates – other genera, and successively species and individuals; they are also principles, since from this plurality Being takes its rise, constituted in its entirety from these its elements" (McKenna translation). Also VI. 2. 7. 16–20: "Being, then, containing many species, is but one genus. Motion, however, is to be classed as neither a subordinate nor a superordinate of Being but as its concomitant; for we have not found Being serving as substrate to Motion. Motion is Being's Act; neither is separated from the other except in thought; the two natures are one; for Being is inevitably actual, not potential" (McKenna translation). Line references are from the Henry–Schwyzer text.

66 *Ennead* VI. 2. 3. 6–10. "It is indeed external: the One is beyond; it cannot, therefore, be included among the genera; it is the [transcendent] source, while they stand side by side as genera. Yet surely the one must somehow be included [among the genera]? No: it is the Existents we are investigating, not that which is beyond Existence" (McKenna translation). Though some other issues are at stake, I would at least amend McKenna's translation to substitute "being(s)" for "existence" or "existents".

67 See Note 65.

68 *Sophist* 248E–249D: "*Str*. But, in heaven's name, shall we really be so easily persuaded that motion, life, soul, and mind are not present in absolute being – that it neither lives nor thinks, but awful and holy, incapable of recognition, fixed and unmoved, it *is*? *The*. That would be a monstrous concession for us to make, Stranger! *Str*. But, if it is capable of cognition, shall we say it lacks life? *The*. How could we? *Str*. But if we say that both of these are in it, shall we then go on to say that it does not have them in a soul? *The*. How else could it have them? *Str*. But if it assuredly has cognition, life, and soul, will it – though remaining ensouled – be at rest and immovable? *The*.

That would be quite unreasonable. *Str.* So then it must be conceded that the moved and the moving are *beings*. *The.* What else could be concluded? *Str.* And so it follows, Theaetetus, that, if *beings* are immovable, there is no cognition anywhere, in anything or of anything. *The.* Exactly so. *Str.* And, on the other hand, if we allow everything to be moved or to move, cognition will – even on that account – be taken away from *beings*. *The.* How so? *Str.* Does it strike you that apart from *rest* anything which stays the same while it is about or of the same thing could ever come to be? *The.* It could not. *Str.* Well, then, without motion and rest can you see cognition among *beings* or coming to be anywhere? *The.* Not in the least. *Str.* Then we must do battle – using every argument – against that man who does away with knowledge, understanding, or cognition and then dogmatizes about anything whatsoever. *The.* Indeed. *Str.* For the philosopher who holds all these in high esteem, it is then imperative not to follow the partisans of the forms (whether they hold to one form to many forms) who insist thal all is at rest and to refuse to listen to those who set being in constant motion. He must, rather, like an importunate child begging for *both*, say, 'Whatever moves and whatever is moved – being and all encompass *both*.' "

69 As we shall see later, the active soul [*pleonasm*!], using the sense impressions of the body, represents an essentially Aristotelian world (a world for which, as we shall see, Plotinus thinks there is adequate ontological ground – though neither it nor any part of it *is* the ground), in so doing, exercising its innate conceptual abilities. I think that Plotinus thinks of such representation as a making or doing which is an appropriate, if low-grade, activity of the soul. The realization that the Aristotelian world is derivative from the great and constant act of *Nous* or intellect is, for Plotinus, all one with the recognition that the representing activity is inseparable from the external exfoliating activity of soul. But see Section 7 below.

70 But this is a general feature of seventeenth-century continental rationalism, though, unlike Spinoza and like medieval neo-Platonists, there is much ingenuity exercised to keep Creator and created separate.

71 Though the term is also used for the exfoliating activity of Intellect, as in VI. 2. 6.

72 As in *Ennead* II. (1).

73 I have in mind, of course, the standard doctrine of thirteenth-century Christian philosophers, echoed in Dante's *Divine Comedy*, and later by Descartes and even Newton.

74 *Ennead* VI. 2. 6–7.

75 Again, see *Ennead* II. (1)., also II. (9)., especially section (7). and (8)., where the *Timaeus* account is defended against the Gnostics.

76 See especially II. (9). Section 8.: "To ask why Soul made the universe is like asking why there is a soul and why the Maker makes. First, it is the question of people who assume a beginning of that which always is: then they think that the cause of the making was a being who turned from one thing to another and changed .... For it is not right to disapprove the management of the All, first of all because it manifests the greatness of the intelligible nature" (Armstrong translation).

77 See especially VI. (4). From Section 5: "We are agreed that diversity within the Authentic depends not upon spatial separation but sheerly upon differentiation; all Being, despite this plurality, is a unity still; 'Being neighbors Being'; 'all holds together'; and thus the Intellectual-Principle [which is Being and the Beings] remains an integral, multiple by differentiai, not by spatial distinction. Souls too? Souls too. That principle distributed over material masses we hold to be in its own nature incapable of distribu-

tion; the magnitude belongs to the masses; when this soul-principle enters into them – or rather they into it – it is thought of as distributable only because, within the discrimination of the corporeal, the animating force is to be recognized at any and every point. For Soul is not articulated, section of Soul to section of body; there is integral omnipresence manifesting the unity of that principle, its veritable partlessness. Now as in Soul unity does not debar variety, so with Being and the Beings; in that order multiplicity does not conflict with unity. Multiplicity. This is not due to the need of flooding the universe with life; nor is the extension of the corporeal the cause of the multiplicity of souls; before body existed, Soul was one and many; the many souls fore-existed in the All not potentially but each effectively; that one collective Soul is no bar to the variety; the variety does not abrogate the unity; the souls are apart without partition, present each to all as never having been set in opposition; they are no more hedged off by boundaries than are the multiple items of knowledge in one mind; the one Soul so exists as to include all souls; the nature of such a principle must be utterly free of boundary" (McKenna translation).

78 Despite Porphyry's claim that Plotinus had studied astronomy (*Life*, 15.22 ff.), albeit "not very mathematically", I can find no evidence that Plotinus was familiar with the *Almagest*. Nor does his refutation of astrology in *Ennead* II. (3). betray any astronomical knowledge beyond Plato's *Timaeus*.

79 That is, for Plato's astronomy in the *Timaeus,* the elapsed time between a particular positioning of *all* of the heavenly bodies and their return to precisely that positioning. *Timaeus* 39D.

80 See especially *Ennead* V(7)., the last few lines of section 2: "Still, admitting the diversity of the Reason Principles, why need there be as many as there are men born in each period, once it is granted that different beings may take external manifestation under the presence of the same principles? Under the presence of all: agreed: but with the dominance of the very same? That is still open to question. May we not take it that there may be identical reproduction from one Period to another but not in the same Period?" (McKenna translation).

81 See Notes 69 and 77 above.

82 A. H. Armstrong, in his Preface to Vol. 1 of the *Plotinus* volumes in the Loeb Classical Library, writes (pp. xix–xx): "When Plotinus concentrates his mind on the inferiority of even this derived reality to its source, any sort of multiplicity to the pure unity to which he aspires, then he comes to think of its production as unfortunate even though necessary, and of the will to separate existence of the intellect and Soul as a sort of illegitimate self-assertion."

83 The use of "mind" as a translation for "*nous*" is, of course, misleading, but it does no harm if one is not tempted thereby to some sort of Cartesianism. The Platonic tradition treats *nous* as an ability (or set of abilities) of soul, not as itself a thing. No Platonist would dream of putting sensations, the results of causal interactions of the physical environment with sense organs, "in the mind" nor for that matter "in" the soul. It is tempting to read Plotinus' taking the soul's appropriation of sensations to represent a world in space and time as inviting the classical attack of skepticism and thus providing the rationale for turning "inward" to get a firm grasp of the principles of "being". As my former student, Michael Wagner, showed me, this is precisely what St. Augustine does in his treatise against the Academic Skeptics. I hope that Wagner's paper on this topic is published soon.

I can find no evidence in Plotinus (or, for that matter, in Plato or Aristotle) of the modern philosophical worry about the "place" of color and other "sense qualities" in an ontology. This is not to say that he ought not to have been worried; there is, after all, a considerable gap between what I have been calling the "plan" and its realization.

[84] See, in particular, *Ennead* II (8), 'How Distant Object Appear Small'.

[85] The doubly-divided line of *Republic* VI, it will be recalled, has in the lower part of main division *eikasia* and *pistis* and in the upper part of the main division *dianoia* and *noesis*. I am, of course, thinking of *eikasia* as extremely naive perceptual acceptance (to use Price's term), thus being taken in by perspective, shadows and other color variations in different lighting, and the like. *Pistis* is full-blooded three-dimensional perception with correction for perspective and the like. The move from naive perceptual acceptance invites a reality appearance distinction which both Plato and Plotinus exploit by turning to principles of mathematics (generally, structure) rather than to the "real" in the material world. Better, they understand what is "real" in the material world only by way of principles of structure. For more detail on the doubly-divided line see my 'The Role of the "Special Sensibles" in the Perceptual Theories of Plato and Aristotle', in Peter Machamer and Robert Turnbull (eds.), *Studies in Perception* (Ohio: Ohio State University Press, 1978), pp. 3–26.

[86] As has been occasionally noted, very lucidly by Robert M. Adams in this volume, Leibniz's *phaenomena bene fundata* are not the sensa or sense-data of Russell and the Vienna Positivists but rather the physical objects in space and time of classical physics.

[87] In invoking the analogy of the line, I am not attempting to suggest that Plotinus uses Plato's distinction between *dianoia* as knowledge of "illustrable" forms (as that of the geometer who can use diagrams) and *noesis* as knowledge of more basic, non-illustrable forms.

FRANK A. LEWIS

FORM AND PREDICATION IN ARISTOTLE'S *METAPHYSICS*

One of the perennial questions in Aristotle's metaphysics concerns the transition from the ontology of the earlier, logical works (the *Organon*), where the individual is primary substance, to the central books of the *Metaphysics*, where pride of place, and the term 'primary substance' now belong to form. The difference between the two viewpoints is centered around the different treatment each gives to the individual. In Aristotle's early theory in the *Categories*, the criteria for determining what is most of all substance are all tied to two canonical forms of predication, relating a subject to its kinds (including, where the subject is a substance, to its substantial kinds), or in the case of a substance, to its accidents. But this analysis offers no technique for inquiring after the inner structure, if any, of the subjects it identifies as basic: individual substances such as the individual man or the individual horse (*Categories* 2, lb4–5). In the *Metaphysics*, the individual substance is treated very differently. An individual substance is an organized, structured entity of a certain sort, and is analyzed accordingly as a compound of matter and form. And of the three, matter, form, and compound, *form* is now primary substance.[1] So a primary substance may *have* a subject or substratum, but it is not itself a compound of form and substratum: no primary substance is "spoken of by way of one thing's being in another, i.e. in something which is its subject as matter" (Z11, 1037b1–4).

Accounts of the progression from the metaphysics of the *Organon* to that of the later works frequently focus on the entry of matter. Important as matter is, however, the mature notion of form which accompanies it is no less novel or complex. The word εἶδος is one of the more overworked in Aristotle's metaphysics. Aristotle can use it for Platonic forms, and for one kind of secondary substance (the species) in the *Categories*, or even, in the biological works, when no distinction is being made between εἶδος and γένος, for any of the kinds under which an individual falls.[2] In the *Metaphysics*, he uses the same word for *form*. Form is, in some sense, a *constituent in* the individual substance, which is thus a *compound of* matter and form. This view of individual substance as a compound dominates Z8, for example, where Aristotle argues that everything that comes to be must be a compound of form and matter. The craftsman, for example, who makes the bronze round

James Bogen and James E. McGuire (eds.), How Things Are, 59–83.

makes neither the bronze nor the round, but something else, viz. this form in something else (1033a32–34, cf. b8–10, 18–19). Again,

what generates is sufficient for the making, and cause of the form's being in the matter. But once we have the whole, such-and-such a form in these flesh and bones, this is Callias or Socrates; and they are different on account of their matter (for that is different), but the same in form (for the form is indivisible). (Z8, 1034a4–8, cf. Mansion, 1971. My translation borrows a phrase from an unpublished version by Montgomery Furth.)

In the same vein, Aristotle uses the phrases *τό σύνολον*, "the composite", or *ἡ σύνολη ούσία*, "the composite substance", to denote the compound of matter and form.[3] The concrete substance can also be described as what is "out of" matter and from, *τὸ ἐκ τούτων*,[4] or as a "this-in-this", *τόδε ἐν τῳδε*,[5] this form in this matter. Hence too the frequent comparisons between natural substances and Aristotle's favourite example, snub nose.[6] But how are we to understand the notion of *compound* that is involved in this language? Sometimes, Aristotle speaks of form as straightforwardly an ingredient or "element" (*στοιχεῖον*) in the compound substance (L4, 1070b10ff). Elsewhere, however, he protests against this view, and calls form a "principle" (*ἀρχή*), as distinct from an element (Z17, 1041b11ff, H3, 1043b4ff [Ross' text]). In this same spirit, he calls matter a *part* of the compound substance (Z10, 1035a3ff, cf. D25, 1023b19–22), but is not apparently willing to talk of form as such as a part. So the precise way in which the concrete substance is a compound of form and matter remains mysterious.

The mystery, though deep, is not perhaps unique to Aristotle. As Lukasiewicz suggested some time ago (Lukasiewicz, 1953, pp. 80–82), Aristotelian form can be regarded as in some respects like a function from matter to a compound, and in the favoured cases, to a compound which is also a substance.[7] If we interpret functions in a Fregean style, similarities open up between Aristotle's views and those of Frege. For Aristotle, matter is itself a material object, but one that lacks the right kind of unity to count as a substance (Z16, 1040b8–10, Z17, 1041b11ff, H3, 1044a2–9, cf. H6). Form, meanwhile, is not (or at least, on most accounts is not)[8] itself a material object, just as for Frege a function is non-material. But in much the way that for Frege a function is "unsaturated", so for Aristotle form is of such a nature[9] as to combine with one material object, namely matter, to make up a second object, which under the right circumstances will be a substance. Now Frege holds that there are functions whose value for an appropriate material object as argument is itself a material object (for example, the function, *the capital of x*). So the structure of a composite substance, which is the compound of a form and a (portion of) matter, can be much like the result of "saturating" a Fregean function with an object as its argument, and the mystery of how one material object can combine with a non-material entity to compose a second material object, is in some respects at least the same in both cases.[10]

In this account, compounding for Aristotle is the counterpart of the operation of application of function to argument in Frege's scheme. If we follow out this comparison, it is perhaps best to regard the operation of compounding as a primitive notion of Aristotle's theory, so that form and matter are constituents of the compound substance, but in a way that defies ordinary comparison or analysis.[11] This way of treating the notion has some support in Aristotle's account in the *Categories*, where he says that non-substances are "present in" individual substances (so that there exist compounds of a substance with an accident), yet immediately warns that they are "in something, *not as a part*" (*Categories* 2, 1a24–25, 5, 3a31–32). Again, in discussing the problem of the "unity of definition" in *Metaphysics* H6, Aristotle says:

> But if, as we say, the one is matter, and the other form, and the one ⟨is⟩ potentially, the other actually, the object of our inquiry would no longer seem to be a puzzle . . . The difficulty disappears, because the one is matter, the other form. What, then, causes this – that which was potentially to be actually – except, in the case of things which are generated, the agent? For there is no other cause of the potential sphere's becoming actually a sphere, but this was the essence of either. . . . But as has been said, the proximate matter and the form are the same and one, the one potentially, and the other actually. Therefore it is like asking what in general is the cause of unity and of a thing's being one; for each thing is a kind of unity, and the potential and the actual are somehow one. Therefore there is no other cause here unless there is something which caused the movement from potency into actuality. (H6, 1045a23–25, 30–33, b17–22, after Ross' translation, cf. L10, 1075b34–37, *De An*. B1, 412b6–9.)

Aristotle here should be taken as saying that there is no *general* account of why matter and form unite to make up a compound, or why the potential so-and-so becomes an actual so-and-so, beyond reference to the existence of some moving cause or other, or the removal of some hindering factor. There are only particular explanations, which refer to the nature of a particular kind and the dispositions to behaviour that constitute that kind. In the same chapter, he dismisses the metaphors of Plato and others as attempts to supply an explanation of a sort that is neither needed nor possible:

> Owing to the difficulty about unity some speak of 'participation', and raise the question, what is the cause of participation and what is it to participate; and others speak of 'communion', as Lycophron says knowledge is a communion of knowing with the soul;

and others say life is a 'composition' or 'connexion' of soul with body. Yet the same account applies to all cases; for being healthy, too, will on this showing be either a 'communion' or a 'connexion' or a 'composition' of soul and health, and the fact that the bronze is a triangle will be a 'composition' of bronze and triangle, and the fact that a thing is white will be a 'composition' of surface and whiteness. (H6, 1045b7–16, Ross' translation.)

Presumably, the last part of this passage means to ridicule the idea that there is anything *explanatory* in any of this talk.[12] A striking difference between Aristotle and Frege, then, is Aristotle's rejection of Frege's tolerant attitude towards the use of metaphor to help us understand the primitive notions in his theory (cf. note 11 above).

II

In the *Metaphysics*, as we have seen, the individual substance is now in some sense a compound of form and matter, and has been supplanted in the role of primary substance by form. Now, it is a key feature of the *Categories* that a primary substance in the *Categories* sense, that is, an individual substance, is above all a subject to both kinds and accidents. It both has things *in* it (its accidents), and has things *said of it* (its kinds). What becomes of this notion of membership in a kind in the *Metaphysics*? And how is the notion related to the new notion of form?

These questions can be sharpened if we look more closely at Aristotle's views on predication as they develop from the *Categories* to the *Metaphysics*. In the *Categories*, we see Aristotle working out the consequences of a basic condition on substancehood:

Further, it is *because the primary substances are subjects for everything else* that they are called substances most strictly. (5, 2b37ff, Ackrill's translation, my italics; cf. 5, 2allff, a34ff, b15ff)

It is this condition that helps Aristotle conclude that the individual substance, Socrates or Callias, is primary among the different grades of substance. Socrates, for example, is a subject to accidents. He is also a member of various kinds, *man* and *animal*, for example. But he is not himself a kind relative to anything else: Socrates is a *this*, one in number, and indivisible (*Categories* 5, 3b10–13). The secondary substances, meanwhile, items like *man* or *animal*, are both the subjects for some kind-predications, but also themselves the kinds *of* other things of which they are predicated. So they are substances to only a lesser degree.

Now the predicability condition on substancehood is only as good as our grip on the notion of predication. The account of predication Aristotle develops in the *Analytics* is, in turn, itself parasitic on the ontological distinctions already drawn in the *Categories*. In the *Analytics*, Aristotle lays down that if a sentence expresses a genuine predication, the subject-entity must be a genuine subject, that is, a substance, and not a combination of a substance with an accident. In all Aristotle's examples in this discussion, a sentence of the form, 'The φ is (a) ψ', where the φ is a compound of a substance with an accident, must be recast as: 'There is a substance a which happens to be φ and is (a) ψ' (cf. *An. Po.* 73b6, 7, 83a2, 10ff, *An. Pr.* 43a35). Recasting the sentence in this form makes clear the role of the substance a as the subject for the accident φ.

This monolithic view of the subject of predication from the *Organon* is replaced in the *Metaphysics* by a two-part system, which recognizes *two* kinds of subject. The principal texts are these:

For, the rest are predicated of the substance, but it [is predicated of] the matter. (Z3, 1029a23–24.)[13]

[The subject] is subject (ὑπόκειται) in two ways, either being a this, as the animal [is subject] to its affections, or as the matter [is subject] to the actuality. (Z13, 1038b4–6.)

For, that of which and the subject (τὸ καθ' οὗ καὶ τό ὑποκείμενον) differ in this, by being either a this or not being [a this]: for example, what is subject to the affections is [a] man, both [his] body and [his] soul, and the affection in the musical or [the] pale.... In all cases of this sort, the subject (τὸ ἔσχατον) [is] substance; but in the cases that are not like this, but what is predicated is a certain form and a this (εἶδός τι καὶ τόδε τι τό κατηγορούμενον), the subject (τὸ ἔσχατον) is matter and matter-like substance. (Θ7, 1049a27ff)[14]

In these passages, the relations "x is predicated of y" and "x is subject to y" are the converses of each other: that is, x is predicated of y if and only if y is subject to x. So there are now *two* kinds of predication, and *two* kinds of subject. Thus we have

(Al) x is predicated of y (y is subject to x) only if *either* x is an accident and y is an individual substance, *or* x is a form and y is a parcel of matter.

(On this latter disjunct: x = οὐσία, Z3, 1029a23–24, x = ἐντελέχεια, Z13, 1038b6, x = ἐνέργεια, H2, 1043a6, x = εἶδος τι καὶ τόδε τι, Θ7, 1049a35.)

In the *Categories*, as we have seen, there is a single basic subject of predication, while in the *Metaphysics* there are two quite distinct kinds of subject. There is also a second difference between the notion of predication in the *Metaphysics*, and the notion from the *Organon* which it replaces. In the

Organon, Aristotle appears to move back and forth freely between a linguistic and a metaphysical notion of predication. Predication is in some places a relation one or both of whose relata are linguistic entities (call this, "linguistic predication"), but in other places, both relata are metaphysical items ("metaphysical predication").[15] In the *Metaphysics* passages, by contrast, predication is represented as exclusively a metaphysical relation between non-linguistic entities.

One result of this shift to an exclusively metaphysical notion of predication is that in the *Metaphysics*, it is no longer obvious in all cases what consequences Aristotle's views on (metaphysical) predication have for language. In the *Categories*, the linguistic implications of metaphysical predication are evident. Metaphysical predication is the union of the *said-of* relation (a relation between higher and lower members of the same category) and the relation of inherence (a relation between a member of a non-substance category and a substance). In Chapter 2 of the *Categories*, Aristotle uses these two sub relations to draw a fundamental ontological distinction between four different kinds of "things that are" (*τῶν ὄντων*, 1a20). The two relations also have immediate and obvious linguistic corollaries in sentences that relate an individual (a substance or an non-substance) to its kinds, or which assign an accident from a non-substance category to a substance as its subject.

In the *Metaphysics*, by contrast, it is no longer obvious what connections there are, if any, between the different cases of metaphysical predication, and linguistic predication. One case envisaged in the *Metaphysics* is clearly continuous with part of the *Categories* account. In the *Categories* and *Metaphysics* alike, to the case where an accident is (metaphysically) predicated of a substance, there corresponds the linguistic predication, for example,

(1) Socrates is pale.

But the remaining case in the *Metaphysics*, where form is (metaphysically) predicated of matter, seems to lack any ready or natural counterpart in ordinary language. Conversely, there is one linguistic case clearly covered in the *Categories* for which no obvious counterpart exists in the new, metaphysical account:

(2) Socrates is [a] man.

By the canons of the *Categories*, (2) asserts that a substance is related to one of its kinds by means of the *said-of* relation. (2), then, is handled easily within Aristotle's earlier theory of predication. But it is not clear how (2) is accomodated by the new apparatus of predication at work in the *Metaphysics*.

(2) clearly does not assert that an accident is predicated of a substance. Nor is it obvious that (2) asserts that a form is predicated of matter. Again, then, how will Aristotle deal with (2) within the scheme of the *Metaphysics* – if we are not to say that that scheme is incomplete as a foundation for the various kinds of linguistic predication?

III

Unexpectedly, the problem of accounting for sentences like (2) surfaces again in connection with a familiar crux in *Metaphysics* Z13.[16] In this section, I outline the familiar crux, and what seems a promising answer to it. But as we shall see, the solution proposed raises again the question of how to interpret the problem sentence (2): I shall return to this question in the sections that follow.

In Z13, then, Aristotle seems committed to the view that

(3) No universal is a substance.

(With (3), Aristotle sets aside the suggestion reported at Z3, 1028b34–35 and again at Z13, 1038b2–8; he repeats his present conclusion in the summaries at Z16, 1041a3–5 and H1, 1042a21, cf. Z16, 1040b23, I2, 1053b16–18, K2, 1060b21, M10, 1087a2.) But Aristotle is also apparently bound to two further points:

(4) Form is universal

(Z11, 1036a29, M8, 1084b5), and

(5) Form is substance

(cf. note 1 above). The familiar difficulty is to say how all three claims can be true.

Aristotle himself is aware of the trouble he faces over the status of the universal (see in particular B6, 1003a5–17, which sets out much of the agenda for Z13–16; cf. K2, 1060b19–23, M10, 1087a10ff), but he does not indicate how he would resolve it. Some conclude that the difficulty is irresoluble (Lesher, 1971, pp. 169–178; Sykes, 1975, pp. 326–328). Others try to resolve it by finding vagueness or ambiguity in one or more of the three propositions that make up the puzzle. One account (Sellars, 1957, pp. 691–699 and 1967, pp. 107–118; cf. Harter, 1975, pp. 11–18; Hartman, 1976, pp. 545–561, and 1977, Chapter 2) finds an ambiguity in the word "form", which can refer either to individual form, that is, a form that is unique to a given individual, or to general form, that is, a form that is or can be shared by a number of different individuals. An individual form is never a universal, but it is the substance of the individual that has it. And a general form, which is, or can be, common to a number of different individuals, is a universal in relation to all of them, but it is the substance of none of them.[17] Alternatively, the formulation of the puzzle in (3)–(5) above equivocates between different senses of the word 'universal'. According to Woods (1967, p. 229), Aristotle's own terminology successfully keeps separate the notions of "things spoken universally" and universals *simpliciter*. All forms are universals, and they are also substances. But no thing spoken universally is a substance. These two claims are consistent, because although all things spoken universally are universals, some universals, namely forms, are not things spoken universally.[18] In a variation on this account, Modrak (1979, pp. 371–381) finds an ambiguity in Aristotle's own word 'universal'. The word in Aristotle covers both what she calls substance-types and universals$_p$. All forms are substance-types, and hence are themselves substances. But no universal$_p$ is a substance. No substance-type, however, is a universal$_p$: so (again) there is no contradiction.[19] Yet another treatment of the puzzle, due to Albritton, fastens on a vagueness rather than an ambiguity in the use of the word 'universal' in the three propositions (3)-(5). Aristotle's thesis (3), that no universal is a substance, must be understood to say that "nothing universal in relation to *species*, . . . , is the substance of any of *them*" (Albritton, 1957, p. 705, his emphasis). A form can be universal in relation to the *individuals* of a species; but if it is the substance of a single *species*, then that species cannot itself be one of the things in relation to which the form is universal.[20]

What seems to be by far the most promising line is in some respects a variation on Albritton's approach. We can begin by recalling Aristotle's definition of 'universal':

> I mean by universal what is such by nature as to be predicted of many. *De Interpretatione* 17a39–40

(cf. Z13, 1038b11–12 and, for example, *An. Pr.* Al, 24a18, *Metaphysics* B3, 999a20–21, D26, 1023b29–32). Now the type-restrictions placed by (A1) above on the relation "x is predicated of y" stipulate that whenever a form is predicated of something, its subject must be some parcel of matter. This fact, combined with Aristotle's definition of 'universal', suggests that form can properly be called a universal only insofar as it can be applied to *many different portions of matter*. These considerations give us this refinement of (4) above:

(4′). A form is universal in relation to the different parcels of matter of which it is predicated.

Secondly, when Aristotle claims that form is substance, at least part of what he seems to mean is that it is in some sense the substance of the concrete individual: compare Z10, 1035b15, Z13, 1038b21–22. Also, just as a primary substance is (essentially) the same as its essence, and hence is the essence of itself, so too a primary substance (that is, a form) is the substance *of itself* (cf. Z6, 1031a17–18). This is a long and difficult story, which I shall not try to develop here. This much, however, is certain: a form is *not* the substance of the portions of matter of which it is predicated. For example, the form of a man is (in some sense) the substance of Socrates, and by knowing his form, we are able to classify Socrates in his appropriate kind. But to say that these flesh and bones have the form of a man is not in any sense to classify the matter in question (as we do, for example, when we say that it is flesh and bones), and does not give its substance. Accordingly, we interpret (5) to say

(5′). A form is the substance of the concrete individuals of which it is a constituent, or of itself.

Finally, instead of (4), we shall understand the conclusion for which Aristotle is arguing in the chapter to stipulate that

(3′) Nothing that is universal to a set of many things is the substance of any of *them*.[21]

This revision of (3) is essential, for it turns out that, contrary to (3), some universals *are* substances.[22] More precisely, a form is universal with respect to various portions of matter, by (4′), while by (5′), a form is a substance (it is the substance of the concrete individual, or of itself). (3), then, is false. (3′), (4′), and (5′), however, are all true for Aristotle, and are all three consistent with one another.[23]

IV

I have argued that, for Aristotle, form is properly predicated only of different parcels of matter, and *not* of individual substances. Hence, if a universal is that which by nature is predicable of many, a form is a universal only with respect to those things of which it is properly predicated – that is, it is universal with respect to matter. At the same time, Aristotle's conclusion in

Z13 is that no universal is the substance of that of which it is the universal. But form is the substance of the concrete individual or of itself, but not of the matter. As concerns the argument of Z13, then, we are not forced to conclude with Lesher that "Aristotle's metaphysics [is] untenable on a fundamental point" (Lesher, 1971, p. 169).

This resolution of the puzzle from Z13, however, leaves one by now familiar loose end untied. If as (4′) claims, form is universal only in relation to different parcels of matter, what becomes of those many cases in which we say of different individual substances that they are members of a given kind? For example (once more).

(2). Socrates is [a] man.

One possibility is that we interpret the predicate of (2) to refer to a form. Then the truth of (2) seems to suggest that a form can be universal with respect to individual substances. This conclusion, however, is incompatible with the resolution to the puzzle of Z13 sketched above, for it seems to allow that a form is both universal to and the substance of one and the same thing.

In fact, this threat to the account of Z13 is readily countered, if we take seriously the point that form is predicated in Aristotle's sense only of matter, and not of the composite substance. I shall try to defend this point in two ways. First, in the remainder of this section, I shall consider the evidence that this is in fact Aristotle's position: that Aristotle would indeed deny that form can be metaphysically predicated of anything but matter. I shall also try to give an account of (2) consistent with this restricted notion of metaphysical predication. In the next section, I shall then try to suggest some independent motivation that Aristotle may have for his restrictions on metaphysical predication.

How confident can we be, then, that for Aristotle, form is predicated only of matter? Part of the evidence for this view is an argument from silence: the passages that support (A1) above say only that form is predicated of matter, and never say that form may be predicated of anything else. (In addition, they do not allow that anything but an accident [*pathos*, Z13, 1038b6, Θ7, 1049a29, 30] may be predicated of a substance.) But there is more evidence than this. Aristotle also has a positive account of how sentences like (2), which someone might take to show that a form can be (metaphysically) predicated of an individual substance, can be understood in terms of a relation between the form and *the matter of* the substance in question. By giving such an account, Artistotle is able to solve simultaneously both of the two problems noted in Section II above. For as we saw, the sentence (2) has no obvious counterpart in Aristotle's notion of metaphysical predication, while the case in which form is metaphysically predicated of matter seems to lack any obvious linguistic counterpart. Aristotle's suggestion that it is the metaphysical relation of predication between a form and some matter that underlies the truth of (2) solves both these problems at a single stroke.

Let us turn, then, to the details of the treatment of (2) that Aristotle proposes.[24] In *Metaphysics* Z17, Aristotle is pursuing the notion of substance as an *αἴτιον* (1041a9–10, cf. A3, 983a26–29, Z13, 1038b7), more particularly, the idea that substance is the *αἴτιον τοῦ εἶναι* for a thing (1041a31–32, b26, 28 cf. D8, 1017b15, H2, 1043a2–4, H3, 1043b13–14 [Ross' text]), that is, it accounts for a thing's being. As an *αἴτιον*, substance supplies the answer to certain kinds of "Why?" questions. But there is work to be done, Aristotle thinks, on the form that the appropriate "Why?" questions must take. Although substance accounts for the being of a thing, for example, it does not answer the question whether the thing *exists*, for this is something we must be able to take for granted (1041b4–5, cf. 1041a15–16, 23–24). Nor are we concerned with answering *why a thing is itself* (1041a14–20, 22–23, b2–4).

This last point helps Aristotle to draw a surprising conclusion. For the expected question, for example.

Why is *this*, viz. this man, Socrates, [a] man?

Aristotle substitutes another:

Why are *these*, viz. these materials, [a] man?

(cf. the example at 1041b6–7). Thus, Aristotle says, "the question 'Why?' is always [meant] in this way: Why does one thing belong to another?" (1041a10–11, cf. 1041a23, 25–26). And in general, the appropriate "Why?" questions ask the *αἴτιον* of the matter – that is, the form – on account of which the matter *is* something (1041b7–9).

What motivates this shift to *matter* as the subject of the question? Suppose someone asks a question in the form, Why is this man a man? Then Aristotle is embarrassed by difficulties concerning the scope of the question word 'why . . . ?' in the question.[25] If 'why' is given widest scope, then the questioner is asking the triviality, Why is he who is a man a man? This treatment cuts all questions from the same cloth, giving short shrift to any differences that may exist between them;[26] to all such questions, there is only a single reply, namely the logical truth that everything is itself. So the question-word must be given narrow scope: Concerning this object which is a man, why is it a man? But Aristotle has no clear way of picking out this interpretation of the question. He sees that the questioner cannot be asking, *de dicto*, about the statement that a man is a man, but he fails to see that the question can be about the man, *de re*, asking why *it*, that object, is a man. So he concludes that the question cannot be about the man at all, but must be about his *matter*, asking why *that* is a man. As Cohen puts it, Aristotle imports matter for lack of a device for referring to the man in a "logically independent" way, as the variables in modern notation are able to do: Concerning this x such that x is a man, why is x a man? (Cohen, 1978, pp. 403–404).[27]

According to Z17, then, the explanation of why a thing belongs to a certain kind runs in terms of a relation between the matter of the thing and its form. In H2, Aristotle explains again that sentences like (2), in which an individual substance is assigned to a kind, must be understood in terms of a relation between form and matter. He argues that a different sense of 'being' is associated with each kind of thing, and suggests that the different senses of 'is' for each kind are to be given by what amounts to a contextual definition. For example, that something *is* a threshold, or *is* ice, is to be understood to say that a certain form (lying thus, or being solidified thus) is predicated of the thing's matter:

> Clearly, then, the word 'is' is used in just as many ways; a thing *is* a threshold because it lies in such and such a position, and its being means its lying in that position, while being ice means having been solidified in such and such a way. (H2, 1042b25–28, after Ross' translation.)[28]

Here, Aristotle is clear on the role of form in explaining why a thing belongs to a kind, but he is less helpful over the role of matter.[29] Later passages, however, confirm that reference to the substance in such paraphrases can be dropped in favour of reference to its matter:

> And as among substances what is predicated of the matter is the actuality itself, so in the other definitions too it is what most resembles the actuality. For example, if we had to define a threshold, we should say, *wood or stone in such and such a position*, or a house, *bricks and timbers in such and such a position*, . . . , and ice: *water frozen or solidified in this way* (H2, 1043a5–10)

(cf. the further examples at 1043a5–11). Hence, we can conclude that "An object *is* a thing of some determinate species if and only if the associated form can be truly predicated of the parcel of matter constituting that object" (Loux, p. 18). Again, ". . . [a] substance's belonging to a kind is not a brute fact. It is something that requires explanation, and the explanation is provided by pointing to the fact that the associated substance-form is truly predicated of the parcel of matter that makes up the substance" (Loux, p. 19).[30]

V

If the account given in the previous section is correct, *Metaphysics* Z17 and H2 show Aristotle attempting to bring his treatment of (2) and sentences like it into line with the official notion of metaphysical predication which he sets out elsewhere in the *Metaphysics*. If the attempt is successful, then Aristotle can defend the *Metaphysics* notion of metaphysical predication, by removing a counter-example to the view that form is predicated only of matter. But *why* is Aristotle so strict about what he will or will not count as a genuine metaphysical predication? Why does the scheme of predication which he offers in the *Metaphysics* allow predication to be a relation between form and matter, or between accidents and an individual substance – but never between form and the individual substance?

These questions concern Aristotle's motivation for the theory of metaphysical predication I am attributing to him. The answer is, in brief, that Aristotle has no *need* to postulate predication as a primitive relation beyond those cases in which either an accident is predicated of a substance, or a form is predicated of matter. Where a compound substance is composed of a given form and a given matter, let us say that that form *supervenes on* the matter in question, but that the form is *constitutive of* the composite substance itself. A form is predicated of a given matter, if and only if it supervenes on that matter. Suppose now that Aristotle were to allow that in addition, the form is (metaphysically) predicated of the *compound substance*. Then the form is predicated of the compound substance just in case it is constitutive of that substance. The trouble is, as we shall see, that in Aristotle's theory, the relation, "y is constitutive of x", can be defined in terms of the relation, "y supervenes on x". So the latter, supervenience, is primitive, and the former, constitution, can be eliminated in its favour. Hence, form is predicated only of the matter, and not of the compound substance.

Let me try to defend this claim in somewhat greater detail. In Aristotle's metaphysics, as we saw in Section I, the operation of compounding is taken as primitive.[31] Two kinds of compounds are fundamental. A substance forms a compound with each of its various accidents, and matter is compounded with form. Let us use the notation, '$x + y$', to denote the compound of x with y. Then Aristotle assumes that

(A2). $x + y$ exists, only if *either* x is a substance and y is an accident, *or* x is matter and y is a form.

Let us use the letter 'a' as a variable ranging over (compound) substances, 'm' as a variable over portions of matter (for reasons that will emerge shortly, 'matter' here will include compounds of matter and form if they fall below the level of substances), and 'x' and 'y' as unrestricted variables, 'φ' will range over accidents, and 'ψ' over forms. Then we can define the relations, "x is an accident of y", and, for the operation of applying form to matter, "x supervenes on y", in terms of the basic operation of compounding. First, we assume that

(A3). x is an accident of y, only if x is an accident and y is a (compound) substance.

(A4). x supervenes on y, only if x is a form and y is a portion of matter.

Then we have the definitions

(D1). φ is *an accident of a* if and only if $a + \varphi$ exists.

(D2). ψ *supervenes on m* if and only if $m + \psi$ exists.

Now according to Aristotle, we can regard a complete substance as composed of a hierarchy of coincident matters, starting from a given portion of prime matter, which is not itself a compound of form and matter, through a sequence of form-matter compounds all the way up to the proximate matter of the finished substance (H4, 1044a15–25, b1–3, cf. D4, 1015a7–11, D6, 1017a5–6, D24, 1023a26–29, *Meteor.* D12, 389b23–28, *PA*. B1, 646a13–24, *GA*. A1, 715a9–12). So an individual is composed by means of successive applications of form to increasingly complex parcels of matter, which (with the exception of prime matter) are themselves compounds of matter and form. So we shall need two notions relating matter to the compound:

(D3). x is *the matter of y* if and only if for some form ψ, $y = x + \psi$.

For example, this bronze is the matter for this statue, but also (given a more complex form) a certain quantity of water is its matter, "if everything meltable is [composed of] water" (D4, 1015a7–10, D24, 1023a28–29).[32] But the proximate matter of a thing is Aristotle's preferred candidate for its matter (H4, 1044b1–3, Θ7, 1049a1–18, *Physics* B3, 195b22–27): thus

(D4). x is *the proximate matter of y* if and only if x is the matter of y and if there is no z such that x is the matter of z and z is the matter of y.

Aristotle's metaphysical relation of predication is the union of the *accident-of* relation and the supervenience relation:

(D5). x is *predicated of* y (y is *subject to* x) if and only if either x is an accident of y or x supervenes on y.

According to (D5), there are two cases of metaphysical predication. These support and explain two kinds of linguistic predication. First, where φ is an accident and a is a substance, we have

(D6). a *is* φ if and only if φ is an accident of a.

For example, Socrates is pale if and only if pallor is one of his accidents. Second, where ψ is a form and m is a portion of matter, and 'k' stands in for some kind-term, we have, as a first approximation, the definition:

(D7). a *is* [a] k if and only if, where ψ is the form of a k and m is the proximate matter of a, ψ supervenes on m.[33]

By (D5), the right-hand sides of (D6) and (D7) give the only two kinds of metaphysical predication that Aristotle allows. Why? In particular, why does Aristotle not allow form to be (metaphysically) predicated also of the individual substance? We can begin an answer to these questions by defining the relation, "x is constitutive of y", which holds between a form and a compound substance:

(*D8). ψ is *constitutive of* a if and only if there is a proximate matter m of a such that ψ supervenes on m.

Constitution suggests a natural way of expanding the notion of metaphysical predication to include a third case, where a form is metaphysically predicated of a compound substance. In place of (D5), we will have:

(*D9) x is *predicated of* y (y is *subject to* x) if and only if either x is an accident of y, or x supervenes on y, or x is constitutive of y.

Given (*D9), finally, we can have this alternative to (D7):

(*D10). a *is* [a] k if and only if, where ψ is the form of a k, ψ is constitutive of a.

The argument against (*D9) is that it is, quite simply, unnecessary. In (*D8), constitution is defined in terms of supervenience. Constitution and supervenience, then, will not *both* be primitive in Aristotle's theory. If his theory already contains supervenience, as suggested here, (*D8) shows how the new relation, "x is constitutive of y", can be eliminated in favour of it.[34] For this same reason, the idea that form can be predicated of a compound substance, as in (*D9), can also be eliminated in favour of the ordinary notion of predication given in (D5). Hence, Aristotle can claim that the scheme of predication given in the *Metaphysics* is adequate as it is to handle sentences like (2) which assign an object to its kind. (D7) rather than (*D10) is the correct rendering of such sentences, and Aristotle's scheme does not require supplementing by a notion of constitutive form in order to take them into account.

VI

In this next to last section, I want to defend once more the basic scheme of predication given in (A1) and (D5) against the charge of incompleteness with respect to the problem sentence (2). The source of the trouble is again the solution sketched to the problem of Z13. As I interpret it, Z13 requires that nothing is universal to that of which it is the substance. Now plausibly, form is the substance of individual substances. But if the predicate of (2) refers to form, then apparently form also is or can be universal to individual substances. One response to this difficulty was sketched in Section IV above. An alternative account appears in Loux (1979). To head off the difficulty, Loux suggests that instead, the predicate of (2) names a *species*. (2), then, expresses a relation between an individual substance and a species. This idea, however, once more calls into question the official account of predication in the *Metaphysics*. For if a species is predicated of an individual substance, this apparently represents a further, basic metaphysical relation that Aristotle's scheme in (D5) fails to take account of.

Before turing to this point, I have one reservation to make, parenthetically, about the notion of a species that appears in Loux' account. His account of (2) takes its start from a passage at Z10, 1035b27–30. Aristotle says:

> But as for man and horse and those things that [apply] to particulars, but universally, they are not substance but a sort of compound from this *logos* and this matter taken universally.

It seems very plausible that, following Ross, a compound of the sort mentioned here is an explication in terms of the *Metaphysics* notions of matter and form of the *Categories* notion of a secondary substance. The examples *man* and *horse*, moreover, are clearly secondary substances that the *Categories* counts as species as opposed to genera. The reservation I want to enter is that the universal compound, or the *Categories*-style species, is not to be identified with the notion of a *γένους εἶδος* (often translated, "species of a genus") in the *Metaphysics*. For reasons not directly relevant to this paper, I hold that a *γένους εἶδος* as it appears in the *Metaphysics* is identical with an *εἶδος* as form. So I shall suppose that in Loux' account, the predicate of (2) refers to a species in the sense of the *Categories* and *not* to a form, and hence, *not* to a *γένους εἶδος* in the sense of the *Metaphysics*.

Let us say, then, that the predicate of (2) names a *Categories*-style species, and let us understand '*k*' as a variable ranging over such species. Then as a beginning, we have the definition (D7) given above:

(D7). *a is* [a] *k* if and only if, where *m* is the proximate matter of *a* and ψ is the form of a *k*, ψ supervenes on *m*.

For example, Socrates is a man only if these flesh-and-bones are the proximate matter of Socrates, and they have an appropriate form (in fact, they are ensouled in an appropriate way). Here, the definition (D7) connects sentences like (2), construed as containing a name of a species as predicate, with predications involving form and matter.

On Loux' account, sentences like (2) represent at least one kind of case in which we apparently refer to species. (D7) suggests that the apparent reference to species in such cases may ultimately be *eliminable* in favour of the slimmer ontology of matter and form. That is, a species in the *Categories* sense may be seen as a derived entity, constructed out of our more basic talk about matter and form. (D7) by itself, however, takes us only part of the way to accomplishing this reduction. What (D7) principally lacks is a means of specifying the form ψ that is independent of the species *k*. If no such means of specifying ψ is available, then the project of eliminating the notion of a *Categories*-style species in favour of that of a form will have failed. In fact, however, it does not seem impossible in principle to find a way of referring to ψ without identifying it through the species *k*. The job will be done, once we have found a specification of the form that runs in terms of its *content*: the structural, functional, behavioural, and whatever other instructions are incorporated into that form. These will be the features that caused us to group *a* and other objects which share its form, as members of the same species in the first place.

(D7), then, points the way to eliminating a *Categories*-style species in favour of the *Metaphysics* notions of form and matter. (D7) also shows, again, that the notion of metaphysical predication as defined in (D5) is adequate after all to handle sentences like (2), in which an individual substance is assigned to a kind.

VII

I have argued that, for Aristotle, the decision on what kinds of predication to include in his metaphysical theory is a matter of what he takes to be primitive within that theory. Predication, defined in terms of the primitive notion of *compounding*, is one of the most nearly basic notions in Aristotle's metaphysics. The notion of predication in the *Organon* directly reflects what in those works Aristotle regards as the primacy of individual substance. In the *Metaphysics*, predication is analyzed differently to reflect a changed view of the metaphysical facts: but it remains one of the most nearly basic notions which Aristotle employs.

In this way, Aristotle preserves a due economy in the primitive base of his metaphysical theory. Two kinds of compounds are primitive, and the two kinds of (metaphysical) predication can be defined in terms of them. In turn, sentences like (2), which at first sight seem to lack any obvious counterpart at the metaphysical level, can in fact be understood in terms of the notion of (metaphysical) predication defined in (D5). This account of (2) requires no addition to the primitive notions of Aristotle's theory, beyond what (D5) already provides. Thus, the moral of Aristotle's treatment of sentences like (2) is in essence a counsel of parsimony. Far from calling for an enlargement in the basic apparatus of his ontology – the addition of a *Categories*- style species, or a notion of the constitutive role of form – the treatment of such sentences is for Aristotle an exercise in staying within the limits set by the primitive notions in his metaphysics.[35]

POSTSCRIPT

In the formalization in Section V in the text, constitutive form is derivative because it is defined in terms of supervenient form:

(*D8) ψ is *constitutive of a* if and only if there is a proximate matter m of a such that ψ supervenes on m.

Might things have gone the other way? That is, might Aristotle have taken constitutive form as primitive, and defined supervenient form in terms of it? For example, might he have offered the following definition?

(*D11) ψ *supervenes on m* if and only if for some compound substance *a*, *m* is the proximate matter of *a* and ψ is constitutive of *a*.

The two halves of (*D11) are without a doubt materially equivalent, but it is clear that (*D11) cannot express a proper definition if the rest of our formalization remains as it is. For, the would-be definiens in (*D11) contains a reference to constitutive form: and how is this notion to be understood? Clearly, not as in (*D8), or the two definitions together take us in a circle. If we accept (*D11) as a definition of supervenient form, then in place of (*D8), we need a definition of constitutive form that is independent of supervenient form. To achieve this, our entire formalization will have to be redone: for example, in broadest outline,

[D1] x is *the matter of y* if and only if for some form ψ, $y = x + \psi$.

[D2] x is *the proximate matter of y* if and only if x is the matter of y and there is no z such that x is the matter of z and z is the matter of y.

[D3] ψ is *constitutive of a* if and only if for some matter *m*, *m* is the proximate matter of *a* and $a = m + \psi$.

[D4] ψ *supervenes on m* if and only if for some compound substance, *a*, *m* is the proximate matter of ***a*** and ψ is constitutive of *a*.

These definitions may be compared with (D3), (D4), (D2), and (*D8) as they appear in the text. In particular, the alternate definition [D4] takes as primitive the *matter-of* relation and the constitutive role of form, while (*D8) in the text is constructed around supervenient form and the *matter-of* relation. It seems evident that the formalization in the text more accurately reflects Aristotle's views. But I am less clear what underlying reasons might have led him to prefer the one formalization over the other.

NOTES

1 *Metaphysics* Z7, 1032b1–2, Z8, 1033b17–18, Z10, 1035b14ff, Z11, 1037a22ff, esp. 29ff, Z17, 1041b7–9 (Ross' text [Ross 1924], II, p. 224). Z3 argues that form is prior to matter, and prior too to the compound of form and matter, 1029a5–7 (1029a29–30 adds that the compound is prior to the matter). The relation between primary substance, that is, form, and the compound substance is explained at Z11, 1037a29–30, cf. Z8, 1033b17–18.

2 Bonitz *Index* 151a57ff, cf. Balme 1962, 1972 (note on *de Partibus Animalium* 639a20).

3 For example, B1, 995b35, B4, 999a33–34, D24, 1023a31–32, b1–2, Z10, 1035a20–21, b19, 22, 32–33, 1036a2, Z11, 1037a26, 29–30, 32, Z17, 1041b11ff, H3, 1043a30, K2, 1060b24–25, L3, 1070a14, cf. *PA*. Al 640b26 (but contrast the use of the word σύνολον for the compound of a substance with its accident, M2, 1077b8, for the compound of genus and final differentia, *An. Po*. B13, 97a39, or for the compound of form and this matter, "taken universally", Z10, 1035b29). Aristotle also uses the expression ἡ συνθετη οὐσία, "the composite substance", or τὸ σύνθετον, "the compound": for example, D24, 1023a31–32, b1–2.

4 Z3, 1029a3–5, Z10, 1035al, 2, b32–33, H2, 1043a18–19, 28, cf. *De An*. B1, 412a9.

5 Z11, 1036b23, cf. τά γε ἐν ὕλῃ, K2, 1060b25, and *De An*. G4, 429b14, *PA*. A1, 640b27, *G. & C.* A5, 321b20–21, τῶν ἐν ὕλῃ εἶδος ἐχόντων.

6 E1, 1025b28–1026a7, Z10, 1035a25–26 (compare the discussion in Z5), *Phys*. B2, 194a5, *De An*. G4, 429b13. For other reference to the combination of matter and form in the compound substance, see Bonitz *Index*, 786a15–23, and Hicks on *De An*. A1, 403b11.

7 This suggestion is developed further in Fine (unpublished). To forestall misunderstanding, I should add the qualification that the comparison between Aristotelian form and the modern notion of a function, especially as it it understood by Frege, is meant to illustrate only one facet of Aristotle's theory. If Aristotelian form shares some of the formal features of a Fregean function, it surely also has many other properties that the comparison with Frege leaves untouched.

8 For the contrary suggestion, see Charlton (1970, pp. 70–73).

9 Cf. H6, esp. 1045a30–33 (quoted in the text immediately below), b17–22, and L10, 1075b34–37.

10 But (as predicted in Note 7 above) there are limitations to how far the analogy will go. For example (I owe the example and the point to Alan Code), the father of John is not a compound of John and the function *the father of x* in precisely the way that according to Aristotle, my table (say) is a compound of certain materials and the appropriate form. Nor is the father of John created by combining function and argument in the way that the craftsman creates the table by imposing a certain form on a given matter (again, the example and the point are Code's).

11 Frege for his part makes no bones about his inability to define a function, or the operation of "saturating" a function with an appropriate argument. In 'What Is a Function?', he writes:

> The peculiarity of functional signs, which we here called 'unsaturatedness,' naturally has something answering to it in the functions themselves. They too may be called 'unsaturated,' and in this way we mark them out as fundamentally different from numbers. Of course this is no definition; but likewise none is here possible. I must confine myself to hinting at what I have in mind by means of a metaphorical expression, and here I rely on my reader's agreeing to meet me half-way. (Geach and Black, p. 115.)

Cf. the similar remarks in 'Function and Concept', p. 32, and 'On Concept and Object', pp. 42–43, 45, 54–55 (references are to the translations in Geach and Black). Aristotle, however, does not share Frege's tolerance for the use of metaphor in these contexts: cf. the remarks in H6 (quoted in the text immediately below), and Note 12 below.

12 See *An. Po.* B13, 97b37–39, *G & C.* A10, 327b13–22, and *Metaphysics* A9, 991a20–

22, 992a26–29, Z14, 1039b5–6, M5, 1079b24–27. Compare also *De Caelo* D3, esp. 310b16–24, together with the comments in Mourelatos (1967, p. 102, esp. Note 23), and Ackrill (1972–3, p. 122).

[13] Z3 offers good evidence (corroborated elsewhere in ZHΘ) for the point that matter is a subject for Aristotle. But thanks to Aristotle's special purposes in that chapter, much of what is said there about matter as a subject is not typical of his conclusions elsewhere. For example, his argument at 1029a20–26 seems to imply that the following inference is valid: accidents are predicated of the *οὐσία*, but the *οὐσία* is predicated of the matter, hence the matter is subject for everything. But '*οὐσία*' here is ambiguous between 'individual substance' and the notion of 'substance of', or form, so that the passage is not evidence that the predication relation is transitive. In fact, it is vacuously both transitive and intransitive, since premisses of the form, 'x is predicated of y' and 'y is predicated of z', for distinct x, y, and z, are never jointly true. (The proof, which depends on the type-restriction (A1) below, is analogous to an argument that shows that the relation, "x is an accident of y", is also vacuously transitive and intransitive: see Lewis (1982, p. 8).) But these reservations about Z3 should not cast doubt on the role of matter as a subject in the *Metaphysics* in general: see Note 14 below.

[14] With the passages quoted, see also B1, 995b35, cf. B4, 999a33–34, H2, 1043a5–6. Some have doubted, in the teeth of the textual evidence, that Aristotle really means to say that matter can be a subject for predication: these doubts are answered in Kung (1978).

[15] In some places in the *Organon*, for example, *An. Po.* A22, 83a4, 5, 9, 15 (*τὸ οὕτω λέγειν*), predication is plainly a relation between expressions, or perhaps between an expression and a non-linguistic item, *Categories* 5, 2a19ff. At other places, it is equally clearly a relation between entities, for example, *Categories* 3, 1b10ff.

[16] A helpful review of the difficulties set by this chapter, together with a careful study of its key argument, appears in Code (1978). An earlier survey of the same subject is in Lesher (1971).

[17] But it is controversial whether Aristotle had a notion of individual form: arguments against the hypothesis are given by Lesher (1971, pp. 174 –176), and Modrak (1979, pp. 376–378). There is also a response to Lesher's discussion in Harter (1975, pp. 15–18).

[18] Against this resolution of the difficulty, Code (1978, pp. 66–67) has argued convincingly that Aristotle's use of the two expressions Woods seeks to keep apart is not in fact consistent, and in any case has no special philosophical significance.

[19] An important premiss in this argument is that form is a substance-type. This claim, however is problematic. Form is not a substance-type in relation to matter, for it does not *classify* Socrates' matter (say) to be told that it is a man (as opposed to flesh, bone, or etc.). In fact, however, as Modrak recognizes (p. 374), matter is the *only* subject of which form can properly be predicated (cf. (A1) above). But if form is not predicated of individuals, we can hardly regard form as universal in relation to individuals, despite Modrak's arguments to the contrary (note 25 and p. 379). Any other conclusion flies in the face of Aristotle's repeated definition of 'universal' (see the paragraph immediately following in the text). How, then, can form be a substance-type?

[20] Albritton's interpretation is controverted by Code (1978, pp. 67–69), cf. also Woods (1967, pp. 221–225) and Sykes (1975, pp. 319–320).

[21] A good defence of (4′) as a correct account of Aristotle's conclusion in Z13 is to

show that it is the conclusion to Aristotle's premisses and assumptions in that chapter. Aristotle's argument at 1038b8–15 depends on the definition of 'universal':

> That is called universal which is such as to belong to many things (1038b11–12, cf. [1] in Code, 1978)
>
> x is a universal if and only if for some y and some z, x is predicated of y & x is predicated of z, and $y \neq z$,

together with two principles:

> A universal is the substance either of all the things to which it belongs, or of none of them (1038b12–13, cf. [3] in Code)

(P1) If x is predicated of y and x is predicated of z and $y \neq z$, then x is the substance of y if and only if x is the substance of z,

secondly:

> Those things of which the substance is one, i.e. the essence is one, are also themselves one (1038b14–15, cf. b9–10, [6] in Code)

(P2) If x is the substance of y and x is the substance of z, then $y = z$.

(P2) deserves more comment than I can give it here. The formulation given here is argued for by Code (1978, pp. 67–69). In this form, (P2) holds only if 'y' and 'z' range not over individual substances, but over *primary* substances or forms; without this restriction, (P2) leads straight to the hypothesis of individual form. Further light on (P2) may come from D6, 1016b8–9, cf. a32ff, perhaps also from B4, 999b20 and Z16, 1040b17. At all events, Aristotle uses the two principles together with the definition of 'universal' to draw the conclusion:

> Nothing universal to a set of many things is the substance of any of them (= (3′) in the text).
>
> If x is universal to y and z, then x is not the substance of y and x is not the substance of z.

Suppose, then, that x is a universal in relation to y and to z: that is, by the definition of 'universal', that x is predicated of y and x is predicated of z, and $y \neq z$. Now let x be the substance of y (cf. 1038b13, ἑνὸς δ' εἰ ἔσται, and Code's [5]). Then by (P1), x is the substance of z, and by (P2), $y = z$. But we know that $y \neq z$. Hence, x is not the substance of y, and by (P1) again, x is not the substance of z. This proves Aristotle's conclusion in the form given in (3′) in the text. (This account of the argument is close to Code's at a number of points [although not at all]; in particular, the use of the principle (P1) and of the premiss at 1038b13 are due to him. Further details are also given in Lewis [unpublished].)

22 The replacement of (3) by (3′) also suggests a similar resolution of a related difficulty. Aristotle says often that a necessary condition for substancehood is not being said of a subject:

(a) Nothing that is said of a subject is a substance

(compare, for example, B5, 1001b30–32, Z3, 1029a8, Z13, 1038b15). Yet as we have seen, form both is the substance of something (cf. (5′) in the text immediately above), and is said of a subject (cf. Section II above). Again, the difficulty is, how can these claims all be true? Aristotle's assertions of (*a*) occur at least twice in close promimity to the observation that there are *two* kinds of subject, insofar as a substance is subject to its accidents and matter is subject to form. Hence, it seems unlikely that by (*a*) Aristotle must mean, but neglects to say, 'Nothing that is said of *a subject of accidents* is a substance' (Modrak, 1979, p. 373). A preferable solution is this refinement of (a):

(b) Nothing that is said of a subject *y* is the substance *of y*.

What form is said of, viz. matter, is not the same as that of which it is the substance. So Aristotle's various pronouncements are not obviously inconsistent after all.

23 The treatment of (3), (4), and (5) sketched in this section appears also in Loux (1979). This way of handling the difficulties of Z13 had occurred to me independently, although not (as it turned out) until after Loux' paper had been published. The viewpoint argued for here is also elaborated in Lewis (unpublished).

24 In what follows, I am indebted to the account in Loux (1979). A similar interpretation of Z17 and H2 was once defended by Montgomery Furth.

25 Here, I follow the account in Cohen (1978, pp. 402–404).

26 *ἀλλὰ τοῦτο κοινόν γε κατὰ πάντων καὶ σύντομον*,1041a19–20. In fact, of course, the two examples Aristotle quotes here – Why is the man a man? and Why is the musical [one] musical? – require quite different treatment, according to Aristotle. The first is answered by specifying a relation between some matter and a form, the second, the same relation between a substance and an accident, cf. (A1) above. In this interpretation, I follow a hint by Montogomery Furth in a note to an unpublished translation of the chapter.

27 On other matters, Cohen and I are in disagreement. Cohen connects the fact that in the *Metaphysics* matter is apparently a subject to form, with the question, Is Aristotle an "Aristotelian" or any other kind of essentialist? He offers this definition of an Aristotelian essence:

> an attribute is an Aristotelian essence if and only if it is essential to every individual having it which is a genuine subject for that attribute (1978, p. 395).

But then

> nothing will count as an Aristotelian essence, so long as matter is held to be the subject for predicates in the category of substance (p. 396).

Cohen concludes that matter is not really a subject for form. My inclination, on the contrary, is to take Aristotle's characterization of matter as a subject for form at face value, and to doubt instead the correctness of the definition of an Aristotelian essence which Cohen gives. For, if matter is subject to form, then – if Aristotle is to be an Aristotelian essentialist at all – our account of an Aristotelian essence must allow for the fact that for Aristotle, an essence is not essential to the very thing of which it is (metaphysically) predicated. Cohen's other attempts to undermine the status of matter as a subject for form seem to me to be unsuccessful. The suggestion (p. 401) that such a subject must itself be a *τόδε τι* (while matter can never be a full-fledged *τόδε τι*), is clearly unwarranted. And his attempt to import the theory of essential predication from the

Posterior Analytics to support his view ("the subject of a substance-predicate has to be *that very thing* that is predicated of it", p. 401, his italics) will not impress those who think that the *Metaphysics*, with its new views about form, matter, and the individual, contains a *different* theory of predication from what is found in the *Organon*. In discussing Aristotle's views in the *Metaphysics*, Cohen suggests that matter-words can be used to refer to the substance, for example, a house, *without implying that it is a house* (pp. 404–405). The words 'these bricks and stones', for example, refer to what is *and* is not the completed substance; for the matter constitutes the substance, but is not actually it. This attempt to have it both ways is courageous but unsuccessful. As Cohen notes, the matter is not identical with the house which it constitutes. If, then, a term refers to the matter, it does not refer to the substance. So Aristotle is not predicating form of the substance when he predicates it of the matter. Elsewhere, Cohen talks of Aristotle's "tendency" to treat form as predicable of matter (p. 402). But what Cohen calls a tendency is in fact settled theory on Aristotle's part (see Section II above). So I cannot accept Cohen's version of the theory of predication found in the *Metaphysics*:

> It is not as if the bit of gold is the individual of which *being a ring* is predicated (as the ring is the individual of which *being round* is predicated. Rather, the ring is the individual of which *being golden* is predicated (pp. 399–400).

[28] Cf. 1042b36–1043a1: 'for still others, to be will be to have been mixed, and not to be [will be] the reverse' (text as in Ross, and disregarding Jaeger's emendation, which is gratuitous); see also *De Anima* B4, 415b13.

[29] In fact, matter is not mentioned in this extract: for there to be a threshold is for *it* (αὐτὸ, 1042b27), the threshold, to lie in such and such a position. With others (for example, Furth in the work cited in note 26 above), I take this to be a solecism on Aristotle's part.

[30] I have argued in the text that the truth of the linguistic predication,

(2). Socrates is [a] man,

is grounded in the fact that the form of a man is metaphysically predicated of Socrates' matter. It may seem a difficulty in this account that it puts so much distance between the linguistic predication and its metaphysical underpinnings. How is it that (2) reflects its underlying ontology so indirectly? An answer to this question requires us to notice a limitation in the aims of Aristotle's inquiry. Aristotle appears to show little interest in setting out a canonical language, in which his ontological theory is directly and immediately reflected in the admissible forms of linguistic predication. In Z17, he contents himself merely with issuing warnings about the inadequacy of ordinary language as a guide to ontology, Z17, 1041a32–b2. (2), then, is not part of any canonical language, and we cannot reasonably expect to read straight off from (2) the ontology that in Aristotle's metaphysical theory underlies it.

[31] What follows is a formalization of only a small fragment of Aristotle's theory of substance. A full-scale formalization is brilliantly done in Fine (unpublished): Fine will recognize some of my principles as descendants of principles in his account. For a somewhat different exploration of the formal principles that underlie Aristotle's theory in the *Metaphysics*, see also Lewis (unpublished).

32 In the claim, "x is the matter of y", therefore, we must not take the 'the' too seriously, for in most cases y will not have a unique matter.

33 Two features of this definition require comment. (i) "where ψ is the form of [a] k": This way of specifying the form ψ, by reference to the class of k's, is only provisional. The definition as stated does not allow us to eliminate the notion of a kind in favour of that of form. Ultimately, however, we will want to replace this specification by one that runs in terms of the content of the form itself, rather than the kind, k. This question is discussed further in the latter part of Section VI. (ii) "where m is the matter of a": Ultimately, if we hope to replace talk of substances belonging to kinds by talk of parcels of matter having a certain form, we will have to specify the matter m independently of the parent substance, a. While I find this unproblematic in the usual case, where m is itself a compound of matter and form, the point is not uncontroversial, as discussion at the Conference showed. Where m is prime matter, and hence not itself a compound of any matter with any form, a different treatment is called for. Again, however, the point is a source of controversy, which I shall not try to resolve here.

34 If supervenience is primitive in Aristotle's theory, constitution will be a derived notion. But could constitution have been primitive, and supervention derived? This possibility is discussed briefly in the Postscript.

35 This paper was first written in Spring 1980, and the final major revision done in Summer 1981. As a result, I have been unable to take into account here the paper, 'EIDĒ in Aristotle's Earlier and Later Theories of Substance' (in *Studies in Aristotle*, Dominic J. O'Meara (ed.), Washington, D. C., 1981, pp. 129–159), by John A. Driscoll, and the monograph, *Form and Universal in Aristotle* (Liverpool, 1981), by A. C. Lloyd.

I am glad to acknowledge a continuing debt to Kit Fine, whose unpublished work on Aristotle's theory of substance has been the motivating force for my own work on that and related topics. I am grateful also to James Bogen, Tyler Burge, Deborah Modrak, and Lynne Spellman for help at different stages in the preparation of this paper. None of these, except perhaps Bogen, is responsible for my remaining mistakes.

D. K. MODRAK

FORMS AND COMPOUNDS

Lewis raises two problems concerning Aristotle's conception of predication in the central books of the *Metaphysics*. The central problem of the paper is drawn from texts describing two kinds of metaphysical predication, viz., the predication of accidents of concrete individuals and the predication of forms of matter. Lewis then asks whether this notion of predication can accommodate sentences which predicate membership in a kind of concrete individuals. Lewis's discussion of this question suggests several different problems which I propose to separate. The first, which I shall label the metaphysical problem, is whether the acceptance of 'Socrates is a man' as a well-formed sentence involves a commitment to a third type of metaphysical predication, the predication of a species or a form of a concrete individual. The second, which I shall call the linguistic problem, concerns the relationship between linguistic predication and metaphysical predication. Can the metaphysical theory of predication serve as the basis for a linguistic theory of predication?

The second problem which Lewis addresses is the perennial problem of the inconsistent triad; namely, how can forms be substances, if forms are universals and no universal is a substance?

In the first section of this paper, I shall consider the metaphysical and linguistic versions of Lewis's central question. Lewis's treatment of the metaphysical problem seems correct in the main, but more needs to be said about why the sentence, 'Socrates is a man', does not predicate form of Socrates. Although Lewis raises the linguistic question, he does not, so far as I can determine, answer it. My answer is: No. Aristotle's metaphysical theory of predication cannot serve as an adequate foundation for a linguistic theory of predication, nor was it intended to perform this function.

The problem of the inconsistent triad is taken up in the second section. I have dealt with this problem elsewhere, proposing a solution to it which invokes a distinction between a broad conception of a universal and a narrower conception.[1] By contrast, Lewis holds that forms are universal in relation to many portions of matter whereas accidents are universal in relation to many concrete individuals. Lewis's solution turns on a notion of matter that is problematic on textual and conceptual grounds, so he fails to

James Bogen and James E. McGuire (eds.), How Things Are, 85–99.

offer a solution which is a viable alternative to mine. In the third and fourth sections, I note some further difficulties with Lewis's treatment of matter and with his definitions.

I

Lewis is concerned to show that sentences which predicate membership in a kind of substantial individuals are consistent with passages in the *Metaphysics* that describe two ways of being a subject, for instance, 1038b4–6:

> [The substratum] underlies in two ways, either being a this, as the animal underlies its affections, or as the matter underlies the actuality.

According to Lewis, this passage and other similar passages (1029a23–24, 1049a27–36) specify the two forms of predication recognized in the central books of the *Metaphysics*. These two forms are summed up in Axiom 1:

> A1. x is predicated of y (y is the subject of x) only if *either* x is an accident and y is an individual substance, or x is form and y is a parcel of matter.

Many sentences of ordinary discourse seem to fit the first disjunct of Al, for instance, (1) Socrates is pale; but other sentences do not seem to satisfy either disjunct, for instance, (2) Socrates is a man.

As Lewis points out, sentence (2) is easily dealt with on the theory of metaphysical predication at work in the *Categories*. On that theory, secondary substances are predicated of primary substances; "Socrates" names a primary substance and "man" refers to a secondary substance. Secondary substances do not figure in the description of predication in the *Metaphysics*. Nonetheless, if "man" is a referring expression, as it seems to be for Aristotle, then its referent would be predicated of Socrates in (2). If the referent of "man" is the species, man, then (2) predicates a species of a concrete individual; if the form, then it predicates form of a concrete individual. In either case, the existence of a type of metaphysical predication not mentioned in A1 would be implied. To resolve this difficulty, Lewis must establish both that (i) species is not a basic ontological category for Aristotle and hence the predication of species is not metaphysically basic and (ii) "man" does not refer to the species form in (2).

Lewis draws much of his argument for (i) from Z17. A passage in this chapter which Lewis does not discuss and which offers further support for his conclusion is 1041a32–b2.[2]

> The object of the inquiry is most easily overlooked where one term is not predicated of another, e.g. when we ask "what is man?"

The form of the question draws our attention away from the formal cause.

Aristotle says that we should ask instead, "Why are these [materials] a man?" This question evokes an answer in terms of the essence or form. If a question about the nature of a species must be answered by citing the species form, then from a metaphysical point of view, form is prior to species.

I agree with Lewis that form is not predicated of concrete individuals, but I do not find his argument for (ii) as persuasive as his argument for (i). Unlike the species interpretation of (2), the form interpretation would not add an additional item to the basic ontology of form, matter, accident and concrete individual; it would add a third type of metaphysical predication. The form interpretation is not without supporters and apparently for good reason.[3]

Aristotle often uses the "what is it?" question to pick out items in the category of substance.[4] "A man" seems the most natural Aristotelian answer to give in response to the question, "What is this thing?" asked while pointing to Socrates (cf. 1028a27). Not surprisingly, "man" shows up repeatedly on lists of natural substances in the *Metaphysics* (e.g. 1032a18–19, 1034b17–18, 1028a17). Since Aristotle holds that form is ontologically prior to the composite (1029a6–7), and since "man" refers to the composite substance, it is arguable that "man" also refers to the form.

Secondly, there are passages that strongly suggest that terms which name artifacts are used homonymously, sometimes referring to the concrete artifact and sometimes to its form (cf. 1033b17–18; 1034a22–23). In H3, Aristotle says that:

> . . . sometimes it is not clear whether a name means the composite substance, or the actuality or form, e.g., whether house is a sign for the composite thing, "a covering consisting of bricks and stones laid thus and thus" or for the actuality or form, "a covering" . . . (1043a29–33).

Elsewhere names of artifacts are used for the conception in the artisan's mind which Aristotle tends to identify with the form (1034a23–24; cf. 1032b1, b22–23; cf. 1035a7). It should be noted that Aristotle does not extend this same treatment to "animal" without qualification –

> "Animal" might even be applied to both [soul and soul in a body], not as something definable by one formula, but as related to a single thing (1043a36–37).

More typically, the names of biological kinds turn up in discussions of terms which cannot be defined without reference to matter of some sort.

Third, the form interpretation of man fits one story that has been told about the transition from the conception of substance in the *Categories* to the conception of substance in the *Metaphysics*:[5] In the *Categories*, species and genera are secondary substance, but species are held to be more truly substances than genera (2b8). In the *Metaphysics*, species are associated with forms. The latter account for the substantiality of concrete individuals and are ontologically prior to them. If, in the *Metaphysics*, species as secondary substance is transformed into form as primary substance, then "man" which once referred to species might now refer to form.

At this point, it may seem that the question is settled in favor of the form interpretation. But as frequently happens in Aristotelian exegesis, there are equally good – indeed better – reasons for rejecting it. There are several passages in *Metaphysics* Zeta where Aristotle explicitly considers the predicate "man" and other species names and asserts that these terms stand for composites of matter and form. For instance, at 1035b27–30, Aristotle says:

> But man and horse and terms which are thus applied to individuals, but universally, are not substance but something composed of this particular formula and this particular matter treated as universal . . . . (Cf. 1033b25–26; 1037a6–7).

Secondly, Aristotle explicitly identifies the form of a living thing with its soul (1035b14–15). If both "soul" and "man" refer to the form of a human being, then the statement that a (human) soul is a man should be unproblematic; but, when Aristotle considers the question whether Socrates is identical to his soul, he maintains the distinction between the composite substance, man, and the soul (1037a6–10; cf. 1036a16–25). Aristotle denies that man and the essence of a man are one and the same in contrast to soul which is one and the same with its essence (1043b2–3; cf. 1036a1–2). The identification of the essence of man with the form of man should be uncontroversial for reasons cited by Lewis, so it seems clear that the referent of "man" should be distinguished from the essence or form of man.

Perhaps the arguments in favour of the form interpretation should be reconsidered. First, we should distinguish between two theses: (a) Aristotle uses "man" to refer to form and (b) Aristotle uses "man" to classify composite substances. The first and third arguments cited above in support of (a) support (b) equally well. The second argument was based on passages which appear to sanction the use of artifact terms for forms, but the support these passages provide is only as good as the evidence for treating natural kind terms as analogous in this respect to artifact terms. As we have seen, Aristotle is cautious about the analogy in *Metaphysics* Eta/3; and elsewhere he expresses reservations about identifying man with soul. His considered judgment is that man cannot be defined without reference to some matter. He likens man to bronze sphere, not to sphere (1033b25–6). Moreover, although Aristotle insists on the priority of form to composite substance in the *Metaphysics*, he does not reject the concrete individual's claim to substancehood; so we should not be surprised at his reserving some substance terms for referring to composite substances.

It is still open to the proponent of the form interpretation to argue that "man" predicates form indirectly of Socrates. After all, Aristotle says that "man" refers "to a compound of this formula and this matter treated as universal (ὡς καθόλου)." To see where this objection goes wrong, we must note the role "ὡς καθόλου" plays in this context. This phrase serves to indicate that species names do not function in quite the same way as other predicates. The species name, like the name of a property, is applied universally (καθόλου); but its ontological correlate is only treated as (ὡς) universal.[6] When we predicate the species name of an individual, we treat as universal what in a strict metaphysical sense we cannot predicate of the individual – its form, for the species name refers to the matter and form of the individual under a general description. Since being a particular composite substance is a precondition for being the subject of properties, being a composite substance cannot be one of Socrates's properties. To assert that Socrates is a man is to assert that Socrates's matter is so organized as to constitute a human being; it is to say that Socrates is a token of a particular substance type.[7]

To sum up, I agree with Lewis that Axiom 1 or, more precisely, the texts it summarizes set out the basic kinds of metaphysical predication. I have tried to bolster Lewis's arguments to show that Aristotle's acceptance of sentence (2) as a legitimate form of linguistic predication need not commit him to adding species to his basic ontology or to extending his notion of metaphysical predication to include the predication of form of concrete individuals.

Lewis sometimes understands the problem posed by sentence (2) in a different and more linguistic manner. He asks, "... how will Aristotle deal with (2) within the scheme of the *Metaphysics* – if we are not to say that that scheme is incomplete as a foundation for the various kinds of linguistic predication?" (p. 65). This question is never explicitly answered by Lewis. I believe that the metaphysical theory of predication is not, nor was it intended to be, a foundation for a linguistic theory of predication.

In the *Metaphysics*, Aristotle is forced to distinguish between two projects – articulating the ontological commitments of the well-formed sentences of ordinary discourse and articulating the ontology implicit in the sophisticated explanatory theory of the *Physics*. The primary subjects of ordinary discourse are concrete particulars; in the *Categories*, these are also the ontologically primary subjects. By the writing of the *Metaphysics*, Aristotle has recognized the importance of a substratum for change; and he has realized that a composite substance cannot serve as the substratum for a substantial change.[8] Aristotle introduces matter to solve this problem, making the form the ultimate determinant of the characteristics of the concrete particular. Aristotle's conception of a metaphysical subject no longer corresponds precisely to the conception of a subject embedded in ordinary discourse.[9] This explains why, in the *Metaphysics*, Aristotle is willing to countenance sentences which do not illustrate either of the basic forms of metaphysical predication. This contrasts with his practice in the *Organon*. For instance, in *Posterior Analytics* A22, he rejects "the white is a log" because it appears to predicate a primary substance of something else (83a10–17). In ordinary discourse, membership in a species may be predicated of a concrete individual and an accident may be predicated of an accident.[10] Neither type of predication falls under A1, and a metaphysical theory of predication which functioned as the foundation for a linguistic theory of predication would have to take these forms of predication into account. But this is not Aristotle's project in the *Metaphysics*; his project is simply to set out the most basic ontological relationships from the perspective of his theory of explanation.

Finally, were Lewis to give a complete account of Aristotle's treatment of predication in the *Metaphysics*, he would have to talk about Aristotle's notion of *per se* predication (*καθ' αὑτό λέγεσθαι*). In *Posterior Analytics* A4, an attribute is said to belong to its bearer *καθ' αὑτό* if the attribute cannot be defined without reference to its bearer or vice versa. This notion of belonging *καθ' αὑτό* is extended in Z4 to cover the relationship between a substance and its essence, and this relationship is contrasted to cases where one thing is said of another (*ἄλλο κατ' ἄλλου λέγεσθαι*). It is possible that Aristotle intends to limit A1 to the latter.[11] If so, A1 is incomplete in a way not discussed by Lewis at all.

II

My major quarrel with Lewis concerns the thesis that form is universal in relation to many portions of matter but the substance of concrete individuals (pp. 66–68).[12] I am sympathetic with Lewis's intent, for I agree that form is predicated of matter and the substance of concrete individuals. However, there are serious textual and conceptual problems confronting the claim that form is predicated universally of many portions of matter. These problems will become obvious when we examine the consequences of Lewis's position for an arbitrary case. Aristotle often uses a bronze sphere to illustrate the relationship between matter and form (e.g. 1033a24–b26). According to Lewis then, the form, sphericity, is (U) universal in relation to some portion of bronze and (S) the substance of the bronze sphere but not of the portion of bronze.[13] (For our purposes, the bronze sphere can be treated as a substance without deciding whether, strictly speaking, a bronze sphere is a substance.[14])

Lewis's conception of constituent matter is somewhat obscure. Lewis might subscribe to and attribute to Aristotle either one of the following conceptions of constituent matter: (i) the matter of a composite is contingently identical with the composite for the lifetime of the composite and (ii) the matter of a composite is not identical to the composite in any way.[15] I am not sure which of these conceptions Lewis holds, so I will examine (*U*) and (*S*) in light of both, beginning with (ii). On the face of it, (ii) seems to be more congenial with Lewis's view than (i). For instance, Lewis talks as if the form of bronze takes water into bronze (p. 72); but presumably, he would not say that the water is contingently identical with the bronze it becomes.

It seems obvious that (*U*) would be false if, by "portion of bronze" is meant any portion of bronze whatsoever, e.g., a rectangular portion. Sphericity is universal in relation to spherical portions of bronze only. This means that on (ii), the acceptance of (*U*) and (*S*) commit one to drawing a distinction between spherical portions of bronze and bronze spheres, for sphericity is the substance of the latter but not of the former. This task may prove impossible. A standard way of distinguishing between the constituent stuff of an object and the object is by recourse to persistence conditions: the bronze which constitutes the sphere might have existed prior to being shaped into a ball, or it might exist after the sphere is destroyed.[16] However, the bronze sphere and the spherical portion of bronze have precisely the same persistence conditions so that the onus is on Lewis to set out their differences.

There is the further problem that, according to (*S*), sphericity is not the substance or form of spherical portions of bronze. What will account for the unique characteristics of spherical portions of bronze? Lewis cannot appeal to the form of bronze, for it is the substance of any and all portions of bronze. It is open to the proponent of the distinction to say that, since the spherical portion of bronze constitutes the bronze sphere, the spherical portion is not a substance, so *ex hypothesi* sphericity does not give the spherical portion of bronze its substance. But does it make sense to say that sphericity is the substance of a bronze sphere but not of a spherical portion of bronze? Traditionally, sphericity is said to be the substance of the bronze sphere because it makes the sphere the kind of object it is; but this would, presumably, also be true of spherical portions of bronze. Indeed, in Lewis's terms, sphericity would seem to be compounded with bronze in both cases.

These difficulties may lead us to abandon (ii) and to interpret Lewis's position in light of (i). Suppose a portion of bronze is contingently identical to a bronze sphere at time *t*. Then according to (*U*) and (*S*): (*U**) sphericity is universal in relation to the portion of bronze at *t* and (*S**) sphericity is the substance of the bronze sphere at *t*. This interpretation avoids the problems mentioned above, but it, too, is open to a variety of objections. First, the textual support for attributing (ii) to Aristotle is greater than the evidence for (i); most importantly, Lewis's treatment of matter as a this conflicts with Aristotle's insistence that matter is not a "that" (ἐκεῖνο), but a "thaten", (ἐκείνινον), e.g., the sphere is not bronze, but brazen. Even on (i), (*U**) requires justification; Lewis should explain why sphericity is not universal in relation to the bronze sphere which is contingently identical to the portion of bronze at *t*. Third, Lewis's argument from A1 for attributing (*U*) to Aristotle is flawed.

A quick survey of Aristotle's description of matter should demonstrate Aristotle's reluctance to identify the matter of a composite with the composite. According to Aristotle, matter is that out of which a thing comes to be (1033a26–27, 1032a17); matter is that into which a thing is dissolved (1035a17–21; 1041b14–16, b31–32). These passages suggest the following picture: we start out with a bronze cube, we melt it down and reshape it producing a bronze sphere. The bronze persists throughout the change from cube to sphere; the cube is resolved into bronze, the sphere comes out of bronze (cf. 1034a10–13). So far, the question whether the bronze is now contingently identical with the bronze sphere remains open. But Aristotle also characterizes matter as: a part of that of which it is the matter (1033al; cf. 1035a4–7); as indeterminate (1049b2; 1037a25); as attribute-like (1041b1–2) and definitely not a this (1042a17–8; 1049a19; *De Anima* 412a8). In Theta 7, Aristotle compares the relationship bewteen matter and the thing it constitutes to the relationship between an attribute and the subject it inheres in (1049a30–b1). The wood of the casket belongs to the casket; it is not identical to the casket; the casket is a this; the wood a thaten. Indeed, "thaten" (ἐκέινινον) is coined by Aristotle as a way of referring to the constituent matter without appearing to sanction the use of nouns for material constituents. To use a noun would be to treat the matter as a concrete particular rather than a constituent of a concrete individual. Letters are the matter of syllables, but syllables are not identical to the letters which make them up (1041b12–13). On balance, these passages favor the (ii) conception of constituent matter for they treat the matter of a composite as a feature of the composite rather than that as the composite itself under a different description.[17]

Lewis's position that form is universal in relation to many portions of matter is tenable only if there is some way to individuate the portions of matter. In our discussion of (ii), the form was used to individuate spherical portions of bronze; but this collapsed the distinction between the matter and the composite. On (i), the matter is contingently identical to the composite, so that the individuation of the matter is much less problematic. But, unfortunately, not only do the passages cited above suggest that Aristotle subscribes to (ii) rather than (i), but so does Aristotle's practice of using plurals when he uses nouns to refer to matter. For instance, Aristotle uses "these flesh and bones" (1034a6) and "these bricks and stones" (1041b6; cf. 1033a15, b20) for the matter of a human and of a house. It is noteworthy that the locution Lewis uses for a material individual – a portion or parcel of matter – has no counterpart in Aristotle's terminology.[18]

According to (U^*), sphericity is universal in relation to a portion of bronze which is contingently identical to a bronze sphere B at t but sphericity is not universal in relation to B at t. What does "being universal in relation to X" mean in this context? The usual explanation that X is an instance of or exemplification of the universal in question must be rejected, because the bronze sphere is an instance of sphericity. Lewis then should give us an account of his notion of universal.

Lewis defends his claim that form is universal in relation to the different parcels of matter of which it is predicated by appeal to Axiom 1, i.e. by appeal to the doctrine that form is predicated only of matter. The crucial question is: does Lewis's conclusion follow from Aristotle's theory of metaphysical predication and his conception of a universal? In the *De Interpretatione*, Aristotle gives the definition of a universal which Lewis cites: "I mean by universal what is such by nature as to be predicated of many." He goes on to say, ". . . and I mean by individual that which is not predicated of many, for example, man is a universal, Callias an individual" (17a38–41).[19] This definition treats the universal as the complement of the individual. Aristotle could have said, instead, that the universal is of such a nature as to be exemplified by many things and maintained the same contrast between the universal and the individual. A bronze sphere exemplifies sphericity so sphericity is universal in relation to the bronze sphere.

At this point, Lewis might well appeal to a theory of predication and insist that *X* is universal in relation to *Y* only if *X* is predicated of *Y*. This seems to be an eminently reasonable principle to adopt in the case of linguistic predication. However, the predication of form of matter is a case of metaphysical predication, as Lewis is at pains to point out. Once a wedge is driven between linguistic and metaphysical predication, there seems to be little reason to suppose that, because form is predicated of matter, it must be universal in relation to matter. For surely, all we mean when we say that a form is universal in relation to *X* is that *X* exemplifies the form in question. Matter *qua* matter does not exemplify the form; the composite individual exemplifies the form.

If Lewis's view that form is universal in relation to many portions of matter is open to serious objections, then an alternative solution to the puzzle of the inconsistent triad is needed. I have argued elsewhere that Aristotle has a circumscribed conception of a universal which he employs in the anti-Platonist arguments of *Metaphysics* Zeta, because he holds that the Platonic metaphysics is committed to treating universals as properties.[20] The latter conception, which I labelled universal$_p$, is inapplicable to Aristotlelian forms which are constitutive of individual substances, but for that reason are not properties of composite substances. Lewis and I agree that Aristotlelian forms are universals, but we opt for different construals of this position. For Lewis, all universals are predicated of many; forms differ from other universals in being predicated of many portions of matter. On my view, universals$_p$ are predicated of many this's, but forms are predicated of matter and matter is not properly described as many or one, since *qua* matter, it is indeterminate. Nonetheless, forms fall under the broad Aristotlelian conception of a universal, according to which anything that has (or could have) a number of exemplifications is a universal. The substantial individual exemplifies the form which is constitutive of it and, thus, the form is universal in relation to the individual. The form is not predicated of the individual, because it is not compounded with the individual.

III

Lewis's commitment to material individuals was found problematic in connection with his interpretation of Z 13. This commitment is also a feature of his notion of compounding; here, too, it is troublesome.[21] According to Lewis (p. 60), "... Aristotelian form can be regarded as in some respects like a function from matter to a compound, and in the favoured cases, to a compound which is also a substance." Or again, "... form is of such a nature as to combine with one material object, namely matter, to make up a second object, which under the right circumstances will be a substance." The operation of compounding is performed on a material object. The individuation of the material object which is compounded with a particular form F cannot be a product of the compounding with F. For instance, the form of bronze does not take a portion of bronze into bronze, it takes a parcel of water into bronze. It is possible to treat a portion of water, for instance, the water in a certain vase, as a material individual, because the water has certain properties which allow us to distinguish it from the vase, from the surrounding air, and so forth. In Aristotle's terminology, the water is a distinct kind of stuff because some matter has the form of water.

However, compounding as expounded by Lewis begins with unenformed bits of matter. These are transformed by successive applications of form into increasingly complex compounds of matter and form. " . . . we can regard a complete substance as composed of a hierarchy of coincident matters, starting from a given portion of prime matter, which is not itself a compound of form and matter . . . ." (p. 72).[22] By positing the existence of primitive material individuals which are not composites of matter and form, Lewis avoids an infinite regress of composite objects, each of which is the result of the application of form to an already enformed material object. On his view, the matters of any composite will form a finite linear sequence terminating in matter which is not combined with form. This, too, must be a material individual (a bit of matter), so Lewis needs some way of individuating unenformed matter. This will not be easy. Indeed, I can see no way to individuate portions of pure matter.

Even such devices as individuating by reference to spatial-temporal location are of little help here. On the not unproblematic assumption that a spatial-temporal matrix can be defined relative to existing complete substances, Lewis could stipulate that whatever stuff occupied a given three-dimensional spatial location during a given temporal period constituted a material individual. Since there would be no way to establish spatial or temporal continuity for featureless portions of matter, these arbitrarily defined individuals could have very peculiar properties. Perhaps, Lewis should avoid this problem by rejecting the notion of prime matter and stipulating that form is always applied to a compound of matter and form.

A further set of problems emerge in connection with the individuation of organic matter. Biological individuals are substances *par excellence* for Aristotle, so a comprehensive account of compounding should be applicable to them. The individuation of organic material seems to require the presence of soul. Aristotle says that "a dead finger is a finger only in name" (1035b25), and he denies that a dead hand is a hand on the grounds that it is not part of an ensouled being (1036b31–32). In the *Generation of Animals*, he extends this treatment to flesh (734b25–27).[23] Ackrill has argued that any name for organic matter is applied homonymously if the matter in question is not part of a living being, i.e. if the matter is not enformed by soul.[24] So Ackrill continues, Aristotle is unable to pick out the matter of an organism by a description which is logically independent of the form of the organism, but such a description is needed in order to make the distinction between matter and form. Irrespective of the correctness of Ackrill's views as a criticism of Aristotle, his argument raises a difficulty for Lewis. Compounding seems to require logically independent descriptions of the material object and the form. Since compounding often involves the successive applications of form to increasingly complex materials, the description of the matter with which the form is compounded will be a description of the proximate matter of the compound. Flesh and bones are Aristotle's stock example of the proximate matter of a human being (1035a19, a33). If these terms contain a tacit reference to the form of a human being, then the phrase, a portion of flesh and bones, cannot be used to pick out the material object with which the form is compounded.

To sum up: compounding presupposes the individuation of the parcels of matter to which the form is applied. Lewis gives us no account of the individuating conditions for matter. Were Lewis to attempt to formulate such a criterion, he would probably be forced to restrict compounding to parcels of enformed matter. Lewis also needs to give an account of the individuation of the matter of biological individuals which avoids Ackrill's objection. These considerations point to the conclusion that a final decision about the adequacy of Lewis's analysis of compounding will not be possible until a criterion for the individuation of matter is added.

IV

I will conclude with a few quibbles about Lewis's definitions and axioms.[25] I am unclear about Lewis's project. Are these axioms and definitions offered simply for the purpose of stating very precisely positions to be discussed? Or, are they intended as an axiomatization of Aristotle's theory of predication? If the latter, I am concerned about definitions 7 through 10. Since the eighth and ninth definitions are preceded by asterisks (to indicate they are not part of the system?), I will confine my remarks to D7 and D10. The items on the left-hand side of these definitions, viz., "*a* is [a] *K*" and "*a* is [a] σ" seem to represent cases of predication. But neither kind-nor species-predication is sanctioned by D5. At this point, Lewis might appeal to the distinction between linguistic and metaphysical predication and say that D7, D10, and D11 define types of linguistic predication in terms of the allowable forms of metaphysical predication. But then, the distinction between linguistic and metaphysical predication should be made explicit within the formal system. For example, D5 should read: *x* is metaphysically predicated of *y* if either *x* is an accident of *y* or *x* supervenes on *y*, and D7 should read: the sentence "*a* is *K*" is true only if for some ψ, ψ supervenes on the matter *m* of *a*. As the definitions stand, there is no indication that "*a* is [a] *K*" or "*a* is [a] σ" does not fall under "*x* is predicated of *y*".

To end with a very tiny quibble: The plus sign is universally used for addition. Aristotle argues vigorously against treating the compounding of matter and form as an additive relation in Z17 (1041b16–26; cf. 1043b5–8). Using the plus sign for compounding suggests the wrong picture of this relation, even though, technically, the usual use of "+" is irrelevant to its use here.

NOTES

[1] D. K. Modrak, 'Forms, Types, and Tokens in Aristotle's *Metaphysics*', *Journal of the History of Philosophy* **17** (1979), 371–81.

[2] I follow Ross in accepting the manuscripts' "*anthropos ti esti*" as the question Aristotle intends to discuss. In this form, it is clearly a question about the species man (cf. 1028a37).

[3] A very clear and persuasive presentation of this position can be found in Michael Loux, 'Form, Species and Predication in *Metaphysics*, *Z*, *H*, and *θ*', in *Mind* (1979), pp. 1–23.

[4] "τί 'εστι" is used in this way at 1028a11–13 and 1034b13. For an interesting discussion of the different roles played by "τί 'εστι" on Aristotle's various lists of categories, see John Malcolm, 'On the Generation and Corruption of the Categories', *Review of Metaphysics* **33** (1981), 662–81.

[5] See, for instance, G. E. L. Owen, 'The Platonism of Aristotle', *Proceedings of the British Academy* **51** (1965), 136–7.

[6] To my knowledge, "ὡς καθόλου" is used in Zeta only at 1035b30 and 1037a7. Thus, it seems very likely that its use at these locations serves to signal the peculiar character of species as predicates.

[7] Modrak, pp. 378–9.

[8] For a clear argument for the persistence of matter through a substantial change, see Alan Code, 'The Persistance of Aristotelian Matter', *Philosophical Studies* **29** (1976), 357–67.

[9] Lewis recognizes the distinction between metaphysical predication and linguistic

predication (pp. 64–65), without recognizing that the reasons for making the distinction call the belief that metaphysical predication serves as a basis for linguistic predication into question.

10 Not only is the sentence, "this white is white", a well-formed sentence in ordinary discourse, but for Aristotle as well (cf. 1031b22–28).

11 Aristotle usually treats the predication of an accident of a composite substance in the *Metaphysics* as an instance of *ἄλλο κατ' ἄλλου* predication (cf. 1030a–6). Since Aristotle says that nothing which is *ἄλλο κατ' ἄλλου λεγέσθαι* is simply a this (1030a4), it might seem that the predication of form of matter which yields a composite substance cannot be an instance of *ἄλλο κατ' ἄλλου* predication. On the other hand, Aristotle describes the essence (*τὸ τί ἦν εἶναι*) as coming to be in another (*ἐν ἄλλῳ*) (1033b7), and the arguments of Z17 assume that the relation between form and matter is that of one thing belonging to another.

12 Cf. Loux (p. 33) who suggests an analysis of Zeta 13 along the lines of Lewis's account.

13 Cf. Lewis, p. 67. Lewis makes form universal in relation to many portions of matter; I prefer the locution; 'portion of matter', to allow for the existence of a universal with only one (actual) instance. Aristotle's favorite examples are artifacts (1033a13–b18, 1034a21–28, 1041a6, 1043a7–20, 1043b7–9), but he expresses reservations about whether artifacts are substances (1033b20, 1043b21, 1060b27, 1070a15, 999b16–20).

14 The form-matter distinction is more easily illustrated with a bronze sphere than a man, and we are in good company using it.

15 A third possibility has been suggested to me by Jim Bogen. On it, 'this portion of bronze is a sphere' is true, although it does not state an identity, since there are no contingent identities. Irrespective of the merit of this position, there is no evidence that Aristotle subscribed to it. Aristotle frequently distinguishes between contingent identity and being the same in essence; an object's being identical with itself is given as an example of the latter at 1019a7–9. But he nowhere makes a distinction between truly saying that *a* is *b* and *a*'s being contingently identical with *b*. I take (i) and (ii) to exhaust the alternatives Lewis might plausibly attribute to Aristotle.

16 Aristotle seems to use the "passing away into" test for human matter as well. Cf. 1035a33: Callias passes away into flesh and bones; cf. 1035a18–19. As Ackrill points out, this is not unproblematic, given Aristotle's insistence upon defining the parts of biological individuals functionally. J. L. Ackrill, 'Aristotle's Definitions of *Psuche*', *Proceedings of the Aristotelian Society*, 1973, pp. 119–33.

17 Cohen argues persuasively against the identification of the matter with the composite (pp. 398–401). S. Marc Cohen, 'Essentialism in Aristotle', *Review of Metaphysics* **31** 387–405. At *Physics* 190b25, Aristotle uses ὁ χρυσός (gold) and ὁ ἄνθρωπος (man) as examples of the matter underlying a change. One might argue that this amounts to the recognition of material individuals. Here it is important to distinguish between being the matter for some change and being the matter of some object. In *Physics* A7, the thing which undergoes a change is the matter, the characteristic which is acquired during the change, the form. Thus, in the *Physics*, but not in the central books of the *Metaphysics*, names of composite substances appear on lists of matter.

18 Cf. Loux (p. 7) who in articulating an interpretation similar to Lewis's says, "Presumably, the following are examples of the kind of predication Aristotle has in mind in H2 and H7:

(1) That pack of flesh and bones is a man
(2) That clump of wood is an oak tree."

We might wonder why if this is Aristotle's position, he did not give us examples of this sort.

[19] Lewis writes as if Aristotle uses "καθόλου" in one sense only. This is clearly not the case. At *Posterior Analytics* 73b26–27, Aristotle says, "I mean by universal whatever belongs to something both in every case and in itself and as such". As John Driscoll points out, form is not universal in relation to matter in this sense. John A. Driscoll, 'EIΔH in Aristotle's Earlier and Later Theories of Substance', Note 75 in ***Studies in Aristotle***, ed. by O'Meara, forthcoming.

[20] Modrak, pp. 273–4.

[21] This criticism was raised in discussion at the Conference. The topic was first broached by Calvin Normore and further elaborated by Alan Code and Jim Bogen.

[22] For the sake of brevity, I will not take up the question of whether Lewis should attribute the notion of prime matter to Aristotle. One passage in *Metaphysics* Z which is often cited as evidence that Aristotle accepts prime matter is 1029a24–26, but see Charlton who argues against this interpretation in his notes on *Aristotle's Physics I, II*, pp. 136–8, as does M. Schofield, '*Metaphysics* Z3: Some Suggestions', *Phronesis* 71, 97–101. H. M. Robinson, on the other hand, finds "Aristotle referring approvingly to prime matter" in this passage (p. 88). H. M. Robinson, 'Prime Matter in Aristotle', *Phronesis* 19, 168–88.

[23] I have some reservations about how wholeheartedly Aristotle endorses this treatment of flesh. At *De Gen. et Corr.* 321b31–32, Aristotle says that flesh and bone would seem to belong more to a corpse than a hand.

[24] Ackrill, pp. 125–30.

[25] Changes in the definitions discussed here were received too late for revision of my comments.

ALAN CODE

ON THE ORIGINS OF SOME ARISTOTELIAN THESES ABOUT PREDICATION

I. G. E. L. OWEN ON THE THIRD MAN ARGUMENT AND METAPHYSICS Z6

The distinction between essential and accidental predication forms the basis of the doctrine of substance and categories that Aristotle had worked out by the time he wrote the *Categories*. In 'The Platonism of Aristotle', G. E. L. Owen presents his own account of the origin of that doctrine.[1] Owen claims that Aristotle first drew the distinction between two sorts of predication when considering which of the following two premises of the Third Man Argument (TMA) should be abandoned in order to avoid an infinite regress:

Non-identity (NI):	What is predicated is always something different from the subjects of which it is predicated
Self-predication (SP):	What is predicated is itself a subject of that same predicate.

While pondering this problem, Aristotle realized (according to Owen) that no *single* answer holds good for all predicates, and hence was led to draw a distinction between two sorts of predicate and two sorts of predication. For the weak predication relation (what is elsewhere called accidental predication), NI holds but SP fails; for the strong predication relation (what is elsewhere called essential predication), SP holds but NI fails. In either case, one of the premises needed to generate a regress is lacking, and Aristotle has a theory immune to the TMA.

He claims that the *Categories* does not face up to such an 'outright rejection' of NI for essential predication and was written when Aristotle was "still at the stage of disarming strong predication by the old plea that 'man' does not stand for any individual thing".[2] However, Owen argues that by the time Aristotle had composed *Metaphysics* Z, he had decided to reject NI outright for *all* essential predications, and came to hold that "if we take any primary subject of discourse and say just what it is, we must be producing a statment of identity ... ".[3] On this view, the subject of a true essential predication is always identical with the predicate. His support for this

James Bogen and James E. McGuire (eds.), How Things Are, 101–131.

interpretation consists of (1) a philosophical argument to the effect that Aristotle must reject NI for essential predications in order to avoid the TMA, and (2) the historical claim that such philosophical reasoning led him to argue, in *Metaphysics* Z6, for the thesis that each primary subject of discourse is identical with its essence.

This motivation for the Z6 thesis involves a serious logical fallacy – that of confusing *all* with *some*. The negation of NI is:

NEG-NI: It is not the case that what is predicated is *always* something different from the subjects of which it is predicated.

To accept NEG-NI for essential predication is to endorse the view that in *some* cases the subject of such a predication is identical with the predicate. NEG-NI does not entail that subject and predicate are *always* identical.[4] Thus, even if it is true that Aristotle rejects NI for essential predication, this fact does not help explain why he adopted the Z6 thesis unless we suppose that he committed a logical fallacy. The main purpose of this paper is to show that the Z6 thesis is not motivated in the way that Owen suggests and that he has not succeeded in demonstrating that essential predications are, according to Aristotle, statements of identity. By developing my own interpretations of the TMA and the Z6 thesis, I will try to show that the former does not provide the motivation for the latter. On my interpretation, the TMA does not involve Owen's NI (though it does involve another version of Non-identity). Where the subject is a primary substance and the predicate the essence of that substance, Aristotle does hold that subject and predicate are identical. However, a particular like Socrates is not, according to the hylomorphic analysis of *Metaphysics* Z, a primary substance, and he is not identical with his essence. Where the subject is a material composite, and the predicate is the essence of that composite, subject and predicate are not identical. Such predications classify a composite according to natural kind but do not identify the composite with the kind that it is.

Furthermore, Aristotle rejects the idea that universals are *separable*. To avoid an infinite regress of Forms of Man, he rejects the separation of the universal man, but does not reject NI. The Form of X, e.g. the Form of Large (or Man), is logically separable from the particulars of which it is predicable *only if* its definition (i.e. the definition of what it is to Be Large, or what it is to Be [a] Man) applies to none of those particulars, though it is involved in the account that gives the conditions for their being large (or men) – an account that neither applies to the Form nor is involved in its definition. Furthermore, a perfectly good reconstruction of one of the main Z6 arguments for the thesis in question can be given without mentioning either the TMA or Owen's NI.

II. A GENERAL CHARACTERIZATION OF THE DISTINCTION BETWEEN ESSENTIAL AND ACCIDENTAL PREDICATION

To facilitate the discussion of the TMA and the Z6 thesis, I begin by stating briefly how the notion of predication figures into Aristotle's thought.[5] Taking the two-place relations *Being* and *Having* as primitive, we may define *essential* and *accidental predication* as follows:

DEF 1: X is essentially predicable of Y iff Y Is X.
DEF 2: X is accidentally predicable of Y iff Y Has X.

Predication is defined in terms of the disjunction of essential and accidental predication; *identity* is simply two-way, or reciprocal, essential predication.[6] A *universal* is an item that can be truly predicated of something distinct from itself; a *particular* is an item that cannot be predicated, either essentially or accidentally, of anything distinct from itself; an *individual* is an item not essentially predicable of anything distinct from itself.[7]

Ontological predication helps us to understand linguistic predication.[8] A universal is essentially predicable of a logical subject X if and only if both the name and the definition of that universal truly apply to X; otherwise, either the universal is not predicable of X, or it is accidentally predicable. One consequence of this, crucial to my assessment of the significance of the TMA is that, since the definition of *man* applies to particular men, the associated universal is an essential predicate of those particulars. Furthermore, since the definition applies to both the universal man and the particular men, the universal is essentially predicable of itself and those particulars in the same way.

According to Aristotelian doctrine, a particular is a logical subject, or subject for predication, in virtue of the fact that it Is something (definable) essentially. The species under which a particular falls is the definable something that it, the particular, must essentially Be if it is to be anything at all.[8a]

For example, Socrates is a particular. He belongs to the species man; and for him to be the particular subject that he is, he must Be a man essentially. The word 'man' is applicable to him because its signification, the species, is predicable of him. Not only does the name of the species apply, its definition applies as well. Suppose that the account that defines man is 'biped animal'. Whatever is a man must be a biped animal, and hence the phrase 'biped animal' applies to Socrates. Consequently, when we use the sentence 'Socrates is a man' to say of Socrates that he is a man, we speak truly *because* the word 'man' signifies what Socrates essentially Is – a man.

Some linguistic predicates, such as 'man', signify universals that are essentially predicable of all the particulars of which they are predicable. These terms may be used to classify particulars according to their natural kinds. In the *Categories*, though not in the middle books of the *Metaphysics*, particulars are primary substances, the natural kinds that are essentially predicable of them (i.e. their species and genera) are secondary substances, and there are no other substances besides these.

In addition to the names of substantial universals, there are also names of the universals that are accidentally predicable of substances. Although, in a few cases, the name of a non-substantial universal ('white', for instance) applies to the substances that Have that item, in general it is a linguistic predicate associated with the name, not the name itself, that applies to all and only those things Having the universal as an accidental property. However, the *definition* of a universal that is accidentally predicable of a subject can never be truly applied to that subject. Thus, although the name 'white' applies to white particulars in virtue of the fact that they all Have the color white, the definition of that color is not linguistically predicable of any of them.

III. ARISTOTLE AND THE THIRD MAN ARGUMENT

Despite the fact that most recent discussions of the 'third man' focus on the argument at *Parmenides* 132a–b,[9] the present discussion will be confined to the infinite regress argument found at 84.21–85.3 of Alexander's commentary on *Metaphysics* A9.[10] Aristotle believed that the logic of a substance-term like 'man' is quite different from that of an incomplete predicate like 'large' – the universals they introduce belong to distinct categories. In particular, only in the case of 'man' does the definition apply to each of the particulars to which the predicate applies. Below, we shall see how this difference helps to explain why Aristotle did not use the TMA to dispute the existence of Forms corresponding to incomplete predicates. The argument in the *Parmenides*, purporting to generate an infinite series of Forms corresponding to the incomplete predicate 'large', does not pose a serious threat to a Platonic theory that denies that the one thing that Is large Has the large.

Suppose that one decides to explain the applicability of the word 'man' to particular men in terms of some universal X and a predication relation R such that not only does the word 'man' apply to X, but also it applies truly to a particular if and only if that particular stands in relation R to that universal

X. This could justify the postulation of a universal *man* predicable of all particular men. Both Aristotle and his Platonist opponent can agree to this much. The latter, however, holds that the item predicable in common of the particulars is *logically separable* from those particulars,[11] while Aristotle believes this separation leads to the Platonist's difficulties, one of which is the TMA.[12] If the Form is separable in this way, Aristotle thinks that it must be an individual *τόδε τι* (a 'this some*what*'), and not essentially predicable of anything but itself. In accordance with the ideas sketched in Section II, we see that since no definition of a separable Form can apply to any of the particulars that participate in it, that Form is not essentially predicable of those particulars. The separability of Forms goes hand in hand with the idea that Forms are participated in *accidentally*. If the Form of Man is separable, then particular men are accidentally, not essentially, men. Aristotle finds this unacceptable. Let us, therefore, consider one consequence of separation – the TMA.

Aristotle took the TMA to presuppose a One Over Many premise:[13]

ALEX-OM: If each member of a plurality of objects is a man, then there is an X such that:
(1) X is a man,
(2) X is separable,
(3) X is predicable of each member of that plurality.

For instance, since each member of the plurality of sensible men is a man, it follows by ALEX-OM that there is an X such that X is a separable man that is predicable of the plurality of sensible men. The Platonist, who adopts this principle, will hold that the term 'man' is applicable to each sensible man in virtue of the fact that this separable man is predicable of each.

Another premise of the TMA is a version of the Non-identity assumption:[14]

ALEX-NI: If what is predicable of each member of a plurality is *separable*, then it is distinct from each member.

If some item is separable (and remember, nothing is separable from itself), it is a *τόδε τι*, and hence an individual (*ἄτομον*). An individual is accidentally predicable of everything (except itself) of which it is predicable at all. Such an item cannot be a member of the plurality of items of which it is accidentally predicable, and thus it must be distinct from those items. Consequently, from ALEX-OM, the principles of Being and Having that underlie ALEX-NI, and the premise that each of the many members of the plurality of sensible men is a man, we may infer the existence of a separable man that is accidentally predicable of, and distinct from, each particular man. What is truly predicated of a plurality of sensible particulars is also something other than the sensibles, and exists in addition to those things of which it is predicable *because* it is separable.[15]

The so-called 'Self-Predication' assumption is not a premise of the TMA. Owen's SP does not play any role at all in this argument. ALEX-OM does have built into it the idea that the postulated item is a man; but, since it does not say that this item is predicable of itself (it says that it is a man, but not how or why), it would be quite misleading to label ALEX-OM1 'Self-Predication'. (ALEX-OM1 is the first of the three conditions given in the antecedent of ALEX-OM.) The Platonist does, in fact, believe that each Form is essentially (not accidentally) self-predicable, but this is not a premise of the present argument.

By an application of ALEX-OM to the group of sensible men we inferred the existence of a separable man, a Form, predicable of each particular man. Since ALEX-NI guarantees the distinctness of this Form from the particulars, we may form a "new" plurality of men – one consisting of the Form together with the particulars. Next, we may apply ALEX-OM to this "new" plurality, and infer the existence of a "new" separable man that is predicable of both the particular men and the man that is truly predicable of the plurality of particular men.[16] Another application of ALEX-NI informs us that this "new" Form that is predicable of this "new" plurality is not identical with any of the members of that plurality, and hence is not identical with the original Form postulated by the first application of ALEX-NI.[17] The "new" Form is accidentally predicable of the original Form and of each sensible man. Repeated applications of ALEX-OM and ALEX-NI generate an infinite regress of Forms.[18]

Relying upon ALEX-OM and ALEX-NI, the infinite regress is generated without relying upon inconsistent premises. The conclusion is, of course, at odds with the claim that there is a *unique* universal that Is itself (a) man, and is that by which all particular men are correctly called men. Furthermore, the infinite regress is at odds with the semantic motivation for the introduction of Forms. To see this, we should now consider some semantic meta-principles.

I have intentionally refrained from the semantic task of disambiguating the copula in the antecedent of ALEX-OM as well as in ALEX-OM1. It is by insisting that the word 'man' applies to the postulated universal and to the particulars *in the same way* (so that the copula is interpreted in the same way for each case) that Aristotle can show the Platonist to be committed to the "new" plurality, and hence to the "new" universal man. Seeing that each particular man is a man, and that the postulated Form is also a man,

Aristotle wants us to form a "new" plurality consisting of that Form together with the particular men, and apply ALEX-OM to that plurality. This move requires:

(A) The term 'man' applies to each member of the plurality of men mentioned in an application of ALEX-OM and to the universal postulated by that application *in the same* way.

This is a semantic meta-principle that sanctions the formation of the "new" plurality, and is not itself a *premise* of the TMA.[19] (The Platonist will want to resist (A) and hold fast to the separation assumption embodied in ALEX-OM2.)

Notice that the application of the term 'man' to the sensible particulars is not to be explained by saying that the original universal is essentially predicable of each of the many men. The separability assumption rules this out. However, it cannot be explained in terms of the Having of this universal unless we also explain the application of the term to the universal itself in the same way. But this option is ruled out as well – nothing is accidentally itself (nothing Has itself). Consequently, the original universal is neither essentially nor accidentally predicable of the particulars; and if we stick to our resolve to explain the application of that term to particulars by reference to a universal and a relation R, there had better be a "new" universal to do the explaining (or a "new" relation – an option not pursued). Thus, it is not the original universal that explains the applicability of the word to particulars, and a "new" universal must be postulated in order to explain the applicability of the word to both the original universal and the particulars. It turns out that this "new" universal cannot, in fact, explain what it is supposed to explain any more than the original did. We are set off on an infinite regress and never do arrive at a satisfactory explanation of the attachment of words like 'man' to things like men.

Using ALEX-NI, we learn that the One, postulated by ALEX-OM, is not identical with any of the Many. That in virtue of which the term 'man' applies to the plurality (i.e. that in virtue of which the members of the plurality are called what they are called) is a separable man that is accidentally predicable of each member of that plurality. Thus:

(B) The term 'man' is applicable to particular men in virtue of the fact that they all Have a (universal) man in common.

Using (A) and (B), one can disambiguate the relevant occurrences of the copula in ALEX-OM. (Unlike (A), principles (B) and (C) – see below – are not used to generate the regress.)

Neither Plato nor Aristotle would accept both (A) and (B), however, for they both agree to the following semantic meta-principle:

(C) The application of the term 'man' to a universal man is not to be explained in terms of something distinct from it that it Has, but rather in terms of what it Is.

The Platonist holds that the term 'man' applies to the Form Man in virtue of what that Form really Is. (This principle is not, however, a premise of the TMA.) The fatal consequence of the conjunction of (A) with (B) is that the term 'man' is applicable to the universal man (postulated by the first application of ALEX-OM) in virtue of something that it accidentally Has, not in virtue of what it Is. Thus (A) and (B) together entail the negation of (C). Since both Aristotle and and the Platonist accept (C), both can agree that one of the other two is false; but they cannot agree as to which is the culprit. The claim (B) that the particulars all Have a universal man in virtue of which they are correctly called men is an integral part of the Platonic theory of Forms, but is actually inconsistent with Aristotle's views on predication. The idea that (a) man is predicable in the same way of both this universal and the particulars (a direct consequence of (A)) should play no role at all in the Platonist's theory, and provides the key to the anti-Platonism of the *Categories*.

Aristotle thinks that Plato's treatment of the item predicable in common as a separable *τόδε τι* rather than a *τοιόνδε* (a 'such') leads to the TMA.[20] If the item predicable in common is a separable *τόδε τι*, then it is an individual and, hence, accidentally predicable of all of the items of which it is predicable (except itself). If the universal man is a *τόδε τι*, it is accidentally predicable of the particular men that participate in it as something they Have in common. Contrary to this, Aristotle holds that each particular man Is (a) man, and cannot participate in or Have (a) man. Aristotle rejects (B).

Since the Platonist accepts (B), he should reject (A) and disown the idea that the term 'man' applies to both a universal man and the particular men in the same way. He will have to maintain that, although the universal man Is (a) man, the particular men merely Have that man. Only in this way could he prevent the formation of a plurality of men consisting of the particulars together the original Form, and hence stop the regress.

Had the argument been concerned with a predicate that is incomplete in its ordinary applications to particulars (such as the predicate 'large'), Aristotle could have been forced to allow such a move. After all, once he grants the existence of a universal *largeness*, he would himself agree that the definition of that universal could not apply to any of the large particulars. Hence by his own lights, that universal would be accidentally predicable of the particular large things, but (since its own definition applies to it) must be essentially predicable of itself. Since largeness would not be predicable of both itself and the large particulars in the same way, there would be no harm (as far as the TMA is concerned) in treating largeness as an individual. Aristotle would have to recognize that the TMA cannot refute the existence of such Forms. In fact, he distinguishes Forms that are supposed to correspond to predicates that are incomplete in their applications to particulars (such as 'large' and 'equal') from Forms that correspond to predicates (such as 'man') that do apply in a complete way, κυρίως, to particulars. Significantly, he does not appeal to the TMA when disputing the existence of Forms that correspond to the former class.[21] According to the Platonist, there must exist a Form of equality that is the bearer of the predicate 'equal' in a complete way, κυρίως and without qualification. Only in this way is the non-equivocal use of that same predicate in its incomplete applications to particulars explicable (or so the Platonist argues). Against this Aristotle can argue that due to the logic of incomplete predicates, nothing can be just equal *simpliciter*, and so there can be no such Form.[22] However, he does *not* invoke the "third man" at this point.

A predicate like 'man', on the other hand, does apply κυρίως to both particular men and to the universal *man*. Not only does the term apply without qualification to the universal, but it also applies without any qualifications or riders to each particular man. Aristotle believes that this supports the claim that the term applies in the same way to both the universal and the particulars, and hence prevents the Platonist from denying (A) in the case of the predicate 'man'. It is when arguing against Forms corresponding to this type of predicate that Aristotle thinks he can rely upon the TMA. This explains why his own regress argument against separable Forms (unlike the regress of *Parmenides* 132a–b) makes use of a predicate that is complete in its ordinary applications to particulars.[23]

The fact that the term 'man' applies κυρίως to both the universal and the particulars requires explanation. I would like to suggest that Aristotle would explain this by appealing to the proposition that the definition of *man* is applicable to both. Since the definition of man is linguistically predicable of both the universal *man* and the particular men, this requires (according to Aristotle) that each Is (a) man, and hence the particulars do not Have a man in common. This is at the heart of Aristotle's early essentialism, and if we cannot accept it, neither can we accept his essentialism. Thus he is not about to concede that (A) is false.

However, the TMA will not be a valid objection to the view that the particular men participate in the one and only thing that really Is (a) man if it would be reasonable for the Platonist to deny (A) while holding firmly to his own metaphysics. Professor Lewis discusses some of the difficulties involved in arguing for (A) in the context of the TMA as I have interpreted it, and at this point I urge the reader to consult his paper ('Plato's Third Man Argument and the "Platonism" of Aristotle', Section 3c, *this volume*).

Since neither NI or SP, as formulated by Owen, are *premises* of the TMA as Aristotle understood it, he could not have hoped to block the regress by denying one of those two principles. Separability is the real issue, and Owen misses the point when he portrays Aristotle as trying to locate the difficulty in either SP or NI, neither of which even mentions separation. The TMA is a *reductio* of ALEX-OM, and it is *that* principle that Aristotle wishes to reject. It is objectionable because it postulates a *separable* Form of Man.[24] He has no quarrel with the idea that there is a universal, itself a man, that is predicable of all particular men, nor does he object to the claim that this universal is not identical with any of the particulars. However, if one goes on to say that this universal is separable, the TMA ensues. To avoid the regress, he denies only separability. Thus, Aristotle did not first draw the distinction between essential and accidental predication when considering which principle, SP or NI, should be rejected in order to prevent the TMA regress.[25] We have thus dispensed with Owen's claim that Aristotle must reject NI for essential predications in order to avoid the TMA. We saw earlier (in Section I) the rejection of NI does not lead to the view that essential predications are statements of identity, and now we see that Owen's NI is not a premise of the TMA. Thus, I submit that the Z6 thesis is not to be understood in terms of an attempt to avoid the TMA by denying NI. The investigation of the true motivation for the Z6 thesis is our next task.

IV. ON THE INTERPRETATION OF THE Z6 THESIS

A. Definition and Essence

The Z6 thesis is the principle that all things *πρῶτα καὶ καθ' αὑτὰ λεγόμενα* (primary and called what they are called in virtue of themselves) are one and the same as their essence.[26] Some of Aristotle's remarks about definition and essence in the *Organon* help to clarify the connection between a definable object and its essence. *Topics* A5 states that a definition (ὅρος; called ὁρισμός at 102a4, 5) is an account (λόγος) signifying (σημαίνων) and essence (*τό τί ἦν* εἶναι).[27] That is to say, an essence is the ontological correlate of a definition. This idea is of the utmost importance for the interpretation of the Z6 thesis.[28] An ὁρισμός is always a λόγος that may be given in place of either an ὄνομα (name) or another λόγος. Aristotle is using 'ὄνομα' to mean a simple name, and contrasts it with a λόγος, which is a complex phrase.[29] A definiendum may be either a syntactically simple name or a complex noun phrase (in which various names are concatenated in grammatically acceptable ways), and a definiens is always a complex phrase. A correct definition contains a definiens that signifies an essence. The essence signified by the definiens is the essence of the item defined. This last assumption is embodied in Aristotle's terminology. Frequently, he picks out an essence using a phrase constructed by attaching 'τὸ τί ἦν εἶναι' to the definiendum in the dative case. Just as the word '*S*' signifies the entity *S*, so the definition of *S* signifies the essence of *S*.

Aristotle never endeavors (in the extant works) to explain the unusual construction of the Greek phrase that we translate as 'essence'. However, there are obvious connections between his use of this phrase and his use of both 'τὸ τί ἐστι' ('the what Is *X*?', or 'the what *X* Is') and 'τὸ εἶναι' ('to be'). Indirect use is made of the former connection in *Metaphysics* Z4 when it is said that the essence of each thing is what it is said to Be in virtue of itself (καθ' αὑτό).[30] An essence is essentially predicable, or predicable καθ' αὑτό, of that which it is the essence. *Posterior Analytics* A4 distinguishes two distinct ways in which some entity *P* might belong to a subject *S* in virtue of itself (in virtue of *S*),[31] and *Metaphysics* Z4 (1029b16–18) indicates that only the first is relevant to the characterization of essence. My discussion will be confined to that case.

In the first of the ways in which something belongs 'in virtue of itself', or *per se*, *P* belongs to *S* in virtue of itself if *P* is in the 'what *S* is' (ἐν τῷ τί ἐστι). The substance of *S* is composed of such *P*s, and such *P*s are in (though, perhaps, he should have said 'mentioned in') the account that says what *S* Is. Thus, *P* belongs to *S per se*, or καθ' αὑτό, just in case *P* is one of the items mentioned in a correct answer to the question 'What Is *S*?'. The *essence* of *S* is the entity signified by the *entire* account saying what *S* Is. Since a definiens is an account signifying an essence, and an account signifying what *S* Is signifies the essence of *S*, it follows that the definiens in the definition of *S* is the account that says what *S* Is.[32] Aristotle includes in the 'what *X* Is' all of the items signified by any part of the appropriate definiens, or signified by any part of the account of a part of the definiens, and so on.

We may now clarify the connection between the ontological predication of an essence and the lingusitic predication of the associated defining expression.

If *X* is a definable object, then its definition is an account that says what it Is. The essence of the definable object is identical with the signification of its definition. The essence of *X* is essentially predicable of *X*, and the definition of *X* is linguistically predicable of *X*. In order to predicate the essence, we may linguistically predicate the definition that signifies that essence.

If *X* is a particular, then it is not a definable object. Nonetheless, in order to be anything at all, it must Be something essentially, and hence the definition of some definable entity must apply to it. Such a definition will be the definition of the essence of *X*, and, consequently, will serve as an account saying what *X* Is. However, since *X* is not itself definable, this definition will not be the definition *of X*. The essence of a particular is not, of course, the signification of the definition of that particular (there is no such definition) – rather, it is the signification of an appropriately related definition. Which definition is 'appropriately related?' The definition that is *an account saying what X Is.*

Whether or not *X* is a definable object, the essence of *X* is identical with the signification of the defining expression that is appropriate for *X*. To predicate an essence of that of which it is the essence, we may linguistically predicate the appropriate associated defining expression. The appropriate definition will be an account saying what *X* Is, and will apply to *X* precisely because it signifies the essence of *X* – the essence of *X* itself being essentially predicable of *X*. In saying what *X* Is, we predicate essentially of *X*; and when we apply the definition appropriate for *X*, we are thereby saying what it Is.

In the case of a particular man like Socrates, his essence is identical with the essence of the species man. Remember the *species* of a particular is what that particular must Be if it is to be anything. The species is what the particular essentially Is, and an account saying what that species Is *ipso facto* is an account saying what that particular Is. The definition of the species applies to each particular member of the species, and hence the defining expression that signifies the essence of a particular just is the definition of its species. In general, the definition appropriate for a particular is identical with the definition of the species to which *X* belongs. In the case of Socrates, the species is man; and Socrates must Be a man in order to be anything at all. Man is what Socrates Is. An account that says what man Is is also an account that says what Socrates Is. The definition of man is an account saying what man Is, and so the definition appropriate for Socrates is the definition of the species man. Thus, the definition appropriate for Socrates is the same as the definition appropriate for the species man. If the appropriate definition is the same for both, then, of course, the signification of the definition appropriate for Socrates is the same as the signification of the definition appropriate for man. Since the essence of X is just the signification of the appropriate definition, it follows that the essence of X is identical with the essence of Y if and only if the signification of the definition appropriate for X is identical with the signification of the definition appropriate for Y. Thus, the essence of Socrates is identical with the essence of the species to which he belong.[33]

Each properly *definable* object is numerically one and the same as its essence. Most discussions about definitions in fact arise from considerations concerning sameness or difference. After all, if we can argue that two items are the same or different, we can, in this fashion, go after definitions – for having shown that the two items are not the same, we shall have destroyed the proposed definition.[34] Whenever some universal is defined, it must be the case that the subject of the definition (the universal) is the same as the essence signified by the definiens. If man is correctly defined as 'biped animal', then *man* and *biped animal* are the same thing. A properly definable item is the same at its essence.

Although it is necessary that in a correct definition the subject and predicate signify the same thing, this is not sufficient. Three types of sameness are distinguished in *Topics* A7. The word 'same' may be correctly applied to cases of numerical sameness, specific sameness, and generic sameness. Everybody agrees that the term 'same' applies most of all to that which is one in number.[35] There is an instance of numerical sameness if there is just numerically one item, but more than one name (or phrase) applied to that item: for instance, 'λώπιον' and 'ἱμάτιον' are two distinct names that may be applied to numerically one and the same thing. This robe is numerically the same as this cloak – I apply two different expressions to just one thing.[36] Things are strictly and in a primary way called 'the same in number' whenever they are the same in virtue of their names or definitions. This man is numerically the same as this biped animal, and this robe is numerically the same as this cloak, in the strictest possible way.[37] Also, definiendum and definiens signify numerically one and the same thing in the strictest possible way. Each properly definable thing is (strictly speaking) numerically the same as its essence. *If* each primary *καθ' αὑτὸ λεγόμενον* is definable, then this last statement implies the Z6 thesis. However, even without the Z6 thesis, we can see that the species man, since it is definable, is (strictly speaking) numerically the same as its essence. Whether or not the species man is a primary *καθ' αὑτό λεγόμενον*, it is numerically the same as the essence signified by its definition. If this kind of sameness entails identity, then it is identical with its essence. In this paper, I will assume that this entailment holds.

B. Woods and the Scope of the Z6 Thesis

Immediately following his discussion of the Z6 thesis, Aristotle adds:

> The sophistical arguments directed towards this thesis, and the question whether Socrates and the essence of Socrates are the same, are obviously resolved in the same way – for there is no difference in either the standpoint from which one might raise the question, or that from which one could successfully resolve it.[38]

One might be tempted to take this as evidence that Aristotle held that Socrates is indeed the same as his essence. However, we have been told no more than that certain unspecified arguments against the Z6 thesis can be unraveled in the same way as the question about Socrates and his essence. This passage does not, by itself, either support or refute the contention that the very same considerations that one would use to vindicate the Z6 thesis in the face of sophistical attacks should lead us to conclude that Socrates is the same as his essence. However, once the question about Socrates has been raised, one would like to know if there are other passages that show us how Aristotle would have answered it.

M. J. Woods, who thinks that the Z6 thesis is meant to apply to particular substances, like Socrates, argues that evidence from *Metaphysics* Δ18, Z4, and Z6 can be put together to show that he was committed to the view that Socrates is, in fact, identical with the essence of the species man.[39] Woods also claims that this "commits him to the position that a statement like 'Socrates is a man' is some sort of identity statement."[40] On his interpretation, (i) true essential predications are identity statements, and (ii) there are true essential predications having an individual substance as the subject. Woods says that Owen has recognized that Aristotle was inclined to accept this consequence of the Z6 thesis,[41] but I suspect that this is not quite Owen's position. Although Owen *does* think that *Metaphysics* Z suggests that essential predications are identities, he does not connect this idea with the claim that an individual like Socrates can be the subject of a genuine essential predication. He says, rather, that this idea "in turn helps to persuade him that the primary subjects of discourse cannot be individuals such as Socrates ... but species such as man".[42] Owen seems to view Aristotle as tempted by the thought that, since the statement that Socrates is a man is *not* an identity, it cannot be a genuine essential predication. Perhaps Owen believes that the sentence 'Socrates is a man' is not in proper, canonical form; and once properly analyzed, it will be apparent that Socrates is not the real subject.

Be that as it may, let us now turn to the argument that Woods gives to show that the statement that Socrates is a man is, in fact, despite appearances, an identity. He argues that, since the definition appropriate for Socrates is the definition of man, and the essence of a thing is the signification of the appropriate definition, it follows that the essence of Socrates is the essence of man. Further, since man is identical with the essence of man, it follows that the essence of Socrates just is the species man. So far, I agree. However, Woods thinks that the Z6 thesis applies to particular substances, and hence Socrates is, on his interpretation, identical with the essence of Socrates. Putting this last claim together with the claim that the essence of Socrates is the species man, one may derive the claim that Socrates is identical with the species man, and Woods thinks that this suffices to show that 'Socrates is a man' is an identity statement.[43] We had better see how Woods supports his application of the Z6 thesis to particulars.

Woods claims that *Metaphysics* Δ18, 1022a24–27, commits Aristotle to the identity of Callias and his essence:

It follows that that which is *by itself* (in its own right) [καθ' ἀυτό] is necessarily also so called in several ways. For in one, (a thing is) in its own right what it is to be each thing, as for instance Callias is in his own right Callias, and what it is to be Callias.[44]

The phrases 'what it is to be each thing' and 'what it is to be Callias' translate 'τὸ τί ἦν εἶναι ἑκάστῳ' and 'τὸ τί ἦν εἶναι 'Καλλίᾳ', respectively. The latter stands for the essence of Callias. Perhaps Woods reasoned as follows: if Callias is καθ' αὑτὸν Callias and *E* (Callias), then (i) Callias is καθ' αὑτὸν Callias and (ii) Callias is καθ' αὑτὸν *E* (Callias); the relation said to hold between Callias and himself in (i) is the same as that said to hold between Callias and *E* (Callias) in (ii); but, since (i) asserts the identity of Callias with himself, (ii) must assert the identity of Callias with his essence. ('*E* (Callias)' abbreviates 'the essence of Callias').[45] However, against this it may be pointed out that the fact that the subject and predicate expression in (i) both pick out the same item does not require that the relation expressed between subject and predicate be identity. If we allow that the Being relation is reflexive, we could interpret (i) as saying that Callias Is Callias. Hence, we would interpret (ii) as saying that Callias Is *E* (Callias). Since the proposition that *X* is *Y* does not entail that *X* Is *Y*, (ii) does not entail that *E* (Callias) Is Callias, and consequently does not commit Aristotle to identifying Callias with his essence.[46] I prefer the latter interpretation to Woods's because Aristotle holds elsewhere in the *Metaphysics* that a material composite is not the same as its essence and indicates that the Z6 thesis is not meant to apply to such objects.[47] Consequently, the material particular that we call 'Callias' is not identical with his essence. In general, no material particular is identified with its essence or with its species. Thus, if we understand the word 'Callias' as picking out a material particular, we must reject the claim that Callias is identical with his essence. On the other hand, the essence of a particular is predicable *καθ' αὑτὸ* of that particular, and thus, although it is not the case that Callias = E (Callias), it is the case that Callias Is E (Callias). Furthermore, the sentence 'Callias is a man' does not express an identity. It expresses the proposition that the species man is essentially predicable of Callias. The sentence may be used to classify him according to kind, but not to identify him with the kind that he is.

According to Woods, Aristotle identifies a *particular* substance with a *universal*, namely, the species to which the particular belongs. We are told that "he seems not to be interested in drawing any distinction, in the case of substances, between particular substances like Socrates, and the form or kind which he shares with Callias."[48] It is quite important to see that Woods does not attribute to Aristotle a doctrine of 'individual essence' according to which the individual essence of a particular is itself particular. E (Callias) = the species man, and as such, is universal. E (Callias) is not an individual essence peculiar to Callias, but is, in fact, identical with E (Socrates). Quite apart from the question as to whether Aristotle held such a view, it is easy to demonstrate that this position is inconsistent. Since a particular cannot be predicated truly of anything else, but a universal can be, it follows that no particular is a universal. Hence, no particular is identical with any universal. This contradicts the claim that a particular substance is identical with the species of which it is a member.

A further difficulty is generated by the fact that on Woods's interpretation, Socrates and Callias are each identical with the form man.[49] Since *identity* is both transitive and reflexive, it follows that Socrates is identical with Callias. In general, on this view, Aristotle would be committed to there being no more than one member of each species. Woods appears to have a way out of this difficulty. He writes: "When we use a proper name like 'Socrates' to pick out an individual man, what we pick out is always the form; though we pick it out as it occurs in a particular piece of matter."[50] This cannot be right. If I use the proper name 'Socrates' to pick out an individual man, then what I pick out is a material particular and not a universal form. Were I to pick out a form, I would emphatically not be *using* the word to pick out a particular. Furthermore, if Aristotle were to believe that his use of a singular term like 'Socrates' picks out a universal form, then the sentences he used containing that word would be about a universal and not a particular. Consequently, even if he were using the word 'Socrates' in this unusual way, this would provide us with no evidence whatsoever that Aristotle intended to identify a *particular* with its essence.

Notice that Woods's qualification that the word 'Socrates' picks out the form *as it occurs in a particular piece of matter* will not block the inference to Callias = Socrates. Although *how* we pick out the form varies, depending upon whether we use the word 'Callias' or the word 'Socrates', nonetheless, *what* we pick out is said to be the same in both cases.

Woods invites us to compare the relation between a word-type and its various occurrences, or tokens, with the relation that holds between a species and its members. Just as the same word may occur in several places on a printed page, so the same species-form may occur in distinct parcels of matter. This analogy does not help to explain how Aristotle could have thought that a particular substance is identical with its essence. Two distinct occurrences of the word 'man' may be described as occurrences of the same *word*, but they are not the same occurrence of that word. Likewise, two distinct particular men may be described as exemplifications of the same species-form, but they are not the *same exemplification* of that form. Furthermore, just as two distinct men may be called the same *animal* (in that each is a man), though, of course, not the same *man*, so also can two distinct occurrences of the word 'man' be described as the same *word* (in that each is a 'man'), but they are not the same 'man'.[51]

Aristotle suggests, but neither explicitly rejects nor endorses, the idea that a singular term like 'Socrates' might have two uses. It could stand for either (i) a particular composite of matter and form, or (ii) just the formal component of such a composite.[52] As we have seen, a material composite is not the same as its essence, and so a word used to pick out such a composite is not being used to pick out something identical with its essence. However, if the word 'Socrates' is used to pick out the form that occurs in Socrates's matter, then it picks out the soul, and this is a primary substance.[53] Since a primary substance is not a material compound (nor is it said to be what it is said to be in virtue of one thing being in another), each primary substance is numerically one and the same as its essence.[54] In such a case, the singular term 'Socrates' is being used to pick out something that is identical with its own essence.

Aristotle also considers the possibility that a word like 'man' has two distinct, but related, uses.[55] In its normal use, it signifies the species man

– a universal composite of matter and form. However, it might also have another use. It might be used to signify the species-form (the formal component of the universal composite) in contradistinction to the species.[56]

To disambiguate, let us use 'Socrates-C' and 'man-C' to pick out the particular and the universal material composite, respectively, and 'Socrates-F' and 'man-F' to pick out the formal component of the material composite and the formal component of the universal composite, respectively. Consider the following:

(i)	Socrates-C = man-C.	FALSE
(ii)	Socrates-C Is man-C.	TRUE
(iii)	Socrates-F = man-F.	TRUE (provided that there are no particular souls)[57]
(iv)	Socrates-F Is man-F.	TRUE

Notice that this list provides four different interpretations of the sentence 'Socrates is a man'. Statements (ii) and (iv) are true under precisely the same conditions, if we take 'Socrates-F' as short for 'the formal component of the material composite Socrates-C'. The material composite, Socrates, is a member of the species man if and only if the formal component of that particular material composite Is essentially (a) human soul. Socrates-C Is (a) man-C because the form man-F is (accidentally) predicable of the material component of Socrates-C.[58]

Woods thinks that a particular material substance is identical with its essence and that 'Socrates is a man' is an identity statement when the subject is taken as referring to a material particular. Thus, he takes the word 'Socrates' in this example to pick out Socrates-C. However, Socrates-C is not identical with either man-C or man-F, and hence is not identical with either his species (which is identical with his essence) or his species-form. Furthermore, in light of these facts, the statement that Socrates is a man, a true statement, cannot identify a material particular with its species.[59]

This should answer the question as to whether the Z6 thesis applies to Socrates. It applies to Socrates-F, but not to Socrates-C. Although the text of Z6 makes it quite clear that the thesis applies to things that are primary and *καθ' αὑτὸ λεγὸμενον*, it is not clear how far it may be extended to cover other cases as well. I would like to suggest that (i) things that are primary and *καθ' αὑτὰ λεγόμενα* are the *primary substances*, and (ii) the thesis may be extended to cover *all* definable objects. A primary substance is the same as its essence,[60] and is *καθ' αὑτὸ λεγόμενον* in that it is definitionally primary and called what it is (essentially) called in virtue of itself (i.e. in virtue of the content of its definition). It is the same as its essence *because* it is a definable substance. The definiendum and the definiens in a correct definition of X always signify the same thing; and since the definiendum signifies X and the definiens signifies the essence of X, it follows that X is the same as its own essence.[61] Although every primary substance is a definable universal, not all definable universals are primary substances. A primary substance is a 'primary definable', for strictly speaking, only substances have definitions,[62] and hence, strictly speaking, only substances are endowed with essences.[63] However, even though there are definitions and essences without qualification only for substances, nonetheless in a derivative (though not a primary) way, items from other categories are also definable and hence are endowed with essences.[64] Thus, since any definable object, whether a primary substance or not, is identical with its essence, the Z6 thesis can be extended to the extent to which the concepts of definition and essence can be extended to accommodate these derivative cases. Furthermore, nondefinable items, such as matter and material particulars,[65] are not identical with essences.[66]

C. *An Argument for the Z6 Thesis*

I now turn to *Metaphysics* Z6, 1031b28–1032a4, a passage that some scholars say appeals to a version of the TMA in order to provide support for the Z6 thesis. According to my reconstruction, this passage supports the idea that a definable substance is the same as the essence signified by its definition, and does so without bringing in the TMA.

The leading ideas employed in my interpretation are (i) that an essence (in the strict sense) is the ontological correlate of a fully analyzed definiens in a correct definition (in the strict sense), and (ii) that the definiendum and the definiens of such a definition must signify one and the same entity.[67] Since each primary substance is definable (in the strict sense), then (by ii) the definiendum and definiens in its definition signify one and the same thing. The definiendum signifies the primary substance, and (by (i)) the definiens signifies the essence of that primary substance. Consequently, the primary substance and its essence must be one and the same. This argument can be extended to cover items that, although definable, are not definable without qualification. So, the Z6 thesis applies to primary substances and it holds good for them *because* they are definable.

Earlier we discussed Owen's view that Aristotle was led to consider the Z6 thesis when he rejected NI in an attempt to disarm the TMA for strong predication. We saw that this involves a logical fallacy. Since the passage that we are about to consider is thought by Woods to confirm Owen's claim,[68] it is important to see that the passage is concerned with definability, not the TMA. Here is the passage in full (*Metaphysics* 1031b28–1032a4).

(1) An absurdity would appear also if one gave a name to each of the essences,
(2) for there will be yet another [essence] besides that one [to which the name was given];
(3) for example, for the essence of horse there will be another essence.
(4) Yet why should not some things be [their] essence straight away,
(5) since [an] essence is [a] substance?[69]
(6) But indeed, not only [are a thing and its essence] one,
(7) but their definition (λόγος) is also the same,
(8) as is clear from what has been said –
(9) for it is not *per accidens* that one and the essence of one are one.[70]
(10) Furthermore, if they are to be distinct it goes to infinity –
(11) for *this* will be the essence of one, and *that* will be the one,
(12) so that also the same argument will apply to the former things.[71]

We learn from (1)–(3) of an absurdity that results from the refusal to identify a primary thing with its essence – the absurdity is that the essence in question must be distinct from its own essence! For instance, if *horse* is a primary thing and it is not identical with E (horse), then E (horse) $\neq$ E (E(horse)). Of course, it is incoherent to suppose that E (horse) is distinct from E (E(horse)). Next, (4) tells us that *some things* just are their own essence,[72] and (5) provides a reason for this – an essence is a substance. I take it that Aristotle means to say that some *essences* are identical with their own essence (and in particular, E (horse) = E (E(horse))). Since the supposition that horse $\neq$ E (horse) conflicts with the fact that some essences are identical with their own essence, the reader should conclude that horse = E (horse).

Given that E (X) = the signification of the definition of X, the absurdity discussed in (1)–(3) follows from a simple application of the general principle that, if two definable objects are nonidentical, then they have distinct definitions with distinct contents. (To put it another way, if the definition of X has exactly the same content as the definition of Y, then X is identical with Y.) The identity conditions for definable objects are given by their definitions.[73] So, given that horse and E (horse) are definable, the content of the definition of horse is not identical with the content of the definition of E (horse), if we assume that horse $\neq$ E (horse). The essence of a thing is just the content expressed by its definition, and so if horse $\neq E$ (horse), then E (horse) $\neq$ E (E(horse)).[74] As (10)–(12) indicate, this argument can be repeated, indefinitely.

Now that we have seen why the refusal to identify a (primary) thing with its essence leads to the absurd conclusion that there is a distinct "new" essence for the original essence, we should ask why Aristotle thought that this conclusion is absurd. The answer is given in (4)–(5) – since an essence is a substance, it follows that some essences are identical with their own essence.

As a first step towards understanding this answer, we should take into consideration the well-known discussion of chains of predicables, in *Posterior Analytics* A22. Where the predicable in question is predicated *ὼσ οὐσία* (as a substance) of the subject, that predicable must be definable. But it would not be definable if infinitely many things are predicable of it *ὡs οὐσία*, for one cannot go through an infinite series in thought. Consequently, if X_1 is predicable of a subject *ὡs οὐσία*, there is no infinite chain of predicables such that X_2 is predicable *ὡs οὐσία* of X_1, and X_3 is predicable *ὡs οὐσία* of X_2, etc.[75]

Since, according to Z6, $E(X)$ is a substance, it is predicable of X *ὡs οὐσία*, and is, in fact, the substance of X (see 1031a18). Likewise, E ($E(X)$) is predicable of $E(X)$ *ὡs οὐσία*, and so on. All such ascending chains are finite, and so the series terminates after a finite number of steps. That is to say, eventually we must reach *some* item in the chain that is identical with its essence. Once we reach that item, the entire chain collapses. Letting E_0 be horse, and E_k be E ($E_{k\text{-}1}$), we may represent the chain as: $E_0, E_1, E_2, \ldots,$ E_n. Suppose the chain terminates at E_n, and hence $E_n = E$ (E_n). Since E_n is the content of the definition of $E_{n\text{-}1}$ and *also* the content of the definition of E_n itself, it follows that the definition of $E_{n\text{-}1}$ and the definition of E_n have exactly the same content. Consequently, since the identity conditions of universals are given by their definitions, $E_{n\text{-}1} = E_n$. In general, for all m such that $0 < m \leq n$, $E_{m\text{-}1} = E_m$. Thus: $E_0 = E_1$, and horse = E (horse).

It is important to notice that in (1)–(3) Aristotle considers a very short chain: {horse, E (horse), E (E(horse))}. He does not *yet* make the general point that an infinite regress results from the claim that horse $\neq E$ (horse). The absurdity mentioned in (3) is simply that E (horse) $\neq E$ (E(horse)), if horse $\neq E$ (horse). (The infinite regress is also absurd, but it is not the problem to which he is immediately calling attention). The argument sketched in the last paragraph makes use of the claim that a chain of essential predications cannot go on indefinitely, but this claim does not by itself contradict the claim that E (horse) $\neq E$ (E(horse)).[76] So, we should see if Aristotle could produce an argument that directly establishes the identity of *E* (horse) with *E* (*E*(horse)), and does so by appealing to the fact that an essence is a substance.

We can see that *E* (horse) = *E* (*E*(horse) if we take into consideration the Z10 claim that the definition of a universal form must be fully analyzed and contain no part capable of further analysis.[77] Z4 and Z5 took great pains to establish that, strictly speaking, only substances are definable.[78] Putting these claims together with the claim that an essence is a substance, we may infer that each essence must have a fully analyzed definition. Now, suppose that the definition of *horse* is 'quadruped animal'. If so, then 'quadruped animal' signifies *E* (horse). Since *E* (horse) is a substance, it must have a fully analyzed definition; and *E* (*E*(horse)) will be the essence signified by that definition. However, if we attempt to define *E* (horse), i.e. *quadruped animal*, we need to find a fully analyzed account that says what *quadruped animal* Is. Thus, *E* (*E*(horse)) will be the signification of the fully analyzed definiens in a definition which contains 'quadruped animal' as the definiendum. Since 'quadruped animal' is the definition of horse, that phrase is already fully analyzed; and, hence, 'quadruped animal' will be the definiens in the definition of *E* (horse). In other words, *horse* and *E* (*horse*) have exactly the same definition. Since the essence of a thing is the item signified by its definition, *E* (horse) = *E* (*E*(horse)).

Since an essence is a substance (as we are told in (5)), *some* essence in the chain is identical with its own essence (as we are told in (4)). Due to the fact that each substance is definable, and *E* (horse) is signified by the definiens in the definition of horse and by the definiens in the definition of *E* (horse), we can be more specific: *E* (horse) = *E* (*E*(horse)). Having established in (1)–(3) that if horse ≠ *E* (horse), then *E* (horse) ≠ *E* (*E*(horse)), this suffices to show the identity of horse with its essence.

According to (8), it is clear from (1)–(5) that not only is it the case that horse and *E* (horse) are one and the same, they are, in fact, *essentially* the same in that they have the very same definition (see (6)–(7) – after all, their unity is not accidental, as (9) points out). We have just seen that this is so. Since horse = *E* (horse), the definition that says what horse essentially Is must also apply to *E* (horse) as what it essentially Is. Otherwise, they would have distinct definitions, and hence *E* (horse) would be distinct from *E* (*E*(horse)). (The definition of horse is applicable to horse because the signification of that definition, *E* (horse), is essentially predicable of horse. If that essence and that of which it is the essence are essentially the same, then the definition that says what it essentially Is must be the same as the definition that says what that of which it is the essence essentially Is.) Furthermore, considerations about definition revealed that horse and E (horse) have the same definition, and this was used to support the claim that they are, in fact, identical.[79]

NOTES

1 Reprinted in J. Barnes, M. Schofield, and R. Sorabji (eds.), *Articles on Aristotle*, Vol. 1 (Duckworth, 1975), pp. 14–34, esp. 21–5.

2 *Ibid.*, p. 23.

3 *Ibid.*

4 NEG-NI follows from SP (provided that something is predicable of something – see Lewis (below), Sec. 1 and n. 7). Since Owen thinks that the *Categories* still accepts NI, but nonetheless contains SP, he must attribute inconsistency to its author.

5 The ideas sketched in this section are given an extended treatment in my 'Aristotle: Essence and Accident', *Philosophical Grounds of Rationality: Intentions, Categories and Ends*, ed. by R. Grandy and R. Warner (Oxford, 1985). The definitions are adapted from unpublished work by H. P. Grice.

6 M. J. Woods, 'Substance and Essence in Aristotle', *Proceedings of the Aristotelian Society* 75 (1974–5), 167–80, claims that in the strongest sense of 'καθ' αὑτό', a καθ' αὑτό, or essential, predication is one in which "that which a predicate is applied to and that in respect of which it is so applied are one and the same" (p. 173). Although I think that, according to *Metaphysics* Z, the most fundamental and basic essential predications all meet the condition that Woods cites, they are, nonetheless, not statements of identity. The subject and predicate of such a predication are identical, but the predication relation in question is *not* the identity relation.

7 Whereas I take an individual (ἄτομον) to be an item that is not *essentially* predicable of anything else, Woods (p. 178) thinks an individual is something that is not *accidentally* predicable of anything. Further, while I allow that some individuals are accidentally predicable of particulars, he claims that some individuals are essentially predicable of particulars. This is connected with the fact that we both argee that the species-form is an individual, but Woods identifies the species-form with the species, while I follow J. L. Ackrill in distinguishing the two (See Ackrill's 'Aristotle's Definitions of *psuche*', reprinted in J. Barnes, M. Schofield, and R. Sorabji (eds.), *Articles of Aristotle*, Vol. 4 (Duckworth, 1979), esp. pp. 67–8. John A. Driscoll ('ΕΙΔΕ in Aristotle's Earlier and Later Theories of Substance', *Studies in Philosophy and the History of Philosophy* 9, ed. by D. J. O'Meara, Catholic University Press (1981), pp. 129–259) argues quite effectively that the middle books of the *Metaph.* (though not the *Cat.*) distinguish the species from the species-form, and treats the latter as primary substance. He discusses in detail the bearing that his findings have on Owen's claims about the TMA.

8 It is actually quite difficult to give an explication of linguistic predication. Roughly: a term is linguistically predicated of an object when one asserts a simple indicative sentence in which (i) that term occurs as (part of) the grammatical predicate, and (ii) the grammatical subject refers to that object. Due to the lack of an indefinite article in Greek, it is rather difficult to translate or paraphrase Aristotle's statements and views about linguistic and ontological predication. He believes, for example, that in the follow-

ing three cases, the word for animal is linguistically predicated of some subject (the subjects being *animal*, *Socrates*, and *man*, respectively):

(a) Animal is animal
(b) Socrates is animal
(c) Man is animal.

Omission of the article leaves one with strange syntax, while supplying it can suggest unwanted readings. (Also, depending upon whether we supply one article or two in (c), we get different readings.) Aristotle further claims that the linguistic predications represented in (a)–(c) are true *because* the item signified by the predicate 'animal' (namely, the genus animal) is essentially predicable of animal, Socrates and man, respectively. (*If* we take the copula as expressing the predication relation, then we may say that it expresses the *same* relation in each of the three cases. Consequently, it is important that the reader does not assume a set-theoretical analysis of predication (for instance, an interpretation which analyzes the copula in (b) in terms of set membership, and the copula in (c) in terms of inclusion).) The reader should note that I am assimilating the distinction in the *Cat.* between 'said of' and 'present in' to the distinction between 'essentially predicable of' and 'accidentally predicable of'.

8a Aristotle may not have articulated the ideas stated in this paragraph at the time he composed the *Categories*. Montgomery Furth ('Transtemporal Stability in Aristotelian Substances', *Journal of Philosophy* 75 (1978), 624–646) points out that in the *Cat.* Aristotle never actually says that "a substantial individual has to belong to a substantial kind" (p. 628), even though this was clearly intended.

9 See Gregory Vlastos, 'Plato's "Third Man" Argument (*Parm.* 132A1–B2): Text and Logic', *Platonic Studies* (Princeton: Princeton University Press, 1973), pp. 342–60. The first appendix to that article (pp. 361–2) lists some recent articles on the TMA.

10 Alexander of Aphrodisias, *In Aristotelis Metaphysica Commentaria*, 85.11–12, indicates that it is this argument, used in the Περὶ ἰδεῶν which is at work in *Metaph.* A9, 990B34, ff. (References to M. Hayduck's text in Volume I of *Commentaria in Aristotelem Graeca*, Berlin (1891).)

11 The present issues are concerned primarily with logical, not existential, separation.

12 *Metaphysics* M9, 1086b6–7, tells us that the difficulties for the Platonic theory of Forms are the result of separation, and Z13, 1039a1–3 states that one of those difficulties, the TMA, is the result of treating the universal as a τόδε τι. Since a Form is separable, it is a τόδε τι; since it is a τόδε τι, there will be a 'third man'.

13 Alex. 84.25–6 attributes to the Platonist the assumption that a *man* is predicable of the plurality of particular men. *If* we further assume that this *man* has been separated from the plurality of which it is predicable (84.22–23), *then* there will be a 'third man' (84.27). Since the basic OM assumption must be general enough to apply to pluralities containing universals as members, it would seem that Aristotle is attacking the assumption that, for each plurality of men, there is a separable man predicable of each member of that plurality. (It is also clear at 84.29–85.1 that the Form in question is a *man*.) This principle has a close affinity to the Sellars's 'at least one' premise (G) – see W. Sellars, 'Vlastos and "The Third Man"', *Phil. Rev.* 64 (1955), 405–37. The 'ὁ ἄνθρωπος' of 84.29–85.1 does not refer to the original Form, but rather to *a* man that is predicable of that Form and the particulars.

14 Nonidentity (though not Owen's NI) is a consequence of separation (84.23), and

the regress is avoided if one denies separation. I take '*καὶ κατ' ἰδίαν ὑφεστώς*' (84.28) to be citing the Form's separability (it Is what it Is in accordance with a substantial nature peculiar to itself – nothing else Is what is essentially Is) as the reason for its distinctness (*ἄλλος*) from the plurality of things of which it is predicable. In order for ALEX-OM and ALEX-NI to be acceptable to the Platonist, one must assume that there cannot be a plurality of men, some of which are essentially, others accidentally, men. All members of a plurality of men are men in the same way. (Alternatively, one might rewrite the principles so as to explicitly exclude consideration of 'hybrid' pluralities.)

Something is a plurality of individual men if and only if there is some universal man X such that either (i) X is essentially predicable of each individual that is a member of the alleged plurality, or (ii) X is accidentally predicable of each such member (but not both). Call this the Plurality Principle. Cindy Edwards and Nicholas J. Sandro recognized that everything that counts as a plurality of individual men in the sense characterized by Lewis (Section 2b, below) is countenanced as a plurality of individual men by this principle (provided, of course, that there are just two ways in which the term 'man' could apply to something – one explained in terms of Being, the other in terms of Having).

15 See Alex. 84.22–23, which is the referent of '*τοῦτο*' in the antecedent of the conditional at 84.27 – the consequent of which states that (if these conditions obtain) there will be a 'third man'. On the distinctness of the Form, cf. *Phaedo* 74a9–12, where *αὐτὸ τὸ ἴσον* is described as *παρα ταυτα παντα ἕτερόν τι*, when contrasting it with sensible equals.

16 84.27–85.1.

17 The 'third man' is *παρά* both the particulars and the Form (85.1).

18 85.2–3.

19 The principle allows one to move from 'the Form of man is a man' and 'each sensible man is a man' to 'there is a plurality of men that consists of the Form together with the sensible men'. Since the word 'man' applies to the sensibles and the Form in the same way, it follows that either there is some X such that they each Have X or there is some X such that each Is X (but not both); hence by the Plurality Principle, there is a plurality consisting of the sensible men together with the Form. (A) need not be considered an additional premise. (In much the same way, one can use the rule of conjunction introduction to infer "P & Q" from "P" and "Q" without treating the rule as one of the *premises* needed to infer "P & Q".) G. E. L. Owen ('Logic and Metaphysics in Some Earlier Works of Aristotle', p. 181) seems to hold that something like (A) is crucial to the TMA, but does not discuss this point in 'The Platonism of Aristotle'.

20 *SE* 178b36–179a10, *Metaphysics* Z13, 1038b34–1039a3. (Compare the language of Alex. 85.1 with *SE* 178b36–37, and 179a34 with *Metaphysics* 1039a1–3.)

21 *Metaphysics* 990b15–17 distinguishes arguments establishing the existence of Ideas '*τῶν πρός τι*' from those asserting the existence of a 'third man.' In his commentary on this, Alexander discusses the *πρός τι* case at 82.11–83.33, and the 'third man' at 83.34–85.12. The former is concerned with linguistic predicates that signify items the definitions of which do not apply *ἀκριβῶς* to particulars (83.10–11), and the example discussed is '*ἴσον*'. The definition of the equal does not apply *ἀκριβῶς* to equal sensibles, and hence an equal particular is not equal *κυρίως*. This fact is appealed to in an argument for the existence of a Form that is *κυρίως* and *ἀληθῶς* equal. No mention is made of the 'third man' at 83.24–33 where Alexander discusses the objections to the existence of such Forms. The 'third man' arguments are introduced at 83.34–84.1

as objections to postulating Ideas corresponding to groups of objects that *are* κυρίως that which is predicable of them in common. See G. E. L. Owen, 'A Proof in the περὶ ἰδεῶν', *Journal of Hellenic Studies* 77 (1957, Part 1), 103–11. See also Owen's 'Logic and Metaphysics in Some Earlier Works of Aristotle', p. 187.

22 "τὰ δὲ πρός τι ἐν τῇ πρὸς ἄλληλα σχέσει τὸ αἶναι ἔχειν (83.25–26), and hence, ὀύ φαμεν εἶναι καθ' αὑτὸ γένος" (*Metaphysics* 990b16–17).

23 83.34–84.1, with 82.13–83.2.

24 Gail Fine, 'Aristotle and the More Accurate Arguments', *Language and Logos*, edited by M. Schofield and M. Nussbaum (Cambridge: 1982), distinguishes a One Over Many principle with a separation assumption from one without it, and stresses that the argument at 84.21ff utilizes the former (p. 164). She does not, however, seem to think that the premise involves logical separation (in my sense). Furthermore, I am in agreement with her claim that the operative principle is not restricted to pluralities of *particulars* (pp. 163–4; see my Note 13).

25 In the paper cited in Note 5, I develop the view that Aristotle's doctrines concerning essential and accidental predication evolve from a Platonic distinction between Being and Having. My interpretation agrees with that of Robert G. Turnbull, 'Aristotle's Debt to the "Natural Philosophy" of the *Phaedo*', *Philosophical Quarterly* 8 (1958), 131–43, in finding in the *Categories* a substantial debt to Plato. Turnbull argues that a number of theses found in the *Categories* have origins in Plato's *Phaedo*. In particular, he claims that the "present in" relation is suggested by the *Phaedo's* "the F in x" locution. He also discusses that dialogue's parallel to "predicate of".

I would like to thank Sally Haslanger for helpful comments on a previous draft of this section.

26 *Metaphysics* Z6, 1032a4–6; see also 1031b11–14, 18–20, and Z11, 1037a33–b4.

27 Topics A5, 101b38; see also 101b21–22, 103b9–10, and 1031a12–13. Note that often Aristotle will use 'λόγος' as his word for 'definition'.

28 See second paragraph of Section IV–C, and paper cited in Note 5.

29 101b38–102a5; see also *De Int.* 16b26–27 (a λόγος must have a significant part – although in that passage a λόγος is a sentence, not a definiens, the point still holds).

30 1029b13–14.

31 73a34–b3.

32 I here ignore the *APo.* B7–10 discussion of different kinds of definition.

33 Consequently, his essence is not peculiar to him alone – it is the same as my essence, for instance. This is not to deny that if Socrates has a definition in an extended sense, then he has an essence in an extended sense as well.

34 These two sentences paraphrase 102a7–9, 11–14; see also 102a16–17.

35 103a23–24.

36 'λώπιον' and 'ἱμάτιον' are here translated as robe and cloak. This account of numerical sameness is not without problems. (Presumably, a robe is a robe whether we use one name or two?) Also, it is not obviously compatiable with *Metaphysics* Δ6, 1016b32–33 and Δ9, 1018a4–5. (Notice that 1016b31–35 distinguishes *four* types of unity, and numerical unity requires matter.)

37 103a25–27.

38 *Metaphysics* Z6, 1032a6–10.

39 Woods, p. 168.

40 Woods, p. 167. He goes on to argue that it is indeed an identity (dropping the quali-

fication "of some sort"). Christopher Kirwan, *Aristotle's Metaphysics*: Books Γ, Δ, and E (Oxford: Clatendon Press, 1971), pp. 100–1, attributes to Aristotle the view that an essential predication is such that, if the statement is true, the subject is identical with the predicate. Kirwan refers to Owen's paper for an account of the origin of the 'dubious theory of predication' according to which essential predications are statements of identity.

41 Woods, p. 167. Woods also endorses (pp. 169–70) Owen's attempt to understand the Z6 thesis against the TMA background, but thinks that the notion of *matter* figures in as well.

42 Owen, p. 23.

43 Using '$E(X)$' to abbreviate 'the essence of X', '$\Delta(X)$' for 'the definition appropriate of X', and '$\Sigma(a)$' for 'the signification of expression a,' I represent his train of thought as follows:

*(1)	Socrates = E (Socrates)	{Metaphysics Δ18}
*(2)	$E(X) = \Sigma(\Delta(X))$	{perhaps Z4, 1030a6–7}
*(3)	Δ(Socrates) = Δ(the species he belongs to)	{perhaps, in part, Z4, 1030a11–13}
(4)	the species he belongs to = man	{assumption}
(5)	Δ(Socrates) = Δ (man)	{from 3, 4, by Substitutivity of identicals (SI)}
(6)	E(Socrates) = Σ(Δ(Socrates))	{instance of 2}
(7)	E(man) = Σ(Δ(man))	{instance of 2}
(8)	Σ(Δ(Socrates)) = Σ(Δ(man))	{from 5, by SI applied to ' Σ(Δ(man)) = Σ(Δ(man))'}
(9)	E(Socrates) = Σ(Δ(man))	{from 6, 8, by Transitivity of Identity (T =)}
*(10)	E(Socrates) = E(man)	{from 7, 9, by T = and Reflexivity of Identity (R =)}
*(11)	man = E(man)	{instance of Z6 thesis}
(12)	E(Socrates) = man	{from 10, 11, by T = , R =}
(13)	Socrates = man	{from 1, 12, by T=}
*(14)	'Socrates is a man' is an identity statement.	{from 13 and ?}

Only the steps marked with an asterisk are explicitly given in his argument on p. 168. To the right of each step, I list the support which I think Woods is prepared to offer for that step. The previous section of the present paper examined the connection between essence and definition, and showed the justification for attributing 2 to Aristotle. Given the ideas sketched in Sections II and IV-A, it is easy to see why E(Socrates) = E(man). As 3 states, the definition appropriate for Socrates is the definition of the species to which he belongs, for the latter is the account that says what Socrates must essentially Be if he is to be anything. His species is, of course, man. Since the definition appropriate for Socrates is the same as the definition appropriate for man (5), and since the essence of a thing is the signification of the appropriately related definition, E(Socrates) =

E(man) as promised in (10). If we admit, as I do, that man = E(man), then E(Socrates) = man (13). Thus, *if* we also grant (1), we are forced to accept (13) – Socrates = man. I do not grant (1).

44 Kirwan's translation, quoted by Woods on p. 168.

45 See Note 43 for abbreviations. David Reier called my attention to the argument just sketched.

46 Καλλίας καθ' αὑτὸν is omitted in EJ, perhaps correctly (cf. Jaeger in OCT). If so, then this passage cites the pair {Callias, E(Callias)} as illustrating the claim that the essence of a thing belongs to it *per se*, and hence it need not be construed as a case of identity of thing and essence.

47 *Metaphysics* Z11, 1037b4–5.

48 Woods, p. 177.

49 *Ibid.*

50 *Ibid.*

51 I would be like to thank David Reier for this point.

52 See *Metpahysics* Z11, 1037a7–10.

53 1037a5; cf. 1035b14–16, 1037a28–29. See Note 57.

54 1037a33–b5.

55 *Metaphysics* H, 1043b2–3.

56 See the Ackrill article cited in Note 7 for the distinction between species and species-form. See Z10, 1035b27–30; 1037a5–10, 1037a25ff.

57 In *Metaphysics* Z Aristotle considers a position according to which (i) each primary substance is definable, (ii) the soul is a primary substance, (iii) only universals, not particulars, are definable, and (iv) Socrates differs from other men in virtue of his matter, not his form. According to such a view, either particular men do not have particular souls (and hence the soul of Socrates would be identical with the soul of Callias), or there are particular souls, but they are indefinable (and hence not primary substances).

The primary substance, *human soul*, is a universal not essentially predicable of anything except itself, and thus cannot be essentially predicable of particular human souls. Thus if there are particular souls, they are not 'souls' in the way that human soul (the primary substance) is a soul. In this paper I intend to use "soul" for the primary substance.

58 The Being of a particular composite man is explained in terms of his body (his matter) Having the substantial form that makes it alive. Thus, the word "man" applies to Socrates in virtue of something his body Has – a substantial form, the definition of which does not apply to Socrates at all. Socrates-C Is (a) man-C because the formal component of Socrates-C Is man-F.

59 Notice that since Socrates-C is not identical with E(Socrates-C), there is no danger that Socrates-C is identical with Callias-C (who is, of course, not identical with E(Callias-C)). E(Socrates-C) = E(Callias-C), but Socrates-C $\neq$ E(Socrates-C), and Callias-C $\neq$ E(Callias-C). However, if Socrates-F = E(Socrates-F), and Callias-F = E(Callias-F), then, unless we attribute to Aristotle a doctrine of 'individual essence', it follows that Socrates-F = Callias-F. That is, the formal component of Socrates-C = the formal component of Callias-G.

60 *Metaphysics* Z11, 1037a33–b2.

61 Z4 paves the way for Z6 by sketching some formal constraints on definition. 1029b19–20 tells us that the account of the essence of X (what I have been calling the definiens)

must satisfy the following two conditions: (i) it must state the same thing as (the name of) *X* (in other words, the definiendum and the definiens must signify the same thing), and (ii) it must not contain as a part the (name of) the thing defined (in other words, the definiendum must not occur in the definiens). He motivates these two conditions (ἄρα) by pointing out in b18–19 that the attempt to define surface by "white surface", a 'definition' that violates both, is not correct. I agree with Ross that the 'αυτό' in b19 refers to surface, but cannot agree that Aristotle rejects the definition because it is tautologous – it is rather obviously false. Against Ross, Woods proposes (pp. 173–4) that the 'αὐτό' refers to *white,* and the definition under consideration in b18–19 is "white surface is white." In favor of his proposal, he correctly points out that it makes "these lines attach smoothly to what precedes, since they answer an objector who tries to make white something which belongs *kathauto* in the relevant sense." However, why should one suppose that Aristotle has such an objector in mind? Furthermore, his proposal does not attach smoothly to what follows, for it leaves the 'ἄρα' of b19 problematic. Woods thinks that Aristotle's objection to the attempt to define white surface by "white" is that the definiens occurs in the definiendum. However, b18–19 neither supports nor is supported by this objection.

62 1030b4–6, 1031a12–14.

63 1030a28–30 with a6–7 and b4–6.

64 1030a30–32, b6–7. Even compounds like white man turn out to have a definition, but not in the way white does, or a substance does (1030b12–13).

65 1036a5, 1037a27, 1039b27–1040a7.

66 1037b4–6. Notice that, since an essence is the ontological correlate of a definition, and only universals are definable, it follows that essences are themselves universal. The universal, man-F, is a primary substance; and, as such, the Z6 thesis is meant to apply to it. Although man-C is not a primary substance, the thesis can be extended to cover it if it is definable. However, since Socrates-C is clearly not definable, the thesis will not, even in an extended form, apply to him. Since each definable item is identical with its essence, and since the essence of a definable item is predicable (essentially) of it, it follows that definable things, whether substantial or not, are essentially self-predicable.

67 See Section IV-A. This interpretation of the Z6 thesis was worked out in collaboration with H. P. Grice, though its subsequent development has been my own. It was originally presented in a paper ('The Relation Between Form, Matter and Composite in Aristotle's *Metaphysics* Z') at a conference held by the University of Victoria in January 1979 (sponsered by the Canada Council). In this section I do not discuss ways to extend the argument to definable universals that are not substances. The argument discussed here presupposes *i*, and the principle (related to *ii*) that the identity conditions for universals are given by their definitions.

The present section benefited from discussion with Alexander Nehamas and Allan Silverman.

68 Woods, p. 179.

69 The 'καίτοι' of b31 is adversative (cf. Denniston, p. 556ff), and the 'καὶ νῦν' has the force "as things actually are," the 'καί' being intensive. The 'εἴπερ' clause makes clear the force of 'καὶ νῦν' by making explicit the condition which supports (6). Line b31 is stating what is actually the case (since an essence is a substance), as opposed to the absurdity cited in (2) and (3). b31 does not begin a new argument as *Notes on Book Zeta of Aristotle's Metaphysics* (recorded by Myles Burnyeat and others, published as

Monograph No. 1, Study Aids, Sub-faculty of Philosophy [Oxford: 1979], pp. 41–2, suggests, taking 'καὶ νῦν' as 'even as it is (sc. without the incentive of the ἄτοπον)'. The adversative "καίτοι" shows that Aristotle is giving his own reason for rejecting the absurd consequence of the failure to identify horse and E(horse), and hence b31 continues the argument initiated at b28.

70 The first 'ἕν' of 1032a2 (as well as the 'ἐνὶ', and also the 'ἐνὸς' and 'ἕν' of a3–4 should not be construed as a reference to the Platonic Form of the One. It functions as a meta-variable ranging over definable substances in general.

71 On 'ἐπ' ἐκείνων' see Ross. I would like to thank Michael Frede for a helpful discussion on the translation of this passage.

72 Note the "ἔνια" in b31.

73 Cf. Owen's claim in 'Particular and General', *Proceedings of the Aristotelian Society* (1978), p. 3, that Z6 considers an "*eidos* which, because its identity is wholly and satisfyingly exhausted by a definition, threatens to become the chief subject of reference in its own right".

74 (1)–(3) generates the absurdity by assigning names to each of the essences. Strictly speaking, the argument does not require new names for the essences. All that is needed is some linguistic expression that (i) signifies the essence in question, and (ii) can function as the grammatical subject of a "What Is X?" question which requests a definition of essence. Since the definiens in the definition of horse (i) signifies E(horse), and (ii) can function as a grammatical subject in such a question, it would suffice for purposes of the argument. However, there are two good reasons Aristotle may have had for assigning new names to the essences. First, his argument should apply whether or not we have definitions of the original items (horse, man, etc.). But we cannot use the definiens to refer to an essence if we do not yet know what the correct definition is. Secondly, as we will see below, if we use the definiens in the definition of horse as the subject of a "What Is X?" question, it is obvious that it will also have to be the definiens in the definition of E(Horse). Hence, it is also obvious that E(horse) = E (E(horse)). However, the logic of the argument is clearer if *first* we see the reason why there must be a new essence distinct from E(horse) (if horse $\neq$ E(horse)), and *then* see why there cannot be an essence of the essence of horse that is distinct from the essence of horse. So, the device of inventing a new name aids in the presentation of the argument.

Of course, if we are to give a name to the essence of horse (on the assumption that horse $\neq$(horse)), it must be a name other than "horse". After all, horse = E(horse) if and only if horse Is E(horse) and E(horse) Is horse. However, we must assume that horse Is E(horse), since an essence is by definition what something (namely, that of which it is the essence) Is; further, E(horse) Is horse if and only if both the name and definition of horse apply to E(horse). Since E(horse) just is the signification of the definition of horse, that definition does apply to E(horse). Thus, if the name "horse" applies to E(horse), it follows that horse = E(horse).

75 See *A Po.* A22, 82b37–a1, 83a39–b8, and the discussion in Jonathan Lear's *Aristotle and Logical Theory* (Cambridge: Cambridge University Press, 1980), pp. 30–3.

76 To derive the contradiction, we also need the premise that their identity conditions are given by their definitions. Notice that the infinite regress point is made quite explicitly later at (10)–(12). If we do not identify a thing with its essence, an infinite regress threatens (see (10)), for once we distinguish $E(X)$ from X, we may apply to $E(X)$,

and to each of the subsequently generated essences, the same argument used in (1)–(3). Of course, such a regress is incompatible with (4)–(5).

77 1034b22–24 poses the question as to whether the account of each part is contained in the account of the whole, and 1035b33–1036a1 answers that only the accounts of the formal parts of the universal being defined occur in the account of that universal. The primary things of Z6 have only formal parts (see Z11 1037a33–b5), and so their definitions contain the definitions of all of their definable parts. (If *horse* turns out to have material parts, it is not a good example of a primary thing.) Since I believe that Z10 continues the discussion of Z6, and that Z7–9 were inserted later, I am glad to find that Z10 helps shed some light on the argument at 1031b28–32 for the Z6 thesis.

78 Z4, 1030b4–6; Z5, 1031a11–14.

79 This paper is an expanded version of parts of my 'Aristotle on the Sameness of Each Thing and its Essence' (Sept. 1980), presented to the December 1980 meeting of the Society for Ancient Greek Philosophy. I am grateful to the National Endowment for the Humanities for their support during the time I did most of the research for that paper. I would also like to thank Frank Lewis for his comments on previous drafts.

A full version of the TMA as I interpret it is included as an appendix immediately following p. 322 of this volume.

FRANK A. LEWIS

PLATO'S THIRD MAN ARGUMENT AND THE 'PLATONISM' OF ARISTOTLE*

In his 'The Platonism of Aristotle', G. E. L. Owen attempts to trace some key theses from Aristotle's *Categories* and *Metaphysics* back to what he argues are their origins in Aristotle's reflections on the Third Man Argument (TMA), first advanced by Plato in the *Parmenides.* [1] Following the testimony of Alexander in his commentary on the critique of the Platonic theory of forms in *Metaphysics* A9, Owen attributes to Aristotle a version of the TMA which is built around these key premisses:

Self-predication (SP).	what is predicated . . . is itself a subject of that same predicate. [2]
Non-identity (NI).	what is predicated is always something different from the subjects of which it is predicated.

According to Owen, these premisses (or their denial) figure prominently both in Aristotle's own theorizing about predication in the *Categories*, and in his theory of essence, especially the thesis that each thing is (essentially) the same as its essence in *Metaphysics* Z6. Owen takes a *literalist* view of Aristotle's version of the TMA: that is, he insists that we can attribute an interest in one or another assumption of the TMA to Aristotle, only if the assumption is attested in the version of the argument Alexander describes as Aristotle's. In Section 1, I sketch the details of Owen's acccount, and suggest a reason for doubting the accuracy of his transcription of the assumptions Alexander takes to be at work in the TMA. In Section 2, accordingly, I consider the revised account of the argument in Alexander given by Code, and add some further refinements of my own. Code also states three semantical principles that bear on Aristotle's controversy with Plato: these will be the subject of Section 3. Code argues that one of the principles, which governs the predicate 'man', but has no analogue for 'large', plays a key role in Aristotle's attempt to revitalize the regress argument against Plato. If Plato can be forced to accept this principle, he will be effectively blocked from applying his way out from the Third *Largeness* Argument in the *Parmenides* in order to evade Aristotle's TMA. The same principle, added to others that we know Plato already accepts, will also lead to an inconsistency in Plato's semantics. I shall

James Bogen and James E. McGuire (eds.), How Things Are, 133–174.

suggest that, in fact, Aristotle has no satisfactory argument to show that Plato must accept the principle in question, and that he does not succeed in showing that Plato is liable to the undesirable consequences described. In Section 4, I return to Owen's historical thesis. On a broader conception of self-predication than Owen's SP allows, it appears that the TMA and the *Categories* are after all concerned with elements in the notion of self-predication to a considerable degree. Contra Owen, however, there is virtually nothing in common between his NI (or its denial) and Aristotle's thesis in *Metaphysics* Z6 that each thing is (essentially) the same as its essence. This last point is developed further in Section 5. I argue that Aristotle's sameness thesis in Z6 is unintelligible outside the new metaphysics of the rest of Book *Zeta*. Aristotle's notion of the *substance of* a thing, first, allows him to construct a variety of anti-Platonism that cannot even be stated within the conceptual framework of the *Categories*. Finally, I discuss the argument Aristotle gives at Z6, 1031b28–1032a2, for the conclusion that each thing is (essentially) the same as its essence. The convolutions of this argument dramatize vividly the way in which the Z6 thesis is linked irretrievably to Aristotle's mature theory of essence, definition, and substance.

1. OWEN'S HISTORICAL THESIS.

In order to avoid the regress generated by the TMA, Owen argues, Aristotle is careful not to assert both NI and SP jointly. But, on Owen's view, Aristotle held that there is no *general* answer to the question, Which of NI and SP is false? Instead, there are *two* kinds of predication to be distinguished, based on a division between two kinds of predicate, and each variety of predication takes a different stand on which of the two premisses from the TMA to assert, and which to deny.

Aristotle's chief weapon in making the distinction in kinds of predicate with which his theory of predication begins, is the notion of synonymy, which he sets out in the *Categories*. On the one side, there are predicates like 'man', which "is used in the same sense whether we use it to describe Socrates or to speak of the kind or species under which Socrates falls" (Owen, p. 135). In support of this, Owen points to Aristotle's view that the definition of man applies equally well to the particular man, Socrates.[3] On the other side, we find the predicate 'white', for example, whose logical behaviour is quite different. For, if Socrates (say) is white, then he is coloured in a certain way. But white is not white in this same sense. White is not coloured in any way; rather, it is a colour of a certain sort.[4]

This difference between predicates, according to Owen, leads Aristotle to posit a corresponding difference in kinds of predication. In one, illustrated by 'Socrates is white', SP is not implied: for "white is not white in the sense in which Socrates is white" (Owen, p. 136). For the remaining kind of predication, however, 'Socrates is (a) man', for example, SP does hold: Socrates and man, presumably, are both men "in the same sense".[5] Ultimately, then, Aristotle will recognize one variety of predication for which SP holds, but not NI (Owen's "strong", or essential, predication), and a second variety for which, conversely, NI holds but SP fails ("weak", or accidental, predication). The *Categories* represents an intermediate stage in the development of this theory, according to Owen. Accidental predication is already in place in the *Categories*. The theory of essential predication, by contrast, is only partly formed, for although Aristotle is already committed to asserting SP for this kind of predication, he has not yet seen his way to denying NI.[6] Instead, he holds that although man is a man, or an animal, in the same sense in which Socrates is, still it is *less of a substance*. By the time that *Metaphysics* Z was written, however, Aristotle has dropped the notion of a secondary substance, in favour of the view that the subject of an essential predication and the essence that is asserted of it are the same: that is, according to Owen, he denies NI for this variety of predication. This move in turn forces a revision in his view of what can be a subject of an essential predication of this canonical sort – not an individual, Socrates or Callias, say, for these apparently share their essence, and two things that are the same as the same essence must be themselves the same. Instead, the appropriate subject within the new theory of essential predication in Owen's view will be the species, for example, man, which has its essence uniquely. In this way, according to Owen, the species threatens to supplant the individual substance of the *Categories* as the primary subject of discourse.

Owen's hypotheses regarding the influence of the TMA on Aristotle's own theories of predication are subject to one evident disadvantage, if we are to take Owen's NI and SP *au pied de la lettre* as transcriptions of what Aristotle took to be the premisses of the TMA. In the reconstruction Owen has in mind, the regress is thought to follow from the assumption that the predication relation is non-empty, that is, that something is predicated of something, call this assumption "P", together with the two principles SP and NI that govern the predication relation. It is easily shown that the premiss-set, {P, SP, NI}, is inconsistent,[7] and hence capable of generating any conclusion whatever. This result trivializes the most significant feature of the TMA, namely the inference to an infinite sequence of forms corresponding to the use of a single, apparently univocal predicate. If the version of the TMA that Aristotle had before him is trivial in this way, then we have philosophical reasons of our own to suppose that he should not find the argument a fit object for philosophical reflection. Clearly, NI, SP, and P cannot jointly be true – not, however, because they permit the inference to an infinite sequence of forms, but because they form an inconsistent set, and so supply an inference to any conclusion whatever.

In fact, however, Aristotle does seem to regard the Third Man regress as a significant result.[8] Arguably, Aristotle himself would have been indifferent to the objection that {P, SP, NI} is an inconsistent set. Even so, there are methodological reasons, other things being equal, to avoid attributing a philosophically problematical position to Aristotle. We have some motive, then, for asking whether he might have had in mind a different premiss-set for the argument. Owen bases his claims about the version of the TMA we should attribute to Aristotle on the version he finds attributed to him in Alexander. But, Code argues, Owen's reading of Alexander is not accurate. Owen's SP is simply not to be found in Alexander's version of the TMA. At the same time, there is an important ingredient in the TMA as it is formulated in Alexander, namely separation, which is given no role at all in Owen's account. Our first task, then, is to establish as best we can the exact form in which the argument is given in Alexander.

2. THE VERSION OF THE ARGUMENT IN ALEXANDER.

In place of Owen's SP and NI, Code finds these premisses for the TMA in Alexander's text:

ALEX-OM. If each member of a plurality of objects is a man, then there is an X such that:
(1) X is a man,
(2) X is separable,[9] and
(3) X is predicable of each member of that plurality.

ALEX-NI. If what is predicable of each member of a plurality is *separable*,[10] then it is distinct from each member.[11]

In this Section, I shall take up three questions in turn about these assumptions. First, are the assumptions logically independent of Owen's SP? Second, are they sufficient to generate the regress, as desired? And finally, what is the novel role that Code assigns to separation in the argument?

(a) *ALEX-OM, ALEX-NI, and Owen's SP*

Clearly, the two assumptions Code supplies do not contain an assertion of SP, or of any instance of SP. It might be thought that, nonetheless, an instance of SP does appear as a part of one of the assumptions of the argument (see clause [1] in the consequent of ALEX-OM), and that it will also appear by detachment as a line in the main body of the argument. Code argues correctly that SP is not present in the argument in even this weaker sense. Suppose there exists a plurality Π of objects, each of which is a man. Then by ALEX-OM, there is an *X* that is a man, is separable, and is predicable of each member of Π. Now the sentence, '*X* is a man', conforms to Owen's SP, only if it expresses a relation of predication *between X and itself*. Nothing in ALEX-OM, however, mandates this interpretation of the sentence. Suppose we make another application of ALEX-OM, this time to the plurality Π *with X*,[12] whose members are all the members of Π together with *X*. Even so, we will conclude only that there is some *Y* that is both a man, and separable, and predicable of each member of Π *with X*. In particular, then, *Y* is predicable of *X*. But this leaves it open whether or not *X* is identical with *Y*; and in fact, ALEX-NI will tell us that they are non-identical. So Code's premisses are consistent with the falsity of Owen's SP.

(b) *Two New Assumptions: ALEX-PL and HOMOGENEITY*

Are the two assumptions Code sets out sufficient to generate the Third Man regress? In fact, the argument is not valid without some further assumptions concerning pluralities. The argument must begin with the assumption that there exists a plurality Π of objects, presumably sensibles, each of which is a man. This permits the first application of ALEX-OM, which generates the existence of *X*. Next, we must go to ALEX-NI to show that *X* is not identical with any member of Π. We must now assume that there exists a new plurality, Π *with X*, in order to clear the way for a second application of ALEX-OM, which will generate the existence of an entity *Y*. Then we go again to ALEX-NI to show that *Y* is not identical with any member of Π *with X*, in particular, that it is not identical with *X*. And so on, to infinity.

If we are to develop the Third Man regress in this way, we must assume that Π is a plurality, that Π *with X* is a plurality, that (Π *with X*) *with Y* is a plurality, and so on. Is there textual warrant in Alexander for these assumptions, as Code's and Owen's literalist programme seems to require? The argument in Alexander assumes the existence of many things,[13] that are alike (ὅμόια). In particular, both the (sensible) men and the forms generated by the regress are alike.[14] And at 93.2ff, there occurs the explicit idea that there is the same εἶδος and λόγος both for the participants and for the form. Accordingly,

just as, in the case of the participants, which are of the same kind (ὁμοειδέσιν), there is something common, so too *in the case of these and the form* from which these <depend>, there will be something common predicated: for it too [sc. the form] is of the same kind (ὁμοειδής) as they.

Alexander goes on to identify this with the argument reported at 84.21–85.3, which both Owen and Code take as the canonical account of the TMA.

These texts allow us to include the following among the assumptions which Alexander assigns to the TMA. The assumption is one governing the formation of pluralities. Its first two clauses together will generate all the pluralities we need; the third clause rules out any other cases:

ALEX-PL. (1) Any collection of sensibles, all of which are alike (in the same respect), is a plurality.
(2) If an object is predicable of each member of a plurality, and it and each member of the plurality are alike (in the same respect), then the result of adding that object to the plurality is itself a plurality.
(3) Nothing else is a plurality.

In general, the *level* of a plurality is equal to the level of its highest-level member, where sensibles are of level 0, a form (immediately) "over" a plurality of sensibles is of level 1, and so on. ALEX-PL, then, countenances two kinds of pluralities: pluralities of level 0 whose members are all sensible, and higher-level pluralities of level 1 or above, with a non-uniform membership of sensibles and forms combined.

ALEX-PL requires that pluralities of any level have members that are all *alike* in a given respect. What does this requirement amount to? One obvious condition on any plurality that is to have a place in a regress argument along the lines of the TMA is a *univocity* condition. In the TMA properly so-called, the members of a given plurality are all men, and the universal that is "over" that plurality is itself a man. The univocity condition requires that the predicate 'man' should be univocal in all these different uses. A similar condition applies to the predicate 'large' in the Third Largeness Argument in Plato's *Parmenides*, and so on.[15]

There are signs in Alexander, however, that more than this may be required. Suppose that there is a collection of sensibles, all of which are men. Then we can use ALEX-PL to form a plurality Π of level zero, all of whose members are men, only if the term 'man' *applies to each member of* Π *in the same way*. More sepcifically, as Alexander points out, if things are alike in a certain respect, they are so *because there is a single form in which they all participate.*[16] Alexander makes it clear that this same requirement holds also for higher-level pluralities. For example, let X be the form in which every member of Π participates. Then we can form a new, higher-level plurality, Π *with* X, by ALEX-PL, only if X and every member of Π are alike in the same respect. That is, again, they must participate in the same form.[17]

Alexander's discussion gives us one possible account of what it is for all the members of a plurality to be alike in a given respect. In the case relevant to the TMA,

> Two things are alike with respect to the term 'man' if and only if 'man' applies to them both in same way.

I shall say that any plurality all of whose members are alike in the sense illustrated here is *homogeneous*. For the TMA to work, we must assume that homogeneous pluralities, in the sense described, are in fact available at each stage in the argument:

HOMOGENEITY. (1) The term 'man' applies to each member of a collection of sensible men *in the same way*.

(2) The term 'man' applies to each member of a plurality of men and to the universal man which is predicable of each member of that plurality *in the same way*.[18]

The same predicative like, then, connects each member of a plurality of men with the universal that is "over" that plurality. That is, as Alexander tells us, they all participate in that universal.

HOMOGENEITY illustrates why, while all the pluralities of the TMA may well be sets, it is clear that not all sets are pluralities in the sense countenanced in the TMA. The homogeneity principle may seem trivial in the case where the men that are members of the plurality are all entities of the same type: that is, they are all *sensibles*. In its full generality, however, HOMOGENEITY is far from trivial, for it places a condition on *all* the pluralities of men that ALEX-PL injects into the argument, including higher-level pluralities, whose membership includes both sensibles and forms. An indication that HOMOGENEITY may not be uncontroversial is the view often attributed to

Plato that very different relations underlie the fact that the term 'man' applies to Socrates, and that it applies to man.[19]

Finally, the TMA requires a single empirical assumption. Where a collection is understood to have at least two members,[20]

There exists a collection of sensibles, each of which is a man.

Once ALEX-PL, HOMOGENEITY, and the empirical assumption are in place, ALEX-OM and ALEX-NI will apply straightforwardly to each of the various pluralities generated at different stages in the argument.

(c) SEPARATION

Aristotle's version of the TMA, as Code understands it, is intended as a *reductio* of the principle ALEX-OM; ALEX-OM in turn is a suitable target because if its antecedent is satisfied, then ALEX-OM introduces a *separable* form of man (Code, p. 105). Aristotle finds the idea that every form is logically separable from the corresponding sensibles objectionable, according to Code, because of his opposition to a further feature of Plato's theory. Plato holds that a form is always predicated *accidentally* of sensible particulars. Aristotle finds this view radically mistaken: for him, Socrates must first of all Be something (in this case, a man), if he is to be a suitable subject for accidents, which he Has. The root cause of Plato's mistake, in Aristotle's diagnosis, is the separation assumption. For Aristotle, the idea that a predicable is predicated accidentally of its subject is a *consequence of* the assumption that it is logically separable from that subject. (This result falls out virtually immediately from the theory of predication in the *Categories*). Accordingly, he builds separation into the premisses of the TMA, so that by attacking separation here, he can also undercut Plato's view that forms are predicated accidentally of sensible particulars.

I begin with the details of *how* separation is meant to fit into the TMA on Code's account. In addition to its appearance in the consequent of ALEX-OM, separation also figures in the antecedent of ALEX-NI, which replaces Owen's NI. One point about the use of separation in these different assumptions is fundamental. A key assumption in the TMA, according to Code, is that a Platonic form is logically separable from the sensibles of which it is predicable. If this is right, the argument must also assume that a form can be logically separable *from other forms*. In the first steps of the argument, where Π is a plurality of sensibles, and the form X is "over" Π, X is logically separable from, and hence by ALEX-NI distinct from, all the members of Π. Where Π is a higher-level plurality, including forms as well as sensibles, and *X* is "over" Π, Aristotle must again suppose that *X* is logically separable from, and by ALEX-NI distinct from, *all* the members of Π. Without this assumption, the regress will not even start. The following principle, then, is fundamental to Aristotle's version of the TMA, on Code's account:

SEPARATION. Where Π is a plurality (of any level) of men, and *X* is "over" Π, *X* is logically separable from all the members of Π.

Further details about SEPARATION concern the claim in ALEX-NI, that if what is predicable of each member of a plurality is separable, then it is distinct from each member of the plurality. The most sustained argument Code gives us in support of ALEX-NI runs as follows.[21] Suppose that (1) *X* is separable. It follows from (1) that (2) *X* is a τόδε τι, and from (2), that (3) *X* is ἄτομον. (3) in turn entails that (4) *X* is accidentally predicable of everything (except itself) of which it is predicable at all.[22] Now by hypothesis, *X* is predicable of all the members of Π. From this and (4), apparently, we know that (5) *X* is accidentally predicable of all the members of Π. (4) and (5) together show that (6) *X* is not identical with any member of Π.

A key move in this argument is the use of (4) and (5) to derive (6). But how sure are we of line (5)? In fact, it seems that (5) follows from (4) and the hypothesis that *X* is predicated of each of the members of Π, only if we already know that *X* is not a member of Π. For, suppose that *X* is predicable of some member *a* of Π. Then *if* $a \neq X$, it follows that *X* is predicable accidentally of *a*. The possibility is left open, however, that, in fact, $a = X$. And if $a = X$, it then follows by Code's principles of Being, that (given that *X* is predicable of *a* at all) *X* is predicable not accidentally but essentially of *a*. Accordingly, we can make the inference to (5), only if we assume that in fact *X* is not identical with any member of Π, which is what the argument was designed to prove.

The argument for ALEX-NI can be repaired, if we add an assumption. One assumption that will do the job, and which Code also accepts, is the homogeneity principle introduced above.[23] HOMOGENEITY ensures that the *same* predicative link connects each member of a plurality of men, with the universal that is "over" that plurality. Now Π is a plurality of (at least two) *distinct* men.[24] Hence, *X* is identical with at most one member of Π, and distinct from all the rest. Now if *X* is predicable essentially only of itself, but predicable of *all* the members of Π, then *X* is predicable accidentally of at least all but one member of Π. Hence by HOMOGENEITY, X is predicable accidentally of *all* the members of Π, as required.

Code has also a second argument from an assumption of separability, to the conclusion that the universal X is distinct from each member of Π of which it is predicated.[25] Suppose that (1′) X is logically separable from the members of Π. From (1′), by the definition of 'logical separation', we know that (2′) the definition of X cannot apply to any member of Π. From (2′) it follows by the definition of 'essential predication' that (3′) X is not essentially predicable of any member of Π. But by hypothesis, as before, X is predicable of each member of Π. If we also assume that accidental and essential predication together *exhaust* the available kinds of predication, we can move from (3′) to the earlier conclusion that (5) X is accidentally predicable of each member of Π, and from here, once more, to the conclusion that X is not identical with any member of Π. This argument is successful, if Code is right that logical separation is a part of Alexander's version of the TMA, and if his definition properly captures the notion of separation Alexander has in mind.

Finally in this Section, I return briefly to the question of *why* in Code's view logical separation as he characterizes it appears as an ingredient in his reconstruction of the TMA. According to Code, Aristotle's motives in opposing logical separation come from what he sees as its association with accidental predication; by attacking separation by way of the TMA, then, Aristotle means to undercut Plato's idea that all predications of sensible particulars are *accidental*. There are a number of flaws in this strategy. It is unclear in general what role, if any, logical separation as Code defines it plays in Plato's own metaphysics. In particular, it is not clear that Aristotle can succeed in undercutting Plato's view that the only predicates true of sensibles are accidentally predicated of them, *unless* he can show that Plato himself must accept Aristotle's use of logical separation as part of the argument for accidental predication. The connection between separation and accidental predication is virtually immediate in the context of Aristotle's own theory of predication in the *Categories*. But Aristotle can hardly *assume* that theory in the context of an argument with Plato designed to settle which of the two rival theories of predication is superior.

Suppose that we waive these difficulties, and grant that the TMA is a successful *reductio* of SEPARATION, and that Plato would accept a connection between a form's being logically separable from a subject of which it is predicable, and its being *accidentally* predicable of that subject. If the TMA demonstrates the falsity of SEPARATION, then it is false that in general for any plurality Π, where the form X is "over" Π, X is logically separable from all the members of Π. It is consistent with this that *X* is after all logically separable from all the *sensibles* that are members of a given plurality, and *not* logically separable only from the forms (if any) that are in that plurality. The fact that *X* is logically separable from all the sensibles in a given plurality, however, is not enough to generate the regress: only wholesale SEPARATION is sufficient to do that, as we have seen. So the idea that a form is logically separable from the sensibles of which it is predicable remains intact. Even if we grant the connection between logical separation and accidental predication, therefore, Aristotle has not damaged the view that the Platonic form man, for example, is predicable accidentally of sensible men.

If this argument is right, Aristotle will be on weak ground if he adopts the strategy Code ascribes to him, and claims to use the TMA to rebut the view that forms are logically separable from, and hence predicable only accidentally of, sensibles. We may prefer to conclude that separation does not appear in the TMA at all. Alternatively, if separation is featured in the argument, either our account of it must be quite different from the one Code provides, or we must find a different account of Aristotle's motivation for including it there.[26]

3. SOME SEMANTICAL PRINCIPLES. THE THIRD LARGE ARGUMENT AND THE TMA.

In the previous Section, we attempted to recover from Alexander the principles that underlie Aristotle's reworking of the TMA. I turn now to a fuller discussion of the anti-Platonic themes Aristotle finds in the TMA. Aristotle's version of the argument points to the particular importance of the principle, HOMOGENEITY, or its near-equivalent, what Code calls the semantic metaprinciple (A):

(A) The term "man" applies to each member of the plurality of men mentioned in an application of ALEX-OM and to the universal postulated by that application *in the same way*.

(A) plays a crucial role in justifying the reasoning of the TMA. (A) is important also, Code argues, because if we add it to other principles that we know Plato accepts, the results will be an inconsistent set. The remaining principles are these:

(B) The term "man" is applicable to particular men in virtue of the fact that they all Have a (universal) man in common.

(C) The application of the term "man" to a universal man is not to be explained in terms of something distinct from it that it Has, but rather in terms of what it Is.[27]

The fact that {(A), (B), (C)} is inconsistent gives Aristotle a second means of arguing against the Platonic idea that forms are always predicable accidentally of sensibles. For, if Plato can be made to accept both (A) and (C), he must give up (B) and the claim that (for example) Socrates is *accidentally* a man. I shall begin by asking more generally how Plato and Aristotle both stand on (A), (B), and (C).

(a) The Principles (A), (B), and (C).

Both Plato and Aristotle are committed to (C). For Plato, the fact that Socrates is a man is grounded in the relation of participation between Socrates and an appropriate form; the fact that man is (a) man is grounded in some *other* relation, which Code identifies with his Being. Aristotle disagrees with Plato over the relation between Socrates and man: participation in Aristotle's view is a more suitable relation between Socrates and his *accidents*, for example, pallor. But Aristotle agrees that 'man' does not apply to man in virtue of a relation of participation. Rather, man Is (a) man. Both philosophers, then, accept (C).

As we have seen, Plato also accepts (B), for he holds that a predicate is true of a sensible in virtue of a relation of participation between the sensible and an appropriate form. Aristotle, meanwhile, is apparently committed to (A), if (A) can be reformulated in properly Aristotelian terms. But (A), (B), and (C) together form an inconsistent set. The fact that he accepts both (B) and (C) means that Plato must deny (A). Aristotle, by contrast, accepts (C) and (A): hence, he must deny (B). Plato and Aristotle, then, disagree over both (A) and (B).

Code's (A), (B), and (C), governing the predicate 'man', should be compared with three analogous principles for the predicate 'large':

(A)-'large'. The term 'large' applies to each member of a plurality of large things mentioned in an application of ALEX-OM and to the universal postulated by that application *in the same way*.

(B)-'large'. The term 'large' is applicable to particular men in virtue of the fact that they all Have a (universal) large(ness) in common.

(C)-'large'. The application of the term 'large' to a universal large(ness) is not to be explained in terms of something distinct from it that it Has, but rather in terms of what it Is.

Plato holds (B) and (C), and he will also hold (B)-'large' and (C)-'large'. Plato denies (A), and he will also deny (A)-'large'. Aristotle implicitly concedes the denial of (A)-'large' to Plato, for his own theory in the *Categories* commits him to the same position. The predicate 'large', in Aristotle's theory, is true of Socrates, for example, by virtue of a cross-categorial relation: Socrates Has largeness. But if nothing can belong in two categories at once, and Having is a *cross*-categorial relation, largeness cannot Have itself. So (A)-'large' is false, on this view.[28]

A similar manoeuvre, however, may *not* be available for the predicate 'man'. That is, it may be possible to show that Plato cannot simply assert (B) and (C), but deny (A). As we have already seen, Aristotle's notion of synonymy indicates one way in which the predicates 'man' and 'large' differ.[29] Perhaps, then, they also differ in that even if Plato denies (A)-'large', he will not be able to deny (A).[30] If this suggestion succeeds, and if Plato persists in holding (B) and (C) as well, his semantics is in this case inconsistent.

The difference in logical behaviour between the predicates 'large' and 'man' also suggests a point about the relation of the TMA, as Aristotle and Alexander refer to it, and the original Third Large Argument in Plato's *Parmenides*. Plato can resist (A)-'large', but perhaps he cannot resist (A) itself. If so, then Aristotle's preference for the TMA over the argument in its original form will not be accidental. For as we shall see, the truth of (A) plays a key role in establishing the validity of the TMA.

(b) (A) and the TMA

Of the three semantical principles given, only (A) is directly involved in the TMA. (A) is the same in intent as our HOMOGENEITY above. Both principles help fill out the requirement in ALEX-PL that for a collection of objects to make a plurality, they must all be *alike* in some single respect. A homogeneity principle interprets this requirement to mean not just, for example, that the term 'man' must apply to every member of a plurality of men, but also that it must apply to them all *in the same way*.

A principle of homogeneity, then, governs the use of ALEX-PL to introduce new, higher-level pluralities as the later steps of the TMA unfold. Code suggests that a homogeneity postulate is crucial to the TMA in yet a second way, by justifying our repeated application of ALEX-OM to the ever-new pluralitites introduced by the use of ALEX-PL. We apply ALEX-OM to a plurality of men, where the plurality is of level n, in order to generate yet another object of level $n+1$, which is predicated of each member of the plurality, and is itself a man. Plausibly, this application of ALEX-OM across-the-board, to pluralities of whatever level, is appropriate only if we can be sure that the pluralities in question are homogeneous in the way described.

We have already seen that Plato denies (A)-'large', and that Aristotle is likely to concede that denial to him. If Plato can successfully deny (A) as well, then we shall no longer be able to use ALEX-PL2 to form higher-level pluralities. And even if higher-level pluralities were after all available, we should no longer be able to apply ALEX-OM to them.[31]

Aristotle's strategy against Plato, as we shall see, is to offer reasons for thinking that (A) is *true*. If Aristotle is right, then our reason for doubting ALEX-PL2 and the application of ALEX-OM to higher-level pluralities is gone, and the TMA is back in place.

(c) 'Man', 'Large', and the Argument for (A)

The difference in logical behavior between the predicates 'man' and 'large' is crucial to Aristotle's attack on Plato. One key way in which the two predicates differ is captured by Aristotle's doctrine of synonymy. So we should be clear first how this difference works.

Aristotle's notion of synonymy is defined in the first chapter of the *Categories*. Schematically,

> x and y are *synonyms with respect to* the predicate 'F', if only if 'F' applies to both x and y, and the definition associated with 'F' applies to both x and y.

For example, if a man and an ox are synonyms with respect to the predicate 'animal', then the predicate 'animal' applies to them both, and the appropriate definition is the same in both cases – that is, they are both (say) animate objects, endowed with sensation (*Categories* 1, 1a6–12).

In *Categories* 5, Aristotle goes on to construct a notion of synonymous predication on this notion of synonymy:

> x is *called synonymously after* y (y is *predicated synonymously of* x), if and only if there is a predicate 'F' "from" y, such that x and y are synonyms with respect to 'F'.[32]

Aristotle argues in the chapter that things are called synonymously after substances or differentiae, and these alone. And by 'substances' here, Aristotle makes it clear that he means *secondary* substances, that is, species or genera. For, only these yield a predicate in the appropriate way (*Categories* 5, 3a36–37). For example, Socrates is called synonymously after his species, man. That is, first, the predicate 'man' is "taken from" the species, man, and the term applies both to Socrates and to his species. Further, the definition associated with the predicate applies both to Socrates and to the species. That is, both Socrates and man are rational animals.

These definitions show that where *a* is a member of a plurality of large things, and large(ness) is the universal that is "over" that plurality, large(ness) and *a* are *not* synonyms with respect to 'large'. For, "*a* is large, but it is not what large(ness) is" (cf. *Metaphysics* G4, 1007a32–33). Equivalently, large(ness) is not predicated synonymously of *a*. But man and each of the members of a plurality of men are synonyms with respect to the predicate 'man': man is predicated synonymously of each member of the plurality.

One possibility is that these very contrasts by themselves are enough to show that, even if Plato can deny (A)-'large', he must acknowledge the truth of (A). Given the notions of synonymy and synonymous predication defined above, Aristotle might argue, Plato must accept this principle:

SYNONYMY. Each member of a plurality Π of men is called synonymously after the universal X that is "over" Π.

It is not immediately evident that the set {SYNONYMY, (B), (C)} is inconsistent, or that SYNONYMY can play precisely the same role as (A) in setting the TMA in motion. Possibly, however, Aristotle can show that SYNONYMY *entails* Code's (A). If so, then his argument against Plato will be complete. Plato will be committed after all to an inconsistent semantics. And to the extent that (A) helps support the TMA, he will be unable after all to resist the TMA.

Does SYNONYMY entail (A)? I have not been able to see that it does. Almost certainly, SYNONYMY and (A) are both true within Aristotle's own theory of predication in the *Categories*. But the fact that Aristotle holds both does not tend to show that the one *entails* the other. Code argues that Aristotle himself accepts the entailment:

> ... since the definition [of man] applies both to the universal man and the particular men, the universal is essentially predicable of itself and those particulars in the same way. (Code, p. 103)

The argument Code appears to have in mind depends on the principle that "a universal is essentially predicable of a logical subject *X* if and only if both the name and the definition of that universal truly apply to *X*" (Code, p. 103). By SYNONYMY, both the name and the definition of man apply to man itself: hence, man is essentially predicable of man. But again by SYNONYMY, both the name and the definition of man apply to particular men; hence, man is also essentially predicable of particular men. This gives the conclusion, that man is essentially predicable of itself and of particular men in the same way.[33]

Plato need have no difficulty with the opening move in this argument. He holds that the form man Is (a) man, and that it Is (a) rational animal. So he can cheerfully concede the intermediate conclusion that man is essentially predicable of itself. The trouble comes with the second step. Aristotle urges that the definition of man applies also to particular men, hence, that man is essentially predicable of particular men. Plato may well jib at this inference. He can accept Aristotle's premiss, that the definition of man applies to particular men, and that it also applies to man itself. But he will contest the sense of "applies to" here. In denying (A), Plato is defending the view that a *different* predicative link grounds the fact that Socrates (say) is a man, and that man is a man. He is not likely to concede, then, that the *same* predicative link is in effect both in the claim that Socrates is a rational animal (if this is the definition of man), and that man is a rational animal. For Plato, the way in which man is either a man or an animal is radically different from the way in which Socrates is a man, or an animal. The form, presumably, has the relevant properties essentially, but Socrates has all his properties accidentally. It is not the case, then, that the universal man is predicable essentially of itself *and of the particular men* in the same way. Accordingly, Plato can accept SYNONYMY, and agree that the form man, for example, Is (a) man, and Is (a) rational animal, while at the same time insisting that Socrates Has all his properties.[34]

I have argued that Plato can consistently acknowledge the truth of SYNONYMY, and continue to deny (A). It is possible, however, that an argument can yet be found to defend the inference from SYNONYMY to (A). But if the TMA is supported by the use of SYNONYMY in this way, then difficulties threaten from a different direction. The trouble lies, as we shall see, with the role Code assigns in the argument to the notion of logical separation.

In Code's reconstruction of the TMA, as we have seen, Aristotle has as his target the idea that a form is logically separable from the particulars of which it is predicable. Separation enters the argument by means of ALEX-OM and

ALEX-NI: its appearances there are based on this assumption (cf. Section 2(c) above):

SEPARATION. Where Π is a plurality (of any level) of men, and *X* is "over" Π, *X* is logically separable from all the members of Π.

Code places the following condition on logical separation: a form *X* is logically separable from the particulars of which it is predicable, only if none of those particulars satisfies the definition of *X* (Code, p. 102). Given this condition, SYNONYMY entails NEG-SEPARATION. For, let Π be a plurality of particulars that are men, and let *X* be the form that is "over" Π. Then by SYNONYMY, each member of Π is called synonymously after *X*. That is, by the definition of synonymous predication, there is a predicate '*F*' from *X* such that *X* and each member of Π are synonyms with respect to '*F*'. By the definition of synonymy, then, the definition associated with '*F*' applies to each member of Π. It follows that the definition of the form *X* applies to the particulars in Π (I assume here that the definition associated with '*F*', where '*F*' is "from" *X*, is just the definition of *X* itself). Accordingly, by the condition on logical separation quoted above, X is *not* logically separable from the particulars in Π. So we have NEG-SEPARATION.

This argument shows that any set that contains SYNONYMY, SEPARATION, and Code's account of logical separation, must be inconsistent. In any version of the TMA built out of these materials, then, the actual generation of an infinite sequence of forms becomes trivial. Any other conclusion would have done equally well, for "from a contradiction anything follows". Again, since SEPARATION and SYNONYMY (and the associated account of logical separation) by themselves form an inconsistent set, we can obtain the conclusion of the TMA directly from these materials alone, without the niceties of ALEX-OM, ALEX-NI, and the rest.[35]

Finally, if SEPARATION is the target of the TMA, as Code suggests, then the mere truth of SYNONYMY alone will suffice to refute SEPARATION, and the TMA itself becomes superfluous.

These results suggest that SYNONYMY and SEPARATION cannot both be included in the TMA. Will the argument make room for either principle? There are independent reasons for thinking that logical separation at least has no place in the argument. On Code's account, Aristotle means the TMA to be, in the short run, a *reductio* of separation. In the long run, the refutation of separation is meant in turn to undercut Plato's theory that forms are invariably predicated accidentally of sensibles. As we have seen in Section

2(c) above, there is a major difficulty with this account of Aristotle's thinking. Suppose we grant that the TMA is a successful *reductio* of SEPARATION, and grant too the connection between separation and accidental predication. Even so, the falsity of SEPARATION is not enough to undercut the view that forms are predicated accidentally of sensibles. The falsity of SEPARATION shows only that there is some plurality Π, not every member of which is logically separable from the universal *X* that is "over" Π. It does not follow from this that any sensible is ever *not* logically separable from a form. But if Aristotle cannot show that the sensibles in a plurality Π are not separable from the universal *X* that is "over" Π, then even assuming the connection between logical separation and accidental predication, he has not succeeded in undercuting the view that *X* is predicated accidentally of all the sensibles in Π. Accordingly, even if the TMA can be ranked as a successful attack on SEPARATION, nothing follows for Plato's theory of accidental predication.

Any account of Aristotle's criticism of Plato must finally come to terms with the notion of separation. But it is not clear that the notion of separation which Code describes plays any useful role in the argument. I suggest that logical separation, as Code explains it, is better dropped from the argument: if separation is at work in the TMA at all, some other account of the notion will have to be found. For all we know, meanwhile, SYNONYMY is true. At all events, we have not yet found any reason to believe that it is threatening to Plato. For all we have seen so far, then, if SYNONYMY does govern the pluralities of the TMA, its effect is totally harmless.

(d) *A Further Argument for (A)*

I have argued that it is at best problematic to use SYNONYMY by itself in order to show that Plato can deny (A)-'large' but not (A) itself. Perhaps, however, there are other means of achieving this same end. One further way in which the predicates 'man' and 'large' differ is, as Code notes, that 'man' is predicated κυρίως both of the universal man, and of each of the members of the plurality of men which that universal is "over". But 'large' is predicated κυρίως only of the universal, large(ness), and not of any sensible large thing, and hence not of every member of any plurality of large things.[36] Code suggests that Aristotle uses *this* difference, which Plato must accept, to explain why Plato must also accept (A), even while he rejects (A)-'large'.

Code argues that Aristotle's strategy falls into two parts. First (1), Aristotle uses SYNONYMY to support the idea that 'man' is predicated κυρίως both of man and of the particular men. He then argues (2) that the notion of κυρίως predication in turn establishes the truth of (A) (Code, p. 109). These suggestions depend for their force on how we interpret the key expression, "x is predicated κυρίως of y". One account on the market is that it singles out complete as opposed to incomplete predicates.[37] If this is right, then the fact that 'man' is predicated κυρίως alike of the universal man, and of each of the members of the plurality of men which that universal is "over", does not by itself seem enough to force Plato to accept (A). On the other hand, Aristotle/Alexander's appeal to the notion of κυρίως predication may mean to invoke some stronger doctrine: the only obvious candidate is SYNONYMY and the notion of synonymous predication, which have already been discussed. On the present showing, then, the notion of κυρίως predication is either too weak to force acceptance of (A), or it is simply *equivalent to* SYNONYMY and the notion of synonymous predication, so that (1) and (2) in the argument Code sketches collapse into one, and the old difficulties of supporting (A) directly by reference to SYNONYMY return.

The fact that there is apparently no knock-down argument in support of (A) that Plato too would find conclusive suggests a broader moral. Behind the dispute over (A), there lies a deeper divergence over ontology and predication that is still less easily resolved. If Plato is to succumb to the version of the TMA discussed here, he must first accept (A). But his opposition to (A) is not based merely on the desire to evade the TMA. His resistance to (A) is embedded in the details of his own ontology and theory of predication, as set out in (B) and (C). If Plato sticks resolutely to (B) and (C), he can resist (A), and remain immune to the TMA. At the same time, if he were after all to concede the truth of (A), he would be forced to give up one or other of (B) and (C). Such a move by itself would create a massive disturbance in the theory of forms – no less a disturbance, perhaps, than the success of the TMA would bring about.

The end-result of the debate over the TMA is apparently a stalemate. For Aristotle, the TMA refutes key features of the Platonic theory of forms. By Plato's lights, however, the TMA shows only that he is right to resist (A), and that he can with impunity retain his theory unaltered, so long as (A) remains safely at bay. So the TMA does not provide the decisive test in the dispute between Aristotelian and Platonic metaphysics for which Aristotle must have hoped.

4. SELF-PREDICATION AND NON-IDENTITY AGAIN

As the discussion in the previous Section shows, some of the assumptions underlying the TMA are the subject of intense controversy between Plato and Aristotle. It is all the stranger, therefore, if there should be no sign in Aristotle's version of the TMA of the self-predication assumption that, in various guises, figures so large in contemporary discussions of the argument. As we saw in Section 2(a) above, the account of the TMA in Alexander shows no trace of the narrow notion of self-predication captured by Owen's SP. If Alexander is to be trusted, then, Aristotle's version of the TMA is innocent of the claim that the form man, for example, *is predicated of itself*. Yet the term "self-predication" is often, perhaps most usually, used for some broader notion than Owen's SP allows. Perhaps, then, there is some broader notion of self-predication that is after all present in the version of the TMA in Alexander, and which also falls under Aristotle's scrutiny in the *Categories*. At the same time, I shall argue, there are good reasons to doubt that a similar connection exists between any non-identity assumption in effect in the TMA and the sameness thesis of *Metaphysics* Z6.

(a) *Self-predication and the Categories*

We should begin by trying to characterize the notion of vicious self-predication often attributed to Plato, and thought to figure in the TMA. Informally, vicious self-predication commits the fault of assimilating the logic appropriate to the universal man to that suitable for particular men: it treats the form man as "just one more for the census-takers to count".[38] It is a little harder to go beyond these general characterizations. It appears that a number of different particular theses go together to make up a notion of vicious self-predication. Some, but not all, may be logically vicious in their own right. And some, perhaps all, are necessary for vicious self-predication. In all likelihood, however, none by itself is sufficient for it.

(i) A minimal assumption that is clearly necessary for vicious self-predication is a *univocity* condition. For example, a single univocal predicate, 'man', should apply both to each member of a plurality Π of men, and to the universal man that is "over" that plurality. This condition seems to be part of what is meant in Alexander by the requirement that all the members of a plurality and the universal that is "over" that plurality should be *alike* in a certain respect: see above, Section 2(a).

(ii) Alexander's condition that the entities mentioned be alike in a given respect may also be taken to support a second condition, that a single predicative link should underlie the application of the term 'man' both to the members of Π, and to the universal man that is "over" Π. This is a *homogen-*

eity assumption, in the sense explained in Section 2 above, and it is captured equally well by HOMOGENEITY (2) and by Code's (A).

HOMOGENEITY is needed both in order to construct the stock of pluralities required in the TMA, and to license the successive applications of ALEXOM in the later stages of the argument. In the *Categories*, Aristotle argues for his own counterpart of HOMOGENEITY, governing the predicate 'man'; at the same time, he would deny that an analogous assumption holds for the predicate 'large'. But as we have seen, it is controversial whether Aristotle can show that Plato himself must accept HOMOGENEITY.

(iii) A third possible ingredient in vicious self-predication is the requirement that each member of Π is called synonymously after the universal man that is "over" Π. That is, both they and the form share the same name, viz. 'man', and the same definition associated with 'man' applies to them all equally. This is just the principle SYNONYMY stated in Section 3(c) above.

In the TMA as reconstructed in Section 2 above, conditions (i) and (ii) are both integral parts of the argument. The role of SYNONYMY is less clear. In the *Categories*, Aristotle argues in favour of SYNONYMY, governing the predicate 'man'. Code suggests that Aristotle puts SYNONYMY to use in the TMA, as part of the argument to show that Plato must accept HOMOGENEITY and hence must be liable to the remainder of the TMA as well. This use of HOMOGENEITY and SYNONYMY, on Code's account, is what accounts for Aristotle's preference for the TMA over Plato's original, the Third *Largeness* Argument. In fact, however, as we have seen, Plato can cheerfully accept SYNONYMY, yet deny that it entails HOMOGENEITY. It may be, then, that SYNONYMY does govern the pluralities of men envisioned in Aristotle's version of the TMA (there is some weak support for including SYNONYMY in the argument in the [admittedly vague] claim by Alexander that a form and its participants must have the same εἶδος and the same λόγος[39]); but its presence in the argument appears to be benign.

Are (i), (ii), and (iii) jointly sufficient for vicious self-predication? Presumably, they are not, since in the *Categories* Aristotle apparently accepts all three, but would not accept vicious self-predication.[40] What must be added to obtain a set of assumptions that are both necessary and jointly sufficient for vicious self-predication? A tempting candidate is the subject of a recent paper by Joan Kung. It is the assumption that (iv) Socrates and man, for example, are *of the same logical type*: that is, they are *both thises*. (i), (ii), and (iii) stipulate in various ways that the term 'man' applies on equal terms both to the form man, and to the particular men, Socrates, for example. But (iv) too may be needed to complete the assimilation of the logic appropriate to the universal man to that suitable for particular men, and so produce a properly vicious notion of self-predication.[41]

Plausibly, we must also add (iv) to the assumptions of the TMA, if we are to be able to count the form man *along with* Socrates, and thus put together the pluralities the TMA requires. On this account, the TMA can be read as a *reductio,* not of logical separation as Code explains it, but of the assumption that the members of each plurality in the argument are all *of the same logical type*. Indeed, it is tempting to think that in one of its uses, the term 'separation' itself answers to just this doctrine, that a Platonic form is a this, and hence, of exactly the same logical type as a particular. In this sense, 'separation' names not a relation between forms and particulars,[42] but a non-relational, "categorial" property of forms.[43]

The idea that Platonic forms are thises is precisely the point Aristotle means to *deny* for his own theory when he says that a secondary substance is a *secondary* substance, or that a secondary substance signifies a *ποιόν τι*, a certain qualification, and not a this (*Categories* 5, 3b13–21).[44]

Of the four conditions given in previous paragraphs, (i) and (ii) are both present in Alexander's version of the TMA, and hence, on the literalist hypothesis, both figure in Aristotle's version of the TMA as well. There is somewhat weaker evidence, as we have seen, that (iii) too is present in the version of the argument in Alexander. Aristotle's own theory in the *Categories* makes room for all three conditions. (iv), finally, may be what is intended by Alexander's phrase, *κατ' ἰδίαν ὑφεστώς*, "existing in its own right", at 84. 28.[45] It is something which Aristotle is apparently concerned to deny in the *Categories*, as we have seen.

There is some reason to believe, then, that the various ingredients (i) – (iv) of a notion of vicious self-predication do figure in Aristotle's version of the TMA as quoted by Alexander, and that they are also objects of reflection by Aristotle in the *Categories*. I conclude that if we adopt a broader notion of self-predication than is allowed by Owen's SP, we may be able after all to find connections of the kind that Owen was looking for between the TMA and Aristotle's *Categories*.

(b) *The Non-identity Assumption and the Thesis of Metaphysics Zeta 6*

I have argued that if our notion of self-predication is suitably broad, then connections open up between the version of the TMA reported by Alexander and Aristotle's own theorizing in the *Categories*. A connection between the TMA and the thesis of *Metaphysics* Z6, however, is far less likely. An initial criticism is due to Code. According to Owen, Aristotle sees that SP and NI cannot be true together. Ultimately, he concludes that there is one kind of predication for which NI is true and SP false, and another for which SP is true, and NI false.[46] This second kind of predication, according to Owen, appears in *Metaphysics* Z6 as the thesis that each thing is the same as its essence. If this account is correct, the denial of NI must be equivalent to the thesis of Z6. But plainly, this is not so, as Code points out. If NI is false, then it is *not* the case that what is predicated is *always* something different from the subjects of which it is predicated. But for the kind of predication which Owen finds in Z6, we need rather the assumption that what is predicated is *never* something different from the subjects of which it is predicated. The negation of NI is simply too weak to express this, and to suppose that Aristotle thought otherwise is to have him confuse 'not . . . always' and 'never'.

Owen's claim can be reformulated relatively easily to avoid this objection. Suppose that, seeing that SP and NI together give the TMA, Aristotle concludes that NI is simply false. He also holds that SP is in some cases false, in others, true (unlike NI, SP contains no explicit quantifiers). As before, then, SP and NI are never jointly true, but now because NI by itself is never true. Aristotle goes on to distinguish the cases in which SP is false and those in which it is true, by supposing that they correspond to two different kinds of predication, which Owen identifies as weak and strong predication respectively. Now, in weak predication, Aristotle supposes that what is predicated is *not* the same as its subject. This by itself forces a decision about strong predication. In strong predication, if what is predicated is again *not the same as* its subject, then (if weak and strong predication together exhaust all of the predications there are) what is predicated is *always* different from its subject, and NI is true after all. In this way, Aristotle is driven to the identity thesis of *Metaphysics* Z6 for strong predication. This is not the way Owen formulates his position. But it preserves a significant portion of his thesis, while being immune to Code's criticism.

Even in this revised form, Owen's thesis is subject to difficulties. For example, the notion of self-predication that according to Owen Aristotle accepts in the *Categories* and elsewhere, is quite different from the notion that on Owen's account Aristotle finds at work in the TMA (cf. n. 41 above). If the thesis of self-predication changes its meaning in this way, then Aristotle himself cannot have the same motivation for denying NI in his own theory that a philosopher would have who accepted the full-strength version of self-predication at work in the TMA, and reasoned that self-predication so understood and NI could not both be true.

Beyond this, the details of Aristotle's own metaphysical theory in the *Metaphysics* give us reasons for doubting that the Z6 thesis can have its origins in the denial of NI, as Owen suggests. By the time of the *Metaphysics*, Aristotle's notion of predication has radically changed. Aristotle's earlier notion of predication in the *Organon* moves back and forth between what is for us the natural notion of predication, where what is predicated is a linguistic item (call this "linguistic predication"), and a notion of *metaphysical* predication, where what is predicated is a *non*-linguistic item. In the *Metaphysics*, by contrast, Aristotle invariably regards predication as a metaphysical relation between entities. And in the theory of the *Metaphysics*, there are only two cases of (metaphysical) predication: either an accident is (metaphysically) predicated of a substance, or a form is (metaphysically) predicated of (a portion of) matter.[47] In either case, it is plausible that what is predicated is *not the same as* that of which it is predicated. Accordingly, in the *Metaphysics* Aristotle is committed to the view that (metaphysical) predication is *aliorelative*:

AL. x is (metaphysically) predicated of y, only if $y \neq x$.

It is important to keep AL, governing (metaphysical) predication, separate from other, superficially similar principles which are not in fact concerned with the same technical notion of (metaphysical) predication. In particular, AL is verbally virtually identical with unnegated NI. But to suppose that the two principles are in fact the same leads immediately to trouble. If AL and NI are the same, then by accepting AL, Aristotle is also committed to NI in the *Metaphysics*. This runs into the difficulty that if Owen is right, in adopting the Z6 thesis Aristotle also sees himself as *falsifying* NI.

This conflict is resolved by noting that AL is *not* equivalent to unnegated NI, and *not* inconsistent with the Z6 thesis. Unnegated NI appears to be inconsistent with the Z6 thesis, if we allow that in the broader sense of 'predicated' at work in NI, an essence is predicated of that of which it is the essence. Unlike NI, however, the principle AL governing metaphysical predication is entirely *consistent* with the thesis that a primary substance is the same as its essence. This is because according to the official notion of (metaphysical) predication at work in the *Metaphysics*, an essence is *not* (metaphysically) predicated at all of that of which it is the essence:

> x is (metaphysically) predicated of y, only if x is not the essence of y.[48]

For, an accident is not the essence of a substance of which it is (metaphysically) predicated, and a form is not the essence of the matter of which it is (metaphysically) predicated. If the essence of a thing is not (metaphysically) predicated of it, then the principle AL has no bearing on whether a thing and its essence are or are not identical.

By accepting AL in the *Metaphysics*, Aristotle does not commit himself to NI, and does not contradict the thesis in Z6 that each thing is (essentially) the same as its essence. This conclusion, however, does not tend to show that the Z6 thesis is in any way based on a reaction against NI, as Owen supposes. The mature notion of (metaphysical) predication at work in the *Metaphysics* is simply *irrelevant to* the relation between a primary substance and its essence. And it is highly implausible that in the middle of an extended treatise governed by this technical notion of (metaphysical) predication, Aristotle would revert without notice to earlier notions of predication to support some of his central theses regarding essence. The fact, then, that the thesis of Z6 appears to falsify NI is of no interest to Aristotle, because he no longer has any interest in the notion of predication that NI contains.

5. THE THESIS OF Z6: "EACH THING IS (ESSENTIALLY) THE SAME AS ITS ESSENCE"

I argued in the previous Section that Aristotle's notion of predication in the *Metaphysics* differs so radically from earlier conceptions, that the earlier principle NI and its negation must both seem simply irrelevant to the theories of the *Metaphysics*. This objection is symptomatic of a general difficulty with Owen's account of the origins of the thesis in *Metaphysics* Z6, that each thing is (essentially) the same as its essence. Far from having its roots in early controversy with Plato, that thesis is unintelligible outside a metaphysical theory that Aristotle does not develop until after the *Categories*, and which he completes only in *Metaphysics Zeta*. Further support for this conclusion comes from two directions: the connections between the Z6 thesis and the anti-Platonic components in the notion of the substance *of* a thing, which Aristotle develops only after the *Categories*, and the wider network of principles governing essence, definition, and substance that Aristotle deploys on behalf of the Z6 thesis in the argument at Z6, 1031b28–1032a2.

(a) *The Anti-Platonism of Z6: Substance and Substance of*

The notion of essence does not appear in the *Categories*, which knows only to ask the question, τι ἐστιν; What is it?, of a thing. Even more significant, the

Categories shows no sign of the notion of the *substance of* a thing. It is a peculiarity of Aristotle's later metaphysical thinking to add to the question, What is substance?, the further question, What is the *substance of* a thing? Indeed, the prominence this second question comes to acquire in Aristotle's later metaphysics leads him to the view that something is a *primary* substance, only if it is also the *substance of* a thing.

Once we reflect on Aristotle's notion of the *substance of* a thing, the real anti-Platonic thrust of the Z6 thesis begins to emerge. At the same time, we shall find that as a response to Platonic metaphysics, the thesis is essentially anachronistic. For as we shall see, the criticism of Plato which it implies is something that Aristotle himself could not have formulated before the mature doctrine of the *Metaphysics*.

At bottom, Aristotle's thesis that each thing is (essentially) the same as its essence is part of the continuing controversy with Plato over the age-old question, What is substance? Plato and Aristotle agree in asking the question, What are the substances? They disagree over Plato's answer: Platonic forms. So far, however, despite this disagreement, Plato and Aristotle at least share the same conceptual ground. But now, Aristotle argues, if Platonic forms are the substances, we must be able to say what they are the substance *of*. Here quite clearly, Aristotle is speaking the language of his own, post-*Categories* theory of substance. The answer he gives on Plato's behalf is that Platonic forms are the substances of their participants. This answer, ever more than the question, is dubiously faithful to Platonic theory. Participation, construed as a relation of Having, is a relation that holds only *accidentally* between a sensible and a form, when it holds at all.[49] But clearly, a predicable cannot be both accidental to and also the substance of the subject on which it supervenes. Even if we modify Plato's theory, however, and say that in certain cases, participation holds *non*-accidentally between a subject and a form,[50] so that a form can after all be the substance of a sensible, this is as far as it is possible to go in adapting Platonic metaphysics to the mould of Aristotle's own later metaphysical theory. Plato cannot go further, and conclude as Aristotle's own theory requires, that a thing and its essence or substance are (essentially) the same. For, it will follow that they are identical, and no form is identical with any sensible. In Aristotle's doctrine of the sameness of a thing with its essence, then, we have a proposition over which the Aristotelian and Platonic theories of substance are in irresoluble disagreement.

(b) The Argument at Metaphysics Z6, 1031b28–1032a2

The thesis of Z6 links the notions of essence and essential sameness:

x is the essence of y, only if y is essentially the same as x.

Closely connected with this is a claim joining Aristotle's notion of essence with the contemporary notion of identity:

x is the essence of y, only if $y = x$.

The evidence in Z6 is that Aristotle is committed to both claims. Almost certainly, his own formulations will involve the notion of essential sameness. But in any case, I take it that the second thesis is entailed by the first, as Code also supposes. One last indication of the rightful place of the Z6 thesis within the apparatus of Aristotle's later metaphysics is the way both claims can be supported by reference to principles at the heart of Aristotle's later theory.

I shall concentrate here on the argument Aristotle gives at Z6, 1031b28–1032a2. For present purposes, it will be sufficient to take the second of the two theses noted above as the conclusion to the argument. (To prove that the first conclusion also follows from Aristotle's argument will require a definition of essential sameness, which I forego here.) The text of the argument is this:

> But it would be obviously absurd, even if one were to give a name to each of the essences: for there will also be, in addition to it [= a given first-level essence] another <essence>, for example, <in addition to > the essence of horse, another essence. Yet what stands in the way of some things even now, right off, being their essences, if indeed essence is substance? But surely not only <are they> one, but they also have the same definition, as is clear also from what has been said: for the essence of one and one are one not accidentally. 1031b28–1032a2.

Suppose that each thing has an essence, from which it is distinct. The absurdity of this hypothesis is evident, if we give each of these essences a name. For, this will reveal the existence of another set of essences.[51] At 1032a2 ff., Aristotle goes on to point out that the argument has enough momentum to go on to infinity. But the argument we have to work with here takes us only two steps into the regress: a set of entities, their essences (which are distinct from them), and the essences of those essences (which are similarly distinct from what they are the essence of).[52]

To follow Aristotle's argument in greater detail, consider a single entity, x, and suppose that there is an essence $E(x)$ of x, where $x \neq E(x)$. Then, Aristotle asks, are we to suppose, by parity of reasoning, that there exists an essence $E(E(x))$ of $E(x)$, where $E(x) \neq E(E(x))$ (cf. ἕτερον, b30)?

Aristotle implies, I think, that he has already rejected the idea that $E(E(x)) \neq E(x)$, and that this constitutes grounds for giving up the initial non-identity of x and $E(x)$ as well.[53] He says:

And yet, what stands in the way of some things[54] even now (καὶ νῦν), right off (εὐθὺς), being their essence, if indeed essence is substance?

Let us put to one side for the moment the relation between x and $E(x)$. The question still remains, Why is Aristotle so sure that it is false that $E(E(x)) \neq E(x)$?

One clue is the equivalence of the notions of essence and substance, which Aristotle expressly remarks on at 1031b31–32 for the third time in this chapter (cf. earlier at a18 and b2–3). Probably, Aristotle's remarks can be taken as a *definition* of the substance-of relation:

(D1) x is *the substance of* y if and only if x is the essence of y.

The connection between essence and substance permits Aristotle to mobilize a principle governing the substance-of relation that gives him the result he needs:

> y is the substance of x, only if y is the substance of y. (*secondary reflexivity*)

(I will return to this principle later in this section, where it will appear as theorem [T3].) Suppose, then, that $E(x)$ is the essence of x, that is, by (D1), that $E(x)$ is the substance of x. It follows by the secondary reflexivity of the substance-of relation, that $E(x)$ is the substance of $E(x)$, and hence by (D1) again, that $E(x)$ is the essence of $E(x)$. That is, $E(E(x)) = E(x)$.

In this argument, the reflexivity principle given can be used to halt the sequence of essences after one step. It remains for Aristotle to show that even one step into the sequence is too many, that is, that $E(x) = x$. This is the aim of the next sentence in the text:

But yet not only <are they> one, but they also have the same definition, as is clear also from what has been said. 1031b32–1032a2.

Two questions need to be answered here. Firstly, how has the previous discussion established that $E(x)$ and x have the same definition? And second, if in fact they do, how does this help obtain Aristotle's conclusion?

Aristotle gives this standard account of definition:

(D2) x is *signified by the definiens in a definition of* y (for short, x is *the definition of* y)[55] if and only if x is the essence of y.[56]

Now as we have already seen, $E(x)$ is the essence of $E(x)$. By (D2), then, $E(x)$ is the definition of $E(x)$. At the same time, $E(x)$ is the essence of x, by hypothesis, so that by (D2) again, $E(x)$ is the definition of x. It follows that $E(x)$ and x have the same definition, as required, namely $E(x)$ itself.

How, second, does this result help Aristotle's major claim, that just as $E(E(x)) = E(x)$, so too, $E(x) = x$? Quite simply, the point will be that things which have the same definition are themselves the same. That is,

> x is the definition of y and x is the definition of z, only if $y = z$. (*partial functionality*)

(Again, discussion of this principle will be deferred to later in this section, where it appears as theorem [T4].) In light of this principle, and since $E(x)$ is the definition of both $E(x)$ itself and of x, we have the result that $E(x) = x$.

The reconstruction of the argument just given contains two principles which have yet to be explained. It takes the addition of a single axiom to the two definitions (D1) and (D2) already given to derive both principles. We begin with the two definitions already stated:

(D1) x is *the substance of* y if and only if x is the essence of y.

(D2) x is *the definition of* y if and only if x is the essence of y.

Trivially, then,

(T1) x is the substance of y if and only if x is the definition of y.

Accordingly, the three relations are equivalent.

Next, I introduce a single axiom:

(A1) x is the definition of y, only if $y = x$.

(A1) asserts the vacuity of the definition-of relation.[57] (A1) and (T1) together give a second theorem:

(T2) x is the substance of y, only if $y = x$.

That is, the substance-of relation too is vacuous. (T2) allows us to prove as a theorem the principle needed for the first half of Aristotle's argument:

(T3) y is the substance of x, only if y is the substance of y.

For, suppose that y is the substance of x. Then by (T2), $x = y$. Putting equals for equals, then, y is the substance of y, which gives (T3).

We are also in a position to prove the second theorem needed in Aristotle's proof:

(T4) x is the definition of y and x is the definition of z, only if $y = z$.[58]

(T4) asserts the partial functionality of the definition-of relation. To prove (T4), suppose that x is the definition of y and x is the definition of z. Then by (A1), $x = y$ and $x = z$, so that $y = z$, as required.

The defense of (T3) and (T4) just given brings to light two oddities in Aristotle's argument. The first rests on a point we have already noticed, that the proof of (T3), needed in the first half of Aristotle's argument, itself requires the truth of (T2). But if (T2) is true, (T2) can be combined with (D1) to give Aristotle's conclusion immediately, without the convolutions of the remainder of the argument as Aristotle actually develops it.

A second oddity is this. The derivations of (T3) and (T4) show that at bottom, Aristotle requires only one basic assumption, namely (A1), in addition to the definitions (D1) and (D2), to obtain the theorems that he appeals to implicitly in the argument. Putting (A1) in place of the derived (T3) and (T4), however, reveals an odd fact about his argument. For, its conclusion follows almost immediately from (D2) and (A1) alone. Thus, let x be the essence of y. Then by (D2), x is the definition of y, and by (A1), $x = y$, as required.[59]

Evidently, there exist simpler arguments to the conclusion for which Aristotle is arguing. Looked at in this light, Aristotle's argument is guilty of considerable redundancy. This very feature of the argument, however, dramatizes once again how deeply embedded its conclusion is in the general *Metaphysics* theory of essence, definition, and substance.[60]

Afterword on Z6, 1031b28–1032a2. As in the discussion in the main text above, Code also finds two stages in Aristotle's argument, the first, establishing that $E(x) = E(E(x))$, the second going on to argue that consequently, $x = E(x)$. Further, many of the principles we use in establishing both conclusions are either the same, or readily interderivable. Beyond this, however, there are also some areas where our two accounts diverge. It may be useful to draw attention here to some of the similarities and dissimilarities in the two accounts, and also to emphasize some of the points of importance in Code's account. I will begin with some details of textual interpretation, and then move on to some points of philosophical interest in Aristotle's argument.

(a) Some Textual Points

I shall concentrate on the interpretation of the key sentence at 1031b31–32 (translated in the main text above):

> *καίτοι τί κωλύει καὶ νῦν εἶναι ἔνια εὐθὺς τί ἦν εἶναι, εἴπερ οὐσία τὸ τί ἦν εἶναι;*
>
> Yet what stands in the way of some things even now, right off, being their essences, if indeed essence is substance?

First, *ἔνια* here is unduly vague for what Code needs. In this part of the argument, Aristotle is concerned with a three-member sequence, $\langle x, E(x), E(E(x))\rangle$, and for Code, *ἔνια* has a very precise reference: it picks out entities at the second position in such sequences, which are said to be the same as their essences (at the third position). For example, $E(x)$ is the same as $E(E(x))$.

Again, *τί κωλύει* is unexpectedly tentative for introducing the claim that $E(x)$ and $E(E(x))$ are the same. The argument that Code gives, based on the full-analysis condition from *Metaphysics* Z10, like the alternate argument sketched in the main text above, seems to necessitate, rather than merely hint at, the sameness result in question.

Accordingly, I take the point of *ἔνια* to be rather to recall the doctrine of Z4, that only things that are suitably primary have an essence (cf. Note 54 above). (The point is amplified at the end of Z11: only things that are primary are [essentially] the same as their essences, 1037a33–b5). So Aristotle is referring to the sameness of entities at the first position in the sequence with their essences; for example, $x = E(x)$.

It follows that Aristotle's words *καὶ νῦν* . . . *εὐθὺς* "even now, right off", peg the discussion not so much to the actual fact of the matter ("as things actually are", Code), set out in the *εἴπερ* clause, as to entities at the first position in the sequence of essences. The *εἴπερ* clause, meanwhile, gives only indirect support for the *ἔνια* clause, concerning the identity of x and $E(x)$. Primarily, the *εἴπερ* clause gives a main premiss in the (omitted) argument for the conclusion that $E(E(x)) = E(x)$. Since Aristotle will use this conclusion to establish that also $E(x) = x$, the *εἴπερ* clause supplies only indirect support for that further conclusion: hence, the less than fully firm *τί κωλύει*. Meanwhile, the primary support for the conclusion that $E(x) = x$ is given in the *ἀλλὰ μὴν* sentence at 1031b32–1032a2, which introduces the connection with definition needed to complete Aristotle's argument.

(b) Three Philosophical Questions

There are three points of philosophical importance concerning Aristotle's sequence of essences which perhaps require more discussion than they are given in the main text above. First, why does Aristotle find an infinite sequence of essences unacceptable? And why is it also unacceptable that $E(x) \neq E(E(x))$? (Code constructs an answer to both these questions around the connection Aristotle makes between essence and substance.) And thirdly, why if $E(x) = E(E(x))$, does it follow that $x = E(x)$, so that Aristotle's regress collapses entirely?

Code uses the connection between essence and substance to show both why an infinite regress of essences is absurd, and also why it is absurd to suppose even that $E(x) \neq E(E(x))$. Code gives an excellent account of Aristotle's philosophical motivation for finding the regress unacceptable. Aristotle rejects out of hand the idea that a definition can be infinitely long, or that a complete explication of a thing's essence can be other than finite in length: *An. Po.* A22, 82b37–83a1, cf. 84a25–26, and see also *Metaphysics* α2 994b16–18 with Alexander *in Met.* 160.30–161.1. And on Aristotle's antipathy in general to infinite regresses, see also Kung 1981, p.225 and Notes 48 and 49.

Why, secondly, does the connection between essence and substance show that it is absurd to suppose even that $E(x) \neq E(E(x))$? I take Aristotle to be appealing here to a network of principles involving the substance-of relation, in particular, (T2) and (T3), where (T3) asserts the secondary reflexivity of the substance-of relation. Code's account relies instead on what appears at first to be a very different set of ideas. He uses the idea that an essence is a substance to show that an essence is therefore a proper object of definition, and hence subject to the full-expansion condition of *Metaphysics* Z10, which stipulates that the definition of a proper definable contain the definition of all its properly definable parts. The argument continues as follows. Suppose that the account $L_{n(n \geqslant 0)}$ of a proper definable E_n is fully-expanded (Code's term: "fully-analyzed"). Then where E_{n+1} is the essence signified by L_n, L_{n+1} is the account of E_{n+1}, only if $L_{n+1} = L_n$. This argument can show that E_n and E_{n+1} have the same definition, and hence, that they also have the same essence. That is, $E(E_n) = E(E_{n+1})$, or equivalently, $E_{n+1} = E_{n+2}$. In particular, it is false that $E(x) \neq E(E(x))$. (This argument will not be able to show, however, that $x = E(x)$.)

This same result can also be obtained independently of Z10 and the full-expansion condition on definition, by taking up again the ever-useful (A1). We know by (A1) that

> If E_{n+1} is signified by the definiens in a definition of $E_{n} (n \geqslant 0)$, then $E_n = E_{n+1}$.

It follows from this that

> If E_{n+1} is signified by the definiens in a definition of E_n, then E_{n+1} is signified by the definiens in a definition of E_{n+1}.

The proof of this reflexivity principle is parallel to that of (T3) from (T2). According to this result, the definiens in a definition of E_n and the definiens in a definition of E_{n+1} *signify the same thing*. That is, $E(E_n) = E(E_{n+1})$, or equivalently, $E_{n+1} = E_{n+2}$.

As the discussion above shows, the appeal to Z10, and the use of the secondary reflexivity of the substance-of relation, come out at much the same place in the end. My inclination, however, is still to regard the reference to substance in the εἴπερ clause at 1031b31–32 as an appeal to the reflexivity principle (or perhaps better, to the network of principles that include that result), and to accept Code's discussion of Z10 as a welcome aid in explaining the philosophical motivation that underlies the truth of the reflexivity principle. Notice, finally, that Aristotle explicitly states the reflexivity principle at Z16, 1040b23–24:

> the substance belongs to nothing but to itself and to what has it, of which it is the substance.

Finally, why if $E(x) = E(E(x))$, does it follow that $x = E(x)$? This question is best combined with another, concerning some minimum materials for generating or collapsing the regress of essences. For as we shall see, a special case of the argument for collapsing the regress will show why, if $E(x) = E(E(x))$, then $x = E(x)$.

I have argued in the penultimate paragraph of the main text that (D2) and (A1) by themselves are strong enough to give Aristotle's sameness conclusion. The importance of these two principles can be dramatized in yet another way. (A1) by itself yields (T4), as we have seen. (D2) and (T4) together allow two arguments, both due to Code, in connection with Aristotle's regress. First, the regress can be generated by the use of (D2) and (T4), if we also assume the initial step of the regress, that $x \neq E(x)$. For, if $x \neq E(x)$, then by (T4), the definiens in the definition of each will signify different things. But by (D2), the essence of a thing is just the entity signified by its definition. Accordingly, if $x \neq E(x)$, then $E(x) \neq E(E(x))$, and in general for any $n \geqslant 0$, if $E_n \neq E_{n+1}$, then $E_{n+1} \neq E_{n+2}$, *ad infinitum*.

The argument just given shows that if the non-identity succeeds at any earlier stage in the regress, it succeeds at every later stage. Trivially, then, if the non-identity fails at any later stage in the regress, then it also fails at every earlier stage. To spell the argument out in greater detail: suppose that $E_n = E_{n+1 (n \geqslant 1)}$. By construction, E_n is the essence of E_{n-1}, and E_{n+1} is the essence of E_n. That is, since by hypothesis, $E_n = E_{n+1}$, the essence of E_{n-1} = the essence of E_n. By (D2), therefore, the definition of E_{n-1} = the definition of E_n. By (T4), therefore, $E_{n-1} = E_n$, as required. This same argument can be repeated for any $E_{m (0<m\leqslant n)}$, descending to the first step of the regress. This, then, is also the argument that shows that if $E(x) = E(E(x))$, then $x = E(x)$.

NOTES

* This paper was written in response to the paper, 'On the Origins of Some Aristotelian Theses About Predication', by Alan Code, which appears elsewhere in this volume. My debt to his paper is extensive. On the TMA, I have absorbed much of his treatment of the argument without comment, and concentrate on points where our accounts diverge. Regarding Aristotle's theory of essence in *Metaphysics* Z6, I have stated my own views first in Section 5 (this part of the paper was written before Code's paper came into my hands), and I have reserved comments on Code's account for the Afterword. Finally, there is one part of Code's paper, on Aristotle's notion of definition and essence, that I do not have the space to comment on here; my own account of these matters is given in Lewis 1984.

1 The TMA exists in many forms. In its original version in the *Parmenides*, the argument is not about men at all, but is a Third *Largeness* Argument, *Parmenides* 132a1b2. Plato also gives a related regress argument at *Parmenides* 132c12–133a4. In his commentary on Aristotle, *Metaphysics* A9, Alexander gives a batch of unrelated arguments that pass under the name 'Third Man Argument' (84.7–21), as well as three separate accounts of the version of the argument to be attributed to Aristotle: one at 83.34–84.7 (said at 85.10–11 to be Eudemus' form of the argument), a second at 84.21–85.3 (this argument is directly attributed to Aristotle at 85.11–12, and at 85.3–5 said to be the same argument as that at 83.34–84.7), and a third at 93.1–7, identified at 93.7 with the argument at 84.21–85.3. On the various cross-references between these three accounts of the argument, including the puzzling δεύτερον at 93.7, see Cherniss (1944), n. 210 and App. IV, pp. 500–501.

2 The context shows that by 'predicate' here, Owen means "property": Owen (1965, p. 134). The formulations of SP and NI given here are taken from Owen *loc. cit.*

3 Although he does not say so, Owen is making two quite distinct points here. For, there are cases in which a term is univocal when applied to two different entities, but the

name and the associated definition do *not* apply equally to both. The term 'white', for example, is univocal when applied to Socrates and to white itself, but the definition of white does not apply to Socrates. The two kinds of principle (one of which is a principle of univocity, and the other, one of synonymy) are discussed further in Sections 3(c) and 4(a) below.

4 Cf. Aristotle *Topics* B2, 109a34–b12.

5 What is it for a kind of predication, or for a true sentence exemplifying a given kind of predication, to imply self-predication? In the true sentence, 'Socrates is a man', we can replace the subject-term, 'Socrates', by a second occurrence of the predicate-term to produce another true sentence: 'Man is a man'. Loosely speaking, the context '... is a man' is to be understood in the same way in both sentences: the exact force of "understood in the same way" will be spelled out later (see Section 4(a) below). Suppose now that we use the same method to obtain from the true sentence, 'Socrates is white', the new sentence, 'White is white'. If this second sentence is true, we will *not* understand the context '... is white' here in the same way as in the original sentence. That is, self-predication is *not* implied.

The notion of self-predication at work in this account requires some comment. Owen seems to suggest that SP holds for the kind of predication illustrated by 'Socrates is a man' (but not that illustrated by 'Socrates is white'), because (in contrast to the other case) Socrates is a man "in the same sense in which" man is a man. That is, perhaps, principles of univocity and synonymy (n. 3 above) are both satisfied. If this reading of Owen is even approximately correct, then he has moved a good distance beyond his official definition of "self-predication", set out in SP. On further shifts in the notion of self-predication, see Note 41 below.

6 As Code points out, then, Owen's account portrays Aristotle in the *Categories* as committed to both SP and NI – even though these are incompatible assumptions about predication, in the way to be explained at the end of this Section. Despite this, however, it is not clear that Owen has exactly the same notion of "self-predication" in mind in talking of the TMA and of Aristotle's own theory in the *Categories*: cf. Note 41 and p. 153 below.

7 Owen 1965, p. 134, says that NI and SP are incompatible; he does not notice that we must also add P to generate an inconsistent set. He also fails to see that if the premiss-set for the TMA is inconsistent in this way, then both the TMA itself and the program of building an account of Aristotle's philosophical development upon it, become pointless, in the sense explained in the text.

8 Aristotle's references to the argument as the "Third Man Argument" seem to suggest that he regarded the infinite sequence of entities, each of which is a man, as a significant result. (See Bonitz *Index* 771a37–39 for Aristotle's references to the TMA.)

9 On the use of 'separable' here, cf. Sections 2(c) and 3(c) below.

10 Cf. Note 9 above.

11 Code cites the authority of Alexander for these two assumptions. In relying on Alexander in this way, Code shares an important assumption with Owen: both are *literalists* in their view of Aristotle's version of the TMA, in the sense explained in the text above. Obviously, any *non*-literalist hypotheses about Aristotle's version of the TMA will lack the virtue of direct evidence, which literalism claims to provide. But there are difficulties, both textual and philosophical, with literalism too. To take the textual arguments first: there are, as we have seen (note 1 above), no less than three separate

descriptions of the TMA in Alexander. (In this connection, notice that the first and second accounts at 83.34–84.7 and 84.821–5.3 are *not* parts of a discontinuous but *single* account.) There are significant differences between the three accounts: yet Alexander apparently regards them as all accounts of the *same* argument. Can we be sure, then, that Alexander is sufficiently strict about what counts as the same argument, to take him as in any sense an authority for the version of the TMA that Aristotle has in mind? A philosophical doubt about literalism concerns the assumption that Aristotle had *just one* version of the TMA consistently in mind throughout his philosophical career. The frequent changes of heart undergone by modern exponents of the argument indicate sufficiently the dubiousness of this hypothesis.

12 This terminology is adapted from a device of Geach's. I avoid the devices of modern set-theory, in order to avoid the suggestion that pluralities can be identified straightforwardly with sets. In fact, not all sets are pluralities, as we shall see later in this Section. Like sets, however, pluralities *have members*.

13 For example, *τινων πλειόνων*, 84.22, cf. *πλείόνων ὄντων*, 84.26. The assumption that our initial plurality Π contains at least two sensibles, that are like each other, begins with Plato, *Parmenides* 132a2, *πολλ' ἄττα μεγάλα*, and passim, although this seems more a consequence of how Plato typically states the One-over-Many assumption, than a feature peculiar (and necessary) to the TMA.

14 *ἔστι δὲ ὁ λόγος οὗτος τῷ πρώτῳ ὁ αὐτός ἐπεὶ ἔθεντο τὰ ὅμοια τοῦ αὐτοῦ τινος μετουσίᾳ ὅμοια εἶναι. ὅμοιοι γὰρ οἵ τε ἄνθρωποι καὶ αἱ ἰδέαι*, 85.4–5, the reference back is to 83.34–84.7, especially 84.1–2, *ἔτι τὰ ὅμοια ἀλλήλοις τοῦ αὐτοῦ τινος μετουσίᾳ ὅμοια ἀλλήλοις εἶναι, ὅ κυρίως ἐστὶ τοῦτο καὶ τοῦτο εἶναι τὴν ἰδέαν.*

15 The univocity condition given imposes a necessary but uncontroversial constraint on the pluralities of the TMA. But it is worth stating separately, if only because it is sometimes confused with other, stronger conditions, cf. Note 3 above. For univocity, see Owen (1975, p. 297 f).

16 See Note 14 above.

17 See Note 14 above.

18 HOMOGENEITY (2) is virtually identical with Code's (A), discussed in Section 3 below. Code notes the need for a principle of this sort governing the membership of the pluralities admissible in the TMA (Code, n. 13), but he does not formally incorporate such a principle into his reconstruction of the TMA. Instead, according to Code, (A) is a semantic meta-principle, and not strictly part of the TMA at all. Meta-principle or not, however, (A) or something like it must be true, if the argument is to be sound: and here, I shall argue, there is trouble ahead (see Section 3(c) below).

The controversial cases of HOMOGENEITY will be where the plurality in question includes not only sensibles but also one or more forms. A number of modern characterizations of "self-predication" in Plato are in effect committed to HOMOGENEITY for such cases: for example, the view that the form man (say) is "logically on a par *with other members of the class*" (Owen 1968, p. 117, my italics), and the similar appeal to class-membership in Vlastos (1973, p. 258, n. 97), cf. also point (iv) in his p. 348, Note 28.

19 The denial of HOMOGENEITY gives Plato one way out of the TMA, and is entailed by other principles that he is apparently committed to (see the discussion of Code's [B] and [C] in Section 3[a] below, and for a different version of the same idea, Cherniss

(1957, p. 370f), and Allen (1960, p. 46). At the same time, however, Code suggests that Aristotle has an argument that, if sound, makes it impossible for Plato to deny HOMOGENEITY. If this is right, then Plato is saddled both with the TMA, and with an inconsistent semantics. These claims are discussed in Section 3 below.

20 The convention that a collection has at least two members corresponds to the fact that accounts of the TMA always begin, "Suppose that two or more sensibles are men (or large, and so on)". And one of Plato's basic principles in the theory of forms is a One-over-*Many* Assumption. At the same time, there seems no reason why there should not be a form corresponding to a *unique* sensible (exactly this hypothesis is made with respect to the visible world, created after the pattern of τὸ παντελὲς ζῷον, in the *Timaeus*), and also no reason why there should not be a *one*-member plurality, from which we can generate a regress of forms along the lines of the TMA. ALEX-PL itself, then, does not rule out the possibility that there exist one-member pluralities.

21 Code, pp. 105–106.

22 Code does not defend these moves in his argument in any detail. Without an analysis of being separable or being a τόδε τι, or of being ἄτομον, why suppose that the second follows from the first, or the third from the second? And without an analysis of being ἄτομον, why suppose that an ἄτομον is predicable accidentally of everything other than itself of which it is predicated, but (as Code suggests elsewhere) predicable essentially of itself? These are more general questions about Code's theory of Being and Having, however, and I shall not pursue them here.

23 Either Code's (A) (Code Section III), or the stipulations regarding pluralities given in his n. 13, will suffice to make the argument valid. Something weaker than HOMOGENEITY will also be sufficient to support ALEX-NI. We will need a principle to govern the members of pluralities of any level, if the regress is not to come to a stop. In general, then,

> For any n, where Π^n is a plurality of level n, if X is predicable of each member of Π^n, then X is predicable accidentally of some member of Π^n only if X is predicable accidentally of all the members of Π^n.

This principle is entaiied by HOMOGENEITY. Notice that like (A) and HOMOGENEITY, the principle avoids the assumption made in Code's n. 13, that "there are just two ways in which the term 'man' could apply to something".

24 Compare the "empirical" assumption above, and the convention that a collection has at least two members.

25 Code, p. 105.

26 For further doubts about the role of logical separation as Code defines it in the argument, see Section 3(c) below.

27 These principles appear in Section III of Code's paper.

28 Code, pp. 109–10, gives a somewhat different argument for this same conclusion (his account uses added materials that I shall discuss later in this Section). Suppose that the definition of the universal does not apply to any of the particulars that fall under it (Code holds that this happens where the corresponding predicate is incomplete – not used κυρίως – in its ordinary application to particulars). It follows, Code argues, that the universal is accidentally predicable of the particulars (this is by his account of essential and accidental predication, given on Code p. 103). But the universal is predicable

essentially of itself (for its own definition does apply to it), so that it is not predicable in the same way of itself and of the corresponding particulars. This result shows that (A)-'large' is false, for 'large' is a predicate which is incomplete, or not used κυρίως, in its ordinary application to particulars. The argument does *not* show, however, that if a predicate has the contrary property, i.e., it is used κυρίως in its application to particulars, it is therefore used in the *same* way, both of particulars and of the relevant universal. I argue later in this Section that Aristotle has no effective way of validating this last inference.

29 Cf. Section 1 above. I discuss the difference synonymy makes further in Section 3(c) below.

30 It is important to see that, for all that has been shown so far, these are two *different* differences. Aristotle's notion of synonymy gives a way in which 'man' and 'large' behave differently (further details on this are given in Section 3(c) below): and even if synonymy is essentially an Aristotelian discovery, it is not implausible that Plato too would accept this difference. But from the fact that the two predicates differ in the way stipulated in Aristotle's doctrine of synonymy, it does not evidently follow that they must also differ in the further way Aristotle claims, that is, that (A)-'large' is false, but (A) must be true. The truth of (A) and the falsity of (A)-'large' falls directly out of Aristotle's theory in the *Categories* (for Socrates is (a) man, and man Is (a) (rational) animal, whereas Socrates Has large(ness), but large(ness) Is a size of a certain sort). But this does not show that Plato, with his very different ontology and very different theory of predication, should also accept these results. I argue in Section 3(c) below that in fact, Plato can consistently accept Aristotle's synonymy doctrine, and at the same time regard (A) and (A)-'large' as both alike *false*.

31 On the account given here, denying (A) constitutes Plato's primary move against the TMA. It is a *consequence* of this (but *not* Plato's first line of defense against the TMA, as some have seen it: Vlastos (1954, pp. 238f), Geach (1956, pp. 274f), Cohen (1971, p. 473, Note 42) that pluralities not of level zero (that is, pluralities whose membership includes forms as well as sensibles) are no longer admissible in the TMA.

32 An example involving the converse relation, "x is predicated synonymously of y", for the case where x is a genus, is given at *Topics* B2, 109a34–b12.

33 Elsewhere, Code argues that for Aristotle, the path from SYNONYMY to (A) leads through the notion of κυρίως predication: the fact that the definition of man applies both to man and to the particular men entails that 'man' is predicated κυρίως of each, while this in turn entails that each Is (a) man, so that (A) is true (Code, p. 109–10). I argue later in this Section and in Section 3(d) below that Plato will not find the move from SYNONYMY to (A) convincing, with or without the mediation of κυρίως predication. Similar reasoning will also lead him to reject the principle quoted, that "a universal is essentially predicable of a logical subject X if and only if both the name and the definition of that universal truly apply to X": see Note 34 below.

34 The reasons given for supposing Plato would resist the argument from SYNONYMY to (A) suggest also that he would reject the principle quoted in the previous paragraph in the main text, that "a universal is essentially predicable of a logical subject X if and only if both the name and the definition of that universal truly apply to X". Specifically, his objection to the inference from SYNONYMY to (A) is an objection to this principle in the right-left direction. For Plato, then, SYNONYMY cannot yield a notion of essential

predication, as perhaps it does for Arisotle. In particular, it does not show that a particular man Is (a) man, contrary to (B).

35 Strictly, these considerations show only that Aristotle *should not* include both SYNONYMY and SEPARATION in his version of the TMA. Aristotle himself may not have been sensitive to the difficulties that follow if the premisses or other principles involved in the TMA form an inconsistent set. Our preference for supposing that, in fact, Aristotle's version of the argument does *not* make room for both SYNONYMY and SEPARATION is based on the methodological bias already noted in Section 1 above.

36 Owen 1957 argues that this difference in logical behaviour between the two kinds of predicate forms the basis of the argument at Alexander *in Met.* 82.9–83.17, cf. Aristotle *Metaphysics* A9, 990b15–17. The same difference also underlies the contrast between the argument concerning relatives Alexander is dealing with at 82.9ff, and the TMA, to which he turns next.

37 Owen (1957), cf. Code Section III.

38 White (1971, p. 117). Cf. Owen (1968, p. 117).

39 Alexander in *Met*. 93.2ff (quoted in Section 2(b) above).

40 (i), (ii), and (iii) are each formulated in terms of the predicate 'man', and in the *Categories*, Aristotle will say that 'man' is univocal as applied both to Socrates and to his species, and that the copula is to be interpreted in the same way both in 'Socrates is a man' and in 'Man is a man' (Code's 'Is'?), and finally, that man is predicated synonymously of Socrates. The same does not hold, however, for the use of the predicate 'white' as applied to Socrates and to white itself. The predicate has the same meaning in both cases, but the use of the copula is not the same, and white is not predicated synonymously of Socrates.

The difference between the two cases lies behind Owen's view that the self-predication assumption does not hold for weak predication, but that it does hold for strong predication (on the sense of "self-predication" here, see Note 41 below). Fine (unpublished) claims to find a difficulty here, for the predicate 'white' figures not only in weak predications, but also in strong predications, for example, 'Della Robia white is white'. She concludes that white is both subject to, and not subject to, self-predication – that is, on her account, it both is and is not a white thing – which is absurd. This objection, however, probably takes too little account of the shifts in Owen's use of the term, 'self-predication' (see Note 41 below). From the fact that 'white' figures in weak predications, for example, 'Socrates is white', Owen means us to conclude only that (roughly speaking) white is *not* white in the way in which Socrates is white (more precisely, neither homogeneity nor synonymy holds). From the fact that the same predicate also appeares in true strong predications, for example, 'Della Robia white is (a) white', we are meant to infer only that (roughly) white *is* white in the same way in which Della Robia white is white (that is, univocity, homogeneity, and synonymy all three hold). There is no absurdity in asserting both of these propositions jointly.

41 There is a gain in clarity from sorting out these various ingredients in vicious self-predication. All four of (i)–(iv) are required to make up a full-blooded notion of vicious self-predication. At the same time, there has been sufficient slack in the use of the term 'self-predication' in the secondary literature, that it may be best to admit a notion of *degrees* of self-predication, according to how many of the criteria (i)–(iv) are in effect. In some places, for example, Owen's notion of self-predication is apparently based on (i)

and (iii) alone (cf. Section 1 above). Sometimes he adds (ii) (in Owen (1968), for example, he describes a "paradeigmatic form" as "logically on a par with *other members of the class*," p. 117, my italics). That is, the same predicative link, namely class-membership, explains both why Socrates is a man, and why man itself is a man. And elsewhere, Owen takes Plato's notion of self-predication to include (iv) as well (for example, Owen (1968, p. 109, Note 1)). When Owen says that the TMA is built around an assumption of self-predication, he presumably means full-strength self-predication, defined by all four criteria (i)–(iv). When, however, he says that Aristotle himself accepts self-predication for strong predication in the *Categories*, he must have a lesser degree of self-predication in mind, based on (i), (ii), and (iii) alone, for Aristotle explicitly denies (iv), as we have seen.

42 The relational notion of separation is expressed by Alexander's κεχωρισμένον αὐτῶν, [ὧν κατηγορεῖται], 84.23. Code, Note 14, takes Alexander to suggest here that non-identity is a *consequence of* separation (in the special sense in which he defines it), with a similar reading of κατ' ἰδίαν ὑψεστώς at 84.28 below. A similar claim appears in Fine (unpublished), who says that separation not only entails non-identity, but even constitutes "the only reason one might have for accepting" non-identity. Like Code, she cites Alexander 84.22–23 in support of this claim, but she does not say what she takes separation to be, or why it should have the consequences she attributes to it. In fact, however, it is not clear that the phrase at 84.23 is any more than a gloss on non-identity, which is asserted immediately before. For the phrase at 84.28, see Note 43 below.

43 This is perhaps the force of Alexander's κατ' ἰδίαν ὑψεστώς at 84.29. If so, then the phrase is not used in exactly the same way as his κεχωρισμένον αὐτῶν at 84.24: see Note 42 above.

44 Further references for the this-such contrast are given in Kung (1981, pp. 207–208) and *passim*.

45 Cf. Notes 42 and 43 above.

46 In the intervening period, however, there is a difficulty, as Code notes. For, according to Owen, Aristotle sees that SP and NI (strictly, with the addition of P) form an inconsistent set, yet on Owen's account, he asserts both for strong predication in the *Categories*.

47 This account of Aristotle's theory of predication in the *Metaphysics* is defended at length in my "Form and Predication in Aristotle's *Metaphysics*", which appears elsewhere in this volume. I argue in that paper that what may appear to be additional cases of (metaphysical) predication, for example, where a species is apparently (metaphysically) predicated of an individual, are at best derivative cases, to be defined in terms of other, more nearly basic notions. A further example is the relation of an essence to that of which it is the essence. In his paper, Code says that a relation of ontological predication underlies the fact that an essence can be linguistically predicated of that of which it is the essence. But *pace* Code, there is no need to introduce this new variety of (metaphysical) predication on Aristotle's behalf: the relation in question is simply the familiar essence-of relation. This may appear to be an argument over terminology rather than substance. Against this, however, if we admit these added cases of metaphysical predication, then the properties of the realtion are radically changed: an example is noted at the beginning of Note 48 below. (The properties of the relation of metaphysical predication are discussed further in Lewis [unpublished].)

48 If Aristotle recognizes a relation of metaphysical predication between an essence and

that of which it is the essence, as Code suggests, then this principle is false. But I have argued that Aristotle does not extend the notion of metaphysical predication in this way (see Note 47 above). Further, the principle given is true, and has important consequences elsewhere in Aristotle's metaphysical theory: I develop this point further in Lewis (unpublished).

It should perhaps be emphasized that in denying that there exists a relation of metaphysical predication between an essence and that of which it is the essence, we are not committed to saying that there are no *linguistic* predications which connect an essence with that of which it is the essence. In the primary case, an essence is linguistically predicated of a primary substance. The essence-of relation appropriate to this case is *vacuous*: that is, an entity and its essence are identical. Aristotle also recognizes secondary cases, in which, for example, an essence is linguistically predicated of a compound material substance: a different essence-of relation is required for this case, with different logical properties. Further details are given in Lewis (1984).

49 The accidental character of the relation is even more evident when it is reinterpreted in the *Timaeus* as a relation between a given region of space, and the likeness of a form that supervenes on it. *Metaphysics* Z4, 1030a13–14, seems to agree that participation is an accidental relation. See also Cohen (1978, p. 82).

50 *Metaphysics* M5, 1080a1 (*αἱ ἰδέαι οὐσίαι τῶν πραγμάτων οὖσαι*), cf. A6, 988a8–11 (*τὰ γὰρ εἴδη τοῦ τί ἐστιν αἴτια τοῖς ἄλλοις*, a10–11), A7, 988b4–5 (*ἀλλὰ τὸ τί ἦν εἶναι ἑκάστῳ τῶν ἄλλων τὰ εἴδη παρέχονται*). Possibly the same point is involved in the controversial passage at A9, 990b27–34, cf. Alexander *ad loc*. and the parallel passage at M4, 1079a24–33. At M5, 1079b15–17, Aristotle argues that the claim that forms are the substances of sensibles cannot be sustained. And some of the arguments in Z14 take Plato to task on the related idea that certain forms are the substance of other *forms* ("the form man is *not accidentally* composed of animal", 1039b7–9, 9–10, 12–13, 15–16); Aristotle remarks at the end of the chapter that similar arguments apply to the relation between forms and *sensibles*.

51 A necessary condition on definition for Aristotle is that it gives a λόγος in place of an ὄνομα (or a longer λόγος in place of a shorter one). So a necessary condition for the existence of a (second-level) definition or essence for a first-level essence $E(x)$ is that $E(x)$ have a name (or a shorter λόγος). Alternatively, Aristotle's remark here may be a variation on the "cloak" gambit practiced at Z4, 1029b27ff, cf. I9, 1058b5: here, the use of an arbitrary name for a first-level essence $E(x)$ in asking for the essence of $E(x)$ will conceal the implausibility of the idea that the essence *of an essence* should be anything but just that essence itself. For an ingenious elaboration of this point, see Code, Note 74.

52 Although Aristotle does not explicitly say so, it is natural to suppose also that any two members of the infinite sequence of essences that results will be distinct from one another: that is, we can rule out any "repeats" or "cycles" in the sequence. With regard to the three-member sequence $\langle x, E(x), E(E(x))\rangle$ Aristotle discusses in our passage, then, we can assume that not only $x \neq E(x)$ and $E(x) \neq E(E(x))$, as Aristotle expressly supposes, but also that $x \neq E(E(x))$.

53 Code also supposes that the argument is divided up into these two stages. Detailed discussion of his account of the argument is reserved for the *Afterword* immediately following this Section.

54 "Some things", ἔνια (b31). Not: somewhere in the sequence of essences, since this

stops at $E(E(x))$. So the point is not that *somewhere* we will get tired of adding new essences. Rather, *in some cases*, we identify a thing and its essence right off, before the sequence of essences ever gets started – that is, when we are giving the essence of things that are properly πρῶτα (Z4, 1031b13ff). A different account appears in Code: see the *Afterword*.

55 Aristotle often uses 'λόγος' not only for a linguistic entity (in the relevant cases, either a linguistic definition or its definiens), but also for form, which is what is signified by the definiens of a linguistic definition. He is sometimes accused of carelessness on this score (see, for example, Kirwan 1971, p. 125 on *Metaphysics* D2, 1013a26, and p. 134 on D6, 1015b24–25), but the different uses are explained as easily by the abbreviation convention adopted in the text.

56 For (D2), see *Topics* A4, 101b21–22, A5, 101b38, A8, 103b9–10, H3, 153a12–22, H5, 154a31–32, *An. Po.* B3, 91a1, B10, 93b29, 94a11, *Metaphysics* D6, 1016a33, 1017a6, D8, 1017b21–22, Z5, 1031a12, H1, 1042a17, and Bonitz *Index*, 764b45–57.

57 (A1) rests on a thesis given in the *Topics*:

x is the definition of y, only if x is numerically the same as y,

Topics A5, 102a7–9, 11–14, 16–17, A7, 103a25–27, cf. Code p. 113. This thesis is crucial, and the credit for drawing attention to its relevance to the argument in Z6 belongs to Code (and also, according to Code, to Grice). By 'numerically the same as' in the thesis stated, I take Aristotle to mean numerical sameness of the strictest kind (*Topics* A7, 103a25–27), that is, essential sameness. Essential sameness is not the same as our identity, but I shall assume without argument here that it entails it. Given this assumption, we can derive (A1) immediately.

58 For this principle, see *Topics* B2, 109b9–11, "the definition <of a thing> belongs <to it and> *to nothing else*" (my italics), cf. A4, 101b19–22, Z1, 139a31–32, b3–4, Z12, 149b22–23, H5, 154b1–3, 10–12.

59 Code p. 119 gives virtually the same argument, again (in effect) using just (A1) and (D2), to give the identity result. (Of Code's two "leading ideas" [p. 119], his [i] is more or less my [D2], and his [ii] is my [A1]. To these add my [D1], which Code also uses in a number of places.) Suppose, then, that a primary substance x has the definition y. Then by (A1), $x = y$, and by (D2), y is the essence of x. Hence, x = its essence, as required. Code suggests that the argument is also good for items that are not objects of definition in the primary sense. If the argument is to be extended in this way, however, we must be able to show that (A1) holds also for things which are not definable in the primary sense. In fact, I believe, (A1) does *not* hold for most if for any of these further cases: this claim is defended in Lewis (1984).

60 Work on this paper was supported in part by an A.C.L.S. Fellowship funded by the National Endowment for the Humanities, hereby gratefully acknowledged. I am grateful also to James Bogen and Alan Code for generous help with earlier versions of the paper: they are not responsible for my remaining mistakes.

MARILYN McCORD ADAMS

THINGS VERSUS "HOWS", OR OCKHAM ON PREDICATION AND ONTOLOGY

Ockham's account of the truth conditions for categorical propositions of the forms '*N* is *B*' and '*A* is *B*' is dictated by his two-fold ontological program: to eliminate universals other than names or concepts and to restrict particular things (*res*) to the categories of substance and quality.[1] For endorsing

(T1) '*N* is *B*' is true if and only if *B*-ness inheres in *N*

and

(T2) '*A* is *B*' is true if and only if something that is *A* has *B*-ness, or *B*-ness inheres in something that is *A*

for all substitutions of proper names for '*N*', concrete general terms for '*B*', and abstract general terms for '*B*-ness' would be tantamount to adopting the ontology of the moderns with a distinct kind of thing for each of the ten Aristotelian categories. According to Ockham's description in the *Summa Logicae*, the moderns saw substances such as Socrates as having a certain layer-like structure: at the core is a particular substance composed of prime matter with several substantial forms; it is made one by the inherence of discrete quantity and extended by the inherence of dimensive quantity; other absolute quality-accidents such as color, flavor, odor, shape, etc. inhere in dimensive quantity immediately and substance mediately. Accidents in the remaining seven categories are relative things (*respectus*) that have their foundations either in the substance, quantity, or qualities and their term in something else. For example, when Socrates and Plato are similar, a relative similarity thing inheres in Socrates' whiteness and is Socrates' similarity to Plato, while another relative similarity-thing inheres in Plato's whiteness and is Plato's similarity to Socrates. For reasons that lie beyond the scope of this paper,[2] Ockham thought this ontology was as prolific of contradiction and absurdity as it was of entities. His own contention is that the truth of a proposition depends not merely on *which* things exist, but on *how* they exist.

In what follows, I shall explain how Ockham sacrifices the simplicity of across-the-board adherence to (T1) and (T2), applying them only to a certain range of case in the category of quality. Then I shall examine a series of objections that his own ontology offers no advantage commensurate with this loss.

James Bogen and James E. McGuire (eds.), How Things Are, 175–188.

1. OCKHAM'S ONTOLOGY AND THE TRUTH-CONDITIONS FOR CATEGORICAL PROPOSITIONS

Whether or not (T1) and (T2) apply depends, according to Ockham, on whether the term substituted for '*B*' is absolute or connotative. (1.1) *Absolute versus Connotative Terms*: According to Ockham, absolute terms do and connotative terms do not signify whatever they signify the same way ("with equal primacy") (*Summa Logicae* I, c. 10; OPH I, 35–6). Realists about universals say that absolute general terms name universal kinds or natures in the same way as proper names name particulars. Loosely speaking, a proper name signifies only one thing per imposition.[3] While the name 'Socrates' is given to many men, each "christening" imposes it to signify or name a single man. Likewise, the term 'man' is conventionally instituted to signify human nature; 'whiteness' the nature of whiteness; etc.[4] Although Ockham agrees that general terms such as 'man' and 'whiteness' signify things and that they are related to their significata the way proper names are related to their bearers, he denies that there are any common or universal things. Hence, he concludes that every such term signifies (or can signify) many particulars, but is related to each just as directly and primarily as a proper name is related to its bearer, and accordingly signifies each exactly the same way as it signifies any other.

Every general term contrasts with a proper name in being able to signify many with equal primacy. But absolute terms signify *whatever* they signify the same way, while connotative terms signify some of their significata one way and others another. Often Ockham says that they signify some *primarily* and others *secondarily*, or that they *signify* some and *consignify* or *connote* others. These connotative terms have a nominal definition (*definitio exprimens quid nominis*) which "explains what is signified by the word – viz., what we ought to understand by the word" (*Summa Logicae* III–3, c. 23; OPH I, 682; III–2, c. 28; OPH I, 556; III–3, c. 26; OPH I, 690). Ockham's favorite example is the concrete quality term 'white' that has as its nominal definition 'something having whiteness' or 'something informed by whiteness' (*Summa Logicae* I, c. 10; OPH I, 367; III–2, c. 28; OPH I, 556). But relative, negative, privative, and infinite terms are genuinely connotative as well (*Summa Logicae* II, c. 12; OPH I, 283). On Ockham's mature logic, the primary significata of a connotative term are those about which it is truly predicated and for which it has personal supposition.[5] Thus, 'white' primarily signifies Socrates and other subjects of whiteness and secondarily signifies the whitenesses. And although he sometimes suggests, he is really in no position to provide definitive grammatical criteria for which are which.[6] In any case, his examples and discussion suggest the view that a fully expanded nominal definition of connotatives would include only absolute terms and primitive syncategorematics not capable of further definition themselves.[7]

Corresponding to each absolute general term, there is a real definition that expresses the nature of things in its extension (*Summa Logicae* III–3, c. 23; OPH I, 682; III–2, c. 28; OPH I, 556). Accordingly, Ockham sometimes refers to absolute concepts as "quidditative" (*Ordinatio* I, d. 3, q. 2; OT I, 419; *Quodlibeta* IV, 2). Strictly speaking, a real definition is "a succinct phrase that expresses the whole nature of the thing and does not mention anything extrinsic to the thing defined" (*Summa Logicae* I, c. 26; OPH I, 85). There are two kinds of real definitions: Metaphysical definitions – such as 'Man is rational animal' – include only nominative-case terms; natural definitions – such as 'Man is a substance composed of body and soul' – include oblique-case expressions but only those that signify the essential parts of the thing (*Summa Logicae* I, c. 26; OPH I, 85). Aristotelian empiricist that he is, Ockham believes that the natures of things can be discovered only by experience. Those who have had an intuitive cognition of a lion will have a simple absolute concept proper to lions or a concept of the differentia needed to formulate the real definition. Others who have never seen lions can "know what the name 'lion' signifies" – i.e. fix the extension of the term – only because they have the absolute concept 'animal' and other non-complex connotative and negative concepts none of which is proper to lions and can put these together to yield a composite concept equivalent to, in the sense of being necessarily co-extensive with, the concept 'lion' (*Summa Logicae* III–2, c. 29; OPH I, 558–60).

1.1. *The Truth-Conditions*

Ockham's theory of universals leads him to say that, from the view-point of unaided natural reason, both concrete and abstract nouns in the category of substance are absolute terms. It follows that (T1) and (T2) cannot apply for substitutions of such terms for '*B*' and '*B*-ness', respectively. The reason is that 'Socrates is a man' is true when 'Socrates has humanity' and 'Humanity exists in Socrates' are false. For either 'humanity' would stand for something extrinsic to Socrates, for Socrates himself, or for one of his parts. Socrates does not have, and is not in, himself. Contrary to the Platonists, nothing extrinsic to Socrates is his humanity. And against the moderate realists, Ockham holds, it is wrong to think of Socrates as composed of parts one of which is his humanity and the other some thisness or individuating principle (see *Ordinatio* I, d. 2, aa. 4–8). Even if humanity is thought of as a particular thing (Ockham's own view), it is not one essential part of Socrates as distinct from others; rather the thing that is Socrates and the thing that is his humanity are altogether one and the same (*Summa Logicae* II, c. 2; OPH I, 250–4; cf. I, c. 7; OPH I, 23–29). Rather, where '*N*' is a singular term (demonstrative pronoun or proper name) and '*A*' and '*B*' are absolute terms, propositions of the form '*N* is *B*' and '*A* is *B*' do not admit of further analysis and are not exponible by a molecular proposition. And he offers a "two-name" account of their truth-conditions:

(T3) '*N* is *B*' is true if and only if there is something for which '*N*' and '*B*' both supposit in the proposition '*N* is *B*'

and

(T4) '*A* is *B*' is true if and only if there is something for which both '*A*' and '*B*' supposit.

Likewise, the truth-conditions for 'This whiteness is a color' and 'Wisdom is a quality', where substitutions for '*A*' and '*B*' are absolute terms in the category of quality, are given by (T3) and (T4), respectively.

Where one or more of the terms is connotative, however, the categorical proposition is equivalent to a molecular proposition. The nominal definition of any connotative term will contain absolute terms signifying substances and/or qualities. But because they will differ with respect to the primitive syncategorematics they contain, no fully general recipe for expounding them is available, and their truth-conditions will have to be assessed on a case-by-case basis. Ockham notes that

(T5) Any proposition that includes a concrete term whose corresponding abstract term signifies a thing that informs another thing, has two exponents – one with both terms in the nominative case and the other with a term in an oblique case

(*Summa Logicae* II, c. 11; OPH I, 281). Letting such concrete inherence terms substitute for '*B*', the corresponding abstract terms for '*B*-ness', singular terms for '*N*', and any general term for '*A*', Ockham stipulates the following:

(T6) '*N* is *B*' is true if and only if '*N* exists and *B*-ness inheres in *N*' is true and

(T7) '*A* is *B*' is true if and only if 'Something is *A*, and *B*-ness inheres in it' is true

(*Summa Logicae* II, c. 11; OPH I, 281). For example, 'Socrates is white' is expounded by 'Socrates exists' and 'Whiteness inheres in Socrates'.

(T6) and (T7) differ from proposals (T1) and (T2) with which we began only in making the existence-conditions explicit, and Ockham thinks they cover propositions of which one or both terms are concrete nouns in the third species of quality, certain concrete disposition or habit terms in the first species of quality, and capacity or incapacity terms from the second (*Summa Logicae* I, c. 55; OPH I, 180–1; *Quodlibeta* I, q. 18; VII, q. 7). But they do not expound the sense where the terms – concrete or abstract – come from other species of quality or from the other eight categories of accident. Thus, (T6) applies to 'Socrates is wise' to yield 'Socrates exists and wisdom inheres in Socrates', but not to 'Socrates is snubnosed' to yield 'Socrates exists and snubnosedness inheres in Socrates' or to 'Socrates is near Plato' to yield 'Socrates exists and Plato exists and proximity inheres in Socrates'. These propositions do not assert the existence of anything over and above Socrates or Socrates and Plato, respectively, but something about *how* they exist: viz., that the parts of Socrates' nose are arranged a certain way, and that Socrates and Plato are positioned a certain way.

2. DOES OCKHAM'S SEMANTICS AFFORD ANY GENUINE ONTOLOGICAL ECONOMY?

Ockham argued that the ontology of unaided natural reason would admit nothing real other than particulars in the categories of substance and qualities. General terms do not signify universal things, but particulars "confusedly", while abstract terms in the last eight categories do not signify distinct quantity-things, relation-things, action-things, etc., but substances and qualities different ways. John Mirecourt was accused of extending the second half of Ockham's program even further, maintaining that

> it is probable in the natural light that there are no accidents but every thing (*res*) is substance; and if it were not for the faith, this should be held and can be held defensibly ....[8]

If it were not for the doctrine of transsubstantiation, Mirecourt allegedly implies, there would have been no need to admit qualities any reality distinct from substance. Similarly, an anonymous *Physics* commentary from the second half of the fourteenth century concludes that "it is indeed possible that neither heat nor cold are things distinct from their subjects." Rather "when fire heats water, it produces nothing, but makes the water to be otherwise (*aliter se habere*) – which is nothing, not any thing or any things but for the same subject to stand in a certain disposition that it did not stand in before."[9] Again, locomotion is identical with the movable object being one way and another (*aliter et aliter se habens*).[10] And had he been bolder, Mirecourt admits, he would have claimed that acts of thought and volitions are only modes or ways things are (*modi se habendi rerum*).[11] By contrast, John Buridan denounces such reasoning, because it drives one to go further still and deny that substantial form is anything really distinct from matter; rather "matter existing one way is fire, existing another way water, air, or stone", which was the "obscure and dangerous" opinion of Melissus and Democritus.[12] Other difficulties aside, are the economies of Ockham's strategy genuine or merely apparent?

2.1. *Do Syncategorematics Name?*

It is useless to treat abstract nouns in categories other than substance and quality as connotatives, if syncategorematics themselves function as names. In the *Summulae Physicorum*, Ockham ridicules the thesis

> that a distinct signification always corresponds to a distinct word, so that the distinction among things signified is as great as that among significant words (*Summulae Physicorum* III, c. 7; Rome 1637, 55–6)

as "the basis of many errors in philosophy" and scoffs that its adherents will have to posit not only wherenesses and whennesses corresponding to place- and time-words, but and-ness and but-ness coresponding to 'and' and 'but' (*ibid.*). Ockham contends that "properly speaking, a syncategorematic signifies nothing" (*Summa Logicae* I, c. 4; OPH I, 15), "because it does not bring any determinate thing to mind" by itself (*Expositio in Librum Perihermenias Aristotelis*, c. 1, §1; OPH II, 378). Rather syncategorematics signify "together with" another, categorematic term and affect the way the latter stands for or supposits for things (*Summa Logicae* I, c. 4; OPH I, 15–16).

This crude semantic thesis apart, however, medieval and modern philosophers have challenged Ockham's strategy for reducing ontological commitments on several further grounds.

2.2 *Modes Versus Things*

According to Ockham, quantity-, relation-, action-, and passion-terms are sorted into distinct categories from each other and from substance- and quality-terms, not because they signify distinct kinds of things, but because they signify substances and qualities in different ways, according to distinct modes of signifying (*modi significandi*). In the grammatical tradition, however, Thomas of Erfurt had insisted that modes of signifying mirror modes of being (*modi essendi*) or the properties of things (*proprietates rerum*).[13] If so, it seems that Ockham's tireless metaphysical polemic results merely in substituting an ontology of modes for one of things. Buridan argued as much in connection with the issue of whether opinions are qualities really distinct from and inhering in the soul:

> Again, 'to be otherwise and otherwise' (*aliter et aliter se habere*) signifies the same as 'to exist in one way and another' (*alio et alio modo se habere*). Therefore, if our intellect existing one way and another is now one opinion and tomorrow will be a contrary opinion, the latter way/mode is not the former way/mode, from which it follows that different modes are posited. Therefore, if there are many modes that are distinct from one another, and the intellect is not and will not be different but is always the same, it is necessary for the intellect to be distinct from those modes and from each of them. And in that case, all of the difficulties that arose about the difference and identity of opinions come back with greater force about modes. Therefore, it is better to stop immediately with the difference of opinions.[14]

Buridan sees no harm in continuing to speak of modes, so long as one acknowledges that they are accidents and hence things (*res*) really distinct from the subject of inherence.

Olivi before and Mirecourt after Ockham did speak about ways or modes of being (*modi se habendi*) in the nominative case[15] and thus invite Buridan's charge that they escape an ontological commitment to things (*res*) only by embracing one to entities of another sort. Later, some philosophers did wish to define a new ontological category distinct from that of Aristotelian substance and accidents. Fourteenth century philosophers had contended that since substance and accidents were distinct things, it was both logically possible for substance to exist without accidents and logically possible for accidents to exist without substance. By contrast, the new ontology held that the subject of a mode could exist without being thus modified, but a mode could not exist without the modified subject.[16]

Ockham would have denied any implicit ontological commitment to modes, however. Favoring the oblique-case expressions '*aliquo modo*' and *'alio et alio modo'*, he rarely uses the nominative case *'modus'* or *'modi'*. He would have agreed with Buridan that *'aliter et aliter se habere'* and *'alio et alio modo se habere'* are synonymous. But whereas Buridan assimilates the former to the latter and treats them both as phrases including expressions that name modes, Ockham takes the adverbial form as canonical. The expression *'alio et alio modo'* is thus misleading in containing terms that merely purport to name something.

2.3. *Quantification Over "Hows"*

Is Ockham's replacement of nouns with adverbs so ontologically decisive? After all, the criterion of ontological commitment seems to be quantification: a theory has an ontological commitment to the entities that must be taken to be values of the variables if the theory is to be true; and a philosopher has an ontological commitment to those entities if he accepts the theory. If Ockham renounces distinct things (*res*) in favor of the same things existing differently, and yet proceeds to quantify over these "hows", he will be as guilty of ontological proliferation as if he had used noun phrases in the nominative case.

Unfortunately, it seems that Ockham and those who extended his ontological program have done just that. (i) This may have been Buridan's thought: to speak of something existing first one way and then another (*alio et alio modo*) in the oblique case, is still to count. Are there not, then, some entities to count? (ii) Again, Ockham seems to quantify over modes explicitly. In *Ordinatio* I, d. 2, q. 1, for example, he asks "whether as much and every mode (*omnis modus*) of identity that obtains between the divine essence and the divine essence, obtains between the divine essence and the attributal perfections in reality" (OT II, 3). He answers that "divine wisdom is the same as the divine essence *in every way* (*omnibus modis*) that the divine essence is the same as the divine essence" (OT II, 17, 18). Again, in *Ordinatio* I, d. 26, q. 2, Ockham sets out to show that "origin and relation are in no way (*nullo modo*) distinct in reality .... Rather everything that is saved by assuming a distinction between generation and the relation can be saved by positing them to be everyway identical (*omnimodam*)" (OT IV, 176). Further, he appeals to the rule that "whenever some thing is absolutely (*simpliciter*) the same and no way distinct, it is impossible to truly affirm and truly deny the same of it" (OT IV, 178). Again, Ockham maintains that "the property that is generation and paternity is the same in every way (*omnibus modis*) nor is it in any way (*aliquo modo*) distinct in reality" (*Ordinatio* I, d. 27, q. 1; OT IV, 193). (iii)

Early in the fifteenth century, Paul of Venice draws inferences that seem tantamount to existential generalization and universal instantiation, respectively. He argues that since x and y are distinct formally, in signification, and in origin, they are *somehow* distinct (*Logica Magna*, Part II, *De Significato Propositionis*; Del Punta and Adams, c. 7, 157–9). And he moves from the conjunction of 'However x-or-y is, x is' and 'x or y is this way' to 'x is this way' (*ibid*., c. 5, 126–8). If Ockham and the others permit such quantification, what credibility can their claims of economy have?

They might defend themselves against this charge of hypocrisy with either of two replies. (iv) First, they could deny, contrary to appearances, that any genuine quantification has taken place. Paul of Venice treats the inference 'x and y are distinct formally; therefore x and y are distinct somehow' as an instance of inferring the superior from the inferior, just as much as 'x is a man; therefore x is an animal' is. He makes the plausible assumption that there is an order of subordination among adverbs analogous to that among nouns in the category of substance. After all, if 'animal' is generic relative to 'man', 'horse', 'cow', and 'pig', why is not 'wickedly' generic relative to 'cruelly' and 'dishonestly' or 'intelligently' to 'insightfully' and 'cleverly'? Paul apparently takes the subordination of 'formally', 'in signification', and 'in origin' to 'somehow' to be of the same sort. Perhaps he would also compare the putative universal instantiation involved in 'x is whatever way y is and y is this way; therefore x is this way' on analogy with 'x is whatever sort of animal y is and y is a man; therefore x is a man'. Still, it looks as if 'man' is inferior to 'animal' or 'sort of animal' and not to 'whatever sort of animal' and hence that the last example does involve quantification over sorts after all. Analogously, 'this way' would be inferior to 'way' and not to 'whatever way' and quantification over ways or modes would remain.

(v) Alternatively, they might insist, contrary to Quine, that it is not mere quantification that carries ontological commitment. "To be" is not merely "to be the value of a variable"; it is to be a property-bearer. But no properties are attributed to ways or "hows" in the propositions 'x is in every way the same as y', 'x is in no way distinct from y', 'x is somehow distinct from y', and 'x is whatever way y is'. Hence, this wording implies no ontological commitment to ways or "hows" after all.

The protest may come back that in an artificial language, one may treat any element of the ordinary-language sentence as the subject and the rest as the predicate. One could hold, therefore, that 'x is whatever . . . y is' is an open sentence satisfied by modes, and hence that the whole sentence attributes a property to a mode.

Yet, even if we allow logicians to stipulate any rules they like for analyzing the sentences in ordinary language, we can assess the ontological commitments of a given author only by reference to the rules he in fact lays down. Obviously, Ockham would reject the above analysis, since he does not permit adverbs to supposit, and insists that expressions such as '*aliquo modo*' and '*nullo modo*' are equivalent to adverbs. His use of these phrases does not saddle him with an unwanted ontological commitment to modes or "hows".

2.4. *The Adequate Significatum of the Whole Proposition*

Even if syncategorematics such as adverbs do not stand for entities on their own, one might still hold that they do so "together with another" – in fact, that the proposition as a whole stands for its own adequate significatum. But what is (*quid sit*) the adequate significatum of the whole proposition?[17]

Paul of Venice records some *modistae* as having argued that the adequate significatum of a whole proposition is the mode of a thing and not a thing. Among their remarkable grounds was the observation that the adequate significatum of the proposition 'You exist' is everywhere, whereas the relevant thing – viz. you – is not everywhere. Similarly, the adequate significatum of 'No chimera exists' cannot be identified with any thing (*res*), because its subject term is necessarily empty. Paul of Venice rejects the opinion as highly illogical. For if the adequate significatum of a proposition were to be identified with a mode of a thing, it would have to be the mode of something signified by the terms of that proposition. And since intrinsic modes are really the same as things of which they are the modes, no mode could exist or be anywhere unless some thing (*res*) signified by the subject or predicate term exists or is thus located (*Logica Magna*, Part II, De *Significato Propositionis*; Del Punta and Adams, 80–85).

Pursuing this question, Ockham's younger contemporary Gregory of Rimini agreed with him that it is not merely *which* things exist, but *how* they exist that determines the truth-value of a proposition:

> . . . The proposition 'A man is a donkey' is not false because a man is not or a donkey is not, but because a man is not a donkey. Nor is the proposition 'A man is white' true because a man exists, or because whiteness exists, or even because a man and whiteness exist, but because a man is white . . . " (*Super Primum et Secundum Sententiarum* I, d. 1, q. 1, a. 1; reprint of 1522 edition, fol. 1Q)

Strictly, the adequate significatum of a proposition is not some thing or things (*res*), but things being this way (*res sic esse*) or existing somehow (*aliqualiter esse*). Nevertheless, Gregory allows that terms such as 'something' (*aliquid*), 'thing' (*res*), or 'being' (*ens*) can be taken three ways: (i) in the broadest sense for whatever can be signified by a proposition or by a term, whether truly or falsely; (ii) less broadly, for whatever can be signified by a proposition or a term, but truly; and (iii) strictly, for an essence or existing thing that can be signified by a complex by itself (*ibid.*, fol. 1 Q–fol. 2 A). It follows that the adequate significatum of a proposition is not something the third way, but may be something the first or second ways. A false proposition such as 'Some chimeras are pink' signifies some thing-(i) – that chimeras are pink, or chimeras being pink – even though there are no things-(iii) for the subject-term 'chimera' to signify. And since some things-(i) and things-(ii) are objects of scientific knowledge, they must be – by Aristotelian standards – necessary and eternal.

But what relation do things-(i) and things-(ii) bear to things-(iii) (things, strictly speaking)? Do the former exist independently of any things-(iii), or do they not? If they do, then some things-(i) and things-(ii) will be necessary and eternal, and hence states of affairs will be necessary and eternal independently of God, the only necessary and eternal thing-(iii) – which runs contrary to the Christian doctrine of creation. On the other hand, if necessary and eternal things-(i) and things-(ii) do depend on God, by what sort of dependence? They cannot be simply identical with God, because different propositions are true about them and about Him (*Logica Magna*, Part II, *De Significato Propositionis*, c. 3; Del Punta and Adams, 102–4). If some alternative reduction is envisaged by Gregory, he needs to specify what it is.

Occasionally, Ockham alludes to what is denoted by a proposition as a whole, but he never enters the debate about the ontological status of its adequate significatum. Had he done so, I suspect he would have concurred with his secretary and student Adam Wodeham in rejecting the question of its quiddity as ill-formed and ridiculous. Adam acknowledges that the proposition 'A man is an animal' signifies that a man is an animal, and turns to consider some objections:

> Suppose you say that for a man to be an animal is either something or nothing. I say that neither should be granted. Instead, it should be held that it is not something, but is for a man to be something, as was said. Thus, I may ask you, whether people is a man or a non-man? Neither should be granted; rather it is not a man, but men.
>
> Suppose you say that if it is not nothing, then it is something. I may argue this way from the other part: if people is not a non-man, then it is a man. You deny both consequences.

> Suppose you say, what then is it (*quid igitur est*)? It must be replied that it is for a rational animal to be a sensitive animate substance. Nevertheless, a more proper response is that for a man to be an animal is not a what (*quid*), but for it to be something (*esse quid*). Therefore, the question is ill-formed the way it would be ungrammatical and perverse to ask, what is a man is an animal? For a man is an animal in reality, every proposition left aside. It should not be granted that a man is an animal is a substance or accident, neither what is something nor what is nothing, because none of these answers would be intelligible or anything to say . . . .[18]

Understood that way, Wodeham's concession – later taken up by Peter of Ailly – that the adequate significatum of a proposition is *res sic esse* or *res sic non esse* is fully compatible with Ockham's ontological program and its contention that only substances and qualities existing somehow are real.

3. THE CHARGE OF ARBITRARINESS AND SUBJECTIVITY

If Ockham steadfastly resists any hypostatization of modes of being or how things are, to mirror the various modes of signifying, does he not thereby jeopardize the correspondence theory of truth and render claims that substances and/or qualities are quantified, related, active, passive, etc. arbitrary and subjective? For his adversaries insist that

(T8) Things are co-specific, co-generic, quantified, related, active, passive, located in space and time, etc. prior to (in the order of explanation) and independently of their being signified, whether naturally or by convention.

But since

(T9) It is only which things (*res*) exist that is thus prior to signification, whether natural or conventional,

Ockham's denial that quantities, relations, actions, passions, etc. are real mind-independent entities distinct from substance and qualities is tantamount to sacrificing (T8).

Ockham would counter that (T9) begs the question against him. In fact, he feels entitled by his putative refutations of the moderns' ontology to the contrary thesis.

(T10) It is not only which things (*res*) exist, but how they exist that may be – in many cases – prior to their being signified, whether naturally or by convention.

Although the facts mentioned in (T8) involve no things other than substances and qualities, (T10) applies to them and guarantees their objectivity.

4. SUMMARY

Ockham's theory of truth-conditions is calculated to accomodate his ontology. He adopts a two-name theory of predication for categories involving absolute terms only, and admits the inherence-analysis for only a restricted range of concrete quality terms. Although he argues extensively that its wider application to the last eight Aristotelian categories of accident leads to contradiction and absurdity, his own positive proposal – to treat all terms in those categories as connotatives that signify particular substances and/or particular qualities existing one way or another – remains essentially programmatic. A full development of it would involve supplementing (T3)–(T7) by stating truth-conditions for statements about how things are related, quantified, active, etc., using absolute terms that name only substances and qualities. Ockham never delivers on the promissory notes he issues. Nevertheless, we must ask how many nominalists have gone further, and by how much, and how recently? In Ockham's view, the realists' program is demonstrably impossible; how much less an evil, if his own is merely incomplete.

NOTES

[1] Although Ockham thinks that unaided natural reason will restrict its ontology to substances and qualities, he himself feels forced by the doctrines of revealed Christian theology – viz., the Trinity, the Incarnation, and transubstantiation – to admit really distinct relations of Paternity, Filiation, and Spiration in the Godhead, a relation of assumption in the human nature of Christ, real relations of inherence between matter and form, substance and accident, and relation of union between parts of the continuum (*Ordinatio* I, d. 30, q. 4; OT IV, 360–374).

[2] See my forthcoming book, *Ockham*, Part One, Chapters 5–9.

[3] Strictly speaking, not; because even Socrates is a composite of many distinct things (*res*) – his prime matter and several substantial forms – although they unite to make one man *per se*. Likewise, nations and cities, which are aggregates of things, received proper names.

[4] See William of Sherwood, *Introduction to Logic*, c. 5; Kretzmann translation, 105–6. See also Walter Burleigh, *De Puritate Artis Logicae, Tractatus Longior*, Prima Pars, cc. 1 & 3, edited by Philotheus Boehner, St. Bonaventure, N.Y., 1955, 1–3, 6–9.

[5] These aren't really criteria for the primary significata, however, since true predication and personal supposition are sometimes explained in terms of primary signification.

[6] Paul Spade, in 'Ockham's Distinction between Absolute and Connotative Terms,'

Vivarium **XIII** (1975), 55–76, offers the following formula for identifying the secondary significata of a term: "(16) A term *t* secondarily signifies a thing *x* if and only if either (a) *t* obliquely signifies *x* or else (b) *t* signifies *x* but not primarily" (72–3). But this is not a criterion for distinguishing the primary from the secondary significata in the first place. Rather he assumes that we have identified all of the significata of a term, and have further determined which ones are the primary significata (by seeing which ones the term is truly predicable of); we derive the secondary significata by subtraction. Ockham's combined remarks suggest that a term *t* signifies a thing *x* primarily, if and only if *x* is signified affirmatively and by a nominative case expression in *t*'s nominal definition, while a term *t* signifies *x* secondarily, if and only if either *x* is signified by an oblique case expression in *t*'s nominal definition or *x* is signified negatively. Since for Ockham a term is truly predicated about something only if the term signifies it directly and affirmatively, Ockham's grammatical criterion includes whatever Spade's includes, but it may take in more besides.

[7] Paul Spade, *op. cit.*, makes this his central thesis. As he puts it, "the paper argues for the claim that connotation-theory, or at least that part of it presented here, requires no new primitive notions beyond those already required for the theory of absolute terms" (*ibid.*, 55). The relation between a connotative term and the thing connoted reduces to the relation between an aboslute term in an oblique case and its significatum.

[8] Hugolino's List, Article 29, quoted in Anneliese Maier, *Zwischen Philosophie und Mechanik*, c. 6, Rome, 1958, 335, note 103.

[9] Maier, *ibid.*, 330.

[10] *Ibid.*

[11] *Ibid.*, 333.

[12] *Ibid.*, 336.

[13] Thomas of Erfurt, *Grammatica Speculativa*, edited with translation and commentary by G. L. Bursill-Hall, Longman, London, 1972; cc. 1–2, 134–40.

[14] Maier, *ibid.*, 337.

[15] *Ibid.*, 328–38.

[16] Paul of Venice, *Logica Magna*, Part II, *De Significato Propositionis*, First Opinion; Del Punta and Adams, Oxford University Press, 1977, 82–85.

[17] This issue forms the bulk of the discussion in Paul of Venice's *De Significato Propositionis, Logica Magna*, Part II.

[18] Gedeon Gál, 'Adam of Wodeham's Question on the "Complete Significabile" as the Immediate Object of Scientific Knowledge', *Franciscan Studies* **XV** (1977), 66–102; esp. 89.

CALVIN NORMORE

BURIDAN'S ONTOLOGY

Here is a recipe for ontology. First divide the expressions of one's language into those which purport to pick things out and those which don't. Then see whether some of those which purport to pick things out can be defined in terms of others. Finally admit into your ontology whatever an undefinable term purports to pick out. This scheme expresses (though vaguely and incompletely) one of the central intuitions behind many ontological programmes. What is admitted by an ontologist operating within this framework will depend, of course, upon how he or she divides expressions, on what resources of definition are available, and, perhaps, on pressures from other theories. What I hope to do in this paper is to show the influence of the intuition behind this sketch on the work of the fourteenth century Parisian master Jean Buridan.

To some extent, I shall compare Buridan's views with those of his contemporary, William Ockham. In doing so, I hope both to present Buridan's own very striking contributions to ontology and to shed a little light on the inner life of 14th century nominalism.

1. LANGUAGE

We think, we speak and we write. For Ockham and Burdian each of these activities is carried out in a language. Both of them postulate a language of thought with a structure common to all men as well as the diversity of natural languages with which we are familiar. Both suppose that written language constitutes a distinct level bearing much the relation to spoken language that spoken language bears to mental. For each of them this relation is not holistic but at the most basic level, piecemeal; particular written words are related to particular spoken words and particular spoken words to particular concepts.[1] But they conceive this relation slightly differently. Ockham calls it subordination and explains that a spoken term is subordinate to a concept if a change in the signification of the concept eo ipso produced a change in the signification of the spoken term.[2] Buridan however calls the relation between term and concept itself signification.

Signification is the most basic term in medieval semantic theory and its

James Bogen and James E. McGuire (eds.), How Things Are, 189–203.

sense varies from author to author.[3] As Ockham understands it a typical term used normally signifies things – those things of which (or proper names of which) it could be truly predicated either in an ordinary assertoric (*de inesse*) sentence (this yields the signification of a term narrowly conceived) or in a sentence *de possibili* (this gives the signification in a wide sense). Thus for Ockham 'man' narrowly signifies all men (all those things of which it would be correct to say 'This is a man') and widely signifies all possible men (all those of which it would be correct to say 'It is possible for this to be a man').[4] Of course not all terms signify ordinary things – 'noun' signifies certain words and 'concept' all concepts.

Ockham's use of 'signify' seems to be a reform of the usage of his contemporaries and not one which found universal acceptance. Buridan in particular seems to use the term more loosely. On his view a term of written language immediately signifies the corresponding spoken term which in turn immediately signifies the concept. Both the written and the spoken term also have an *ultimate* signification which is determined as Ockham suggests and which is what is 'conceived by' the concept. Thus Buridan's 'immediate signification' is Ockham's 'subordination' and his ultimate signification is what Ockham calls 'signification' *tout court*.

Not all terms have (ultimate) signification. Both Buridan and Ockham accept the traditional distinction between categorematic terms (like 'man', 'word') which have signification and syncategorematic terms (like 'and', 'not', 'all' and 'is') which do not. Syncategorematic terms are sometimes described as modifying the signification of categorematic terms. But this is, as well shall see, something of an over simplification.[5] There are two kinds of categorematic terms. Ockham calls them absolute and connotative, Buridan sometimes uses 'absolute' as Ockham does but calls appellative those terms which Ockham calls connotative. On both views absolute spoken terms correspond to simple concepts (concepts without internal logical structure) while for Ockham but not for Buridan appellative/connotative terms invariably correspond to complex concepts or complexes of concepts.

Terms are so-called etymologically because they are the ends of sentences. Sentences are distinguished from terms by being either true or false and an adequate theory of language has to give an account of this. Central to the account Buridan and Ockham share is the notion of *supposition*.

Supposition, on the view Buridan and Ockham share, is properly speaking a property of the subject or predicate of a sentence and is in Buridan's words "the taking of a term in a sentence for some thing or some things, of which the term is truly affirmed." For example in this sentence "a horse runs" this term 'horse' stands or supposits for every horse which exists, because picking out any of them it would be true to say 'This is a horse'."[6] Similarly, Buridan claims, in "a horse ran" 'horse' stands for everything of which it could truly be said "This is or was a horse" and in "a horse will run" 'horse' stands for every horse which is or will be, and in "a horse can run" for every horse which is or can be. I should emphasize that when Buridan speaks of every horse which was or will be or can be he is referring *inter alia* to horses which do not exist. In the terminology of his time tenses and modalities 'ampliate' the term. Thus if an ontology includes whatever is picked out, Buridan's ontology embraces *possibilia*.

On Buridan's view a term can have just two kinds of supposition: when it stands for what it ultimately signifies it supposits personally, when it does not it supposits materially.[7] It is in terms of relations among the (virtual) sets of supposita of their terms that the truth conditions for categorical sentences are given.

Like Ockham's, Buridan's analysis of categorical sentences has a presupposition. It presupposes that such sentences have two terms (a subject and a predicate) which are joined by a copula. In presupposing this Buridan is in the mainstream of medieval work on predication – though there were significant counter-currents.[8] Most of the fourteenth century discussions take the two term approach for granted, and debate centres instead on two points – the first is whether the copula expresses some relation of inherence, the second is whether what is picked out by the predicate term is something (e.g. a form) of a different ontological type from what is picked out by the subject term.

Buridan's answer to both these questions is an unequivocal 'no' and this is supported by the alternative account of the truth conditions of categorical sentences which he and Ockham share. On this account the predicate term of a sentence always has personal supposition and so in typical cases doesn't stand for anything ontologically odd; it is of horses and not of stranger entities that 'This is a horse' is true. And on this account the copula and associated syncategorematic terms tell us which relations of inclusion or intersection must hold between the sets of *supposita* of the subject and predicate terms for the sentence to be true. Thus – "Nestook is a man", a singular sentence, is true if and only if the supposition of 'Nestook' is one of the things for which 'man' stands, "Some men are Rastiferians" is true if 'man' and 'Rastiferian' have at least one *suppositum* in common and "All men are automata" is true just if all of the things for which 'man' stands are among the things for which 'automata' stands. If we keep in mind that medieval theorists thought that affirmative sentences had existential import and negative sentences did not, we can now generate the truth conditions for negative sentences and for various molecular (or in Buridan's terminology 'hypothetical') sentences in the way familiar from elementary logic. Thus "Some man is not a Rastiferian" is true just if "All men are Rastiferians" is false, i.e. if either there are no men (or no Rastiferians) or if something for which 'man' stands is not among the things for which 'Rastiferian' stands; "All men are Rastiferians and Socrates is a man" is true just in case each of its conjuncts is, "Socrates is a body or (*vel*) Buridan loved the Queen of France" is true just if at least one disjunct is.[9]

In general the articulation of the truth conditions for grammatically complex sentences is for Buridan the province of logic and detailing it is beyond the scope of this paper. Of more immediate interest is another phenomenon – the reduction of the truth conditions for grammatically categorical sentences which seem to have unfortunate ontological commitments to the truth conditions for hypothetical sentences without those commitments – the process called by fourteenth century theorists expounding or exposing (*expositio, exponere*). *Expositio* is a natural result of the recognition that the surface grammar of a sentence is not always a reliable guide either to its truth conditions or its inferential connections to other sentences. Once the distinction between surface grammar and deep or 'logical' grammar is recognized it isn't surprising to find philosophers, particularly logicians, attempting to make the logical form of sentences explicit. Fourteenth century theorists were further encouraged in this by their realism about the language of thought. For example, Ockham, who discusses the matter at some length, seems to have thought that mental language contains no sentences whose surface grammar and logical form differ. Thus expounding or exposing a spoken or written sentence is just finding a second equivalent complex spoken or written sentence whose structure mirrors that of the mental sentence to which the first is subordinated.[10]

We can bring all of this into much sharper focus, and prepare for what is to come by considering some elements of the theory of definition.

There are, within the medieval tradition, two competing demands on a definition. First it should be synonymous with the term defined. Second, it should necessarily express the essence of the things to which the defined term applies. I shall follow medieval practice in calling a definition which satisfies the first condition a nominal definition and a definition which satisfies the second, a real definition. A claim crucial to Buridan's (and Ockham's) semantics and metaphysics is

Absolute terms have no nominal definition.

The reason why absolute terms have no nominal definition, Buridan explains, is that for two expressions to be synonymous they must signify the same things and appellate the same things. But an absolute term does not appellate anything, while in a definition by genus and differentia (the medieval paradigm) there is always something appellated. Apparently Buridan believes this because he believes that a complex expression appellates what its parts signify. An expression like "rational animal" appellates all animals other than men while "man" appellates nothing at all. Thus they are not synonyms.

From this it follows at once that absolute terms are indefinable and that an ontologist who follows the schema which I outlined at the beginning will admit whatever an absolute term purports to pick out. And this I claim both Buridan and Ockham do.

The natural analogue of the claim that absolute terms have no nominal definition would be the claim that

Connotative terms have no real definition.

If this were true then there would be a clear sense in which absolute terms "named" things with essences and in which connotative terms did not. But that, on Buridan's view, is too strong because there are several different kinds of connotative terms.

In q. 4 of Bk VII of his *Quaestiones in Libros metaphysicam aristotelis* Buridan takes up the question of which terms express an essence by asking whether for grammatically substantive terms "*sit idem ipsum et esse ipsum*" that is whether for any substantive term x, x is what it is to be x. He concludes that the equation holds for just those terms x which do not appellate anything for which they do not supposit. Thus it holds for absolute terms. It also holds for connotative terms like 'creative' which on Buridan's view have no 'external' connotation because they connote only powers not really distinct from the subject which has them. Thus the ontological commitment carried by a term can be discovered by discovering whether it is absolute or connotative and then determining whether its connotation is external. Just how, in practice, is this done?

2. ONTOLOGY

Perhaps the most widely held ontological assumption of the fourteenth century was that there were substances. Buridan also accepted the existence of particular forms and particular parcels of matter.[11] These are not substances in the fullest sense but they fit this category more naturally than any other so there they lodged as substances *per reductionem*.

Buridan believed that on Aristotle's view there are only substances and not also entities in the accidental categories. This leaves him with the task of explaining Aristotle's account of essential predication. He takes Aristotle to agree with his own view about predications in which both subject and predicate are substance terms. For example he thinks that in "Socrates is (a) man", 'Socrates' and 'man' stand for what they signify (Socrates and all the men there are respectively) and that the singular sentence form indicates that the sentence is true just if what 'Socrates' stands for is among the things 'man' stands for. Because Latin blurs the distinction between adjectives and concrete nouns, a sentence like "*Sortes est albus*" could be as well translated by "Socrates is a white thing" as by "Socrates is white". There is no *special* difficulty in extending the truth conditions for sentences involving substance terms to sentences predicating *concrete* terms in other categories. On Buridan's view problems appear only when we consider abstract terms in the accidental categories. What, for example, are we to make of "The whiteness of Socrates exists"?

In his *Questiones in Metaphysicam Aristotelis* IV q. 6 Buridan considers whether accidents can be called beings (*entia*) in the same sense as substances. He claims that on Aristotle's view they cannot, because on that view to say the whiteness of Socrates exists is just to say that Socrates is (a) white (thing). Thus "The whiteness of Socrates exists" has no ontological commitments above those of "Socrates is a white thing" and that, as we have already seen, commits one only to substances.

The metaphysical core of Buridan's interpretation is his attribution to Aristotle of the view that it is absolutely impossible that there be a whiteness without something being white. Attributing to Aristotle the view that an accident is inseparable from the substance in which it inheres is hardly revolutionary. Indeed those who attribute to him the view that there are *particular* accidents usually claim that they are individuated by the substance in which they inhere. But, for Buridan, it is precisely the capacity to exist separately – as a subject – which marks a being.[12] What cannot be a subject can only be said to be derivatively (*secundum quid*). Thus, for Buridan, Aristotle's particular accidents don't exist properly speaking.

This metaphysical thesis is supported by the semantic thesis that where '*x*ness' is an abstract noun in an accidental category ('whiteness' say) *x*ness is what it is for something to be *x* (whiteness is what it is for something to be white). Note that this is not the view I ascribed to Buridan himself above. On Buridan's view whiteness may be what it is to be white*ness* but it is not what it is for something to be white. But it is the latter position he ascribes to Aristotle.

To say that whiteness is what it is for something to be white implies that a particular whiteness exists just in case a particular thing is white. Buridan takes Aristotle to be claiming that these are just two ways of saying the same thing. We are given a recipe for eliminating 'whiteness' (*albedo*) from the language in favour of 'is a white thing' (*est album*) which latter expression stands only for the substance which is white. Thus the temptation to include accidents in the ontology is eliminated. There are only substances – though substances are in many ways.

I think that so far Buridan and Ockham agree. They agree also in thinking that this simple picture cannot be maintained. The most pressing problem is that an ontology which does not allow for accidents to be subjects seems unable to explain the phenomena of the Eucharist. As Buridan and Ockham understand it, faith requires that one believe that in the consecration of the Eucharist the bread and wine are destroyed to be replaced by the body and blood of Christ. But the colour, taste, etc. of the bread and wine apparently remain, and moreover, on the view Buridan and Ockham share these do not inhere in the body and blood of Christ. The straightforward conclusion is that they do not inhere at all and thus that accidents can exist without inhering and so can be subjects. Thus Buridan and Ockham are both lead to accept a larger ontology than the one they ascribe to Aristotle.

Here it seems, as Professor Adams has argued, Ockham digs in his metaphysical heels; he will allow substances and two species of quality and nothing else.[13] Buridan is more liberal. What I shall argue is that his liberality is principled and comes from a view about what constitutes a satisfactory account of change.

Aristotle allows three kinds of change: generation/corruption, alteration, and motion. We can conceive of these in two fundamentally different ways. The first way (Aristotle's way!) is to think of them as different kinds of processes which a single thing, Socrates say, may in some sense suffer: Socrates was born, can move, can change size, can die. A second way is to conceive of the different types of change as involving the creation and destruction of different kinds of things – in generation and corruption substances are created and destroyed, in movement, motions, in alteration, qualities, quantities, etc.

This second picture does in a sense unify change by bringing all its species under the description "creation or destruction of something". Professor Adams has argued that Ockham understands the project of accounting for change in accord with the first picture. I want to argue that Buridan, on the other hand, is guided by the second and that this leads him to multiply entities but reduce modes.

Unfortunately, Buridan is no readier than Ockham to give a systematic account of the ontology of change. Nevertheless he seems to think that every alteration can be accounted for in terms of changes of quality, quantity, place, and relation. So it is on these that I will focus.

While Buridan thinks that it is the Eucharist which shows us that the existence of a whiteness is different from something's being white, his arguments for the existence of magnitude and motion are physical. In his *Questiones in Octo Libros Physicorum* Bk. I. q. 8 he canvasses the available explanations for the behaviour of a gas when heated or cooled and concludes that only by supposing a "*magnitudo*" distinct from the thing can we save the phenomena. He then proposes that "just as a thing already white is not able to become whiter except by the generation in it of whiteness together with the existing whiteness . . . so a thing already extended is not able to become greater without some generation of magnitude together with the existing magnitude" (f. xi r. b.). Thus an increase of size is reduced to the generation of bits of magnitude – and decreases are handled similarly. Later in the same question Buridan makes it clear that he thinks that magnitudes like whitenesses are in principle separable from the subjects in which they inhere; indeed he claims that "were the magnitude removed from the matter through the divine power still that matter would have mutually distinct parts though it would not have parts located outside nor inside because the location which is the ground (*ratio*) of magnitude would have been removed" (f. xi v. a.).

Buridan's arguments for the reality of motion, like those for magnitude, are physical, though the key here is not a fact of experience but a thought-experiment. His opponents believed that motion is relative and so can be eliminated with relations. In Bk III q. 7 of his *Questiones in Octo Libros Physicorum* Buridan argues that this is inadequate. He considers the question whether the universe as a whole could be moved, something which a relativist would have to deny for lack of something relative to which it could move. Buridan takes it as uncontroversial that if there *were* something outside the outermost sphere of the universe God could move the outermost sphere with respect to it, and he takes it that God can do now whatever he would then do to move it. He thinks it would be odd to claim that *were* God to do that now the outermost sphere would be unaffected and so he concludes that the motion of the sphere is not a relation but an additional thing. And, one might add, it is a strange thing, for Buridan's sixth and ultimate conclusion in Bk III q. 7 is that "the motion of the outermost sphere is a process (*res pure successiva*) namely something of which there is a prior part and a posterior part not enduring at the same time". (f. 1. v. b.)

Buridan doesn't explain why he thinks magnitudes and motions are as separable by the divine power from their subjects as whitenesses are but I think one need not really look further than considerations of parity of reasoning. The same phenomenal facts which require one to admit that the colour of the bread remains after consecration would require one to admit that its size and motion do too. So if size and motion are not eliminable in favour of something else (and Buridan has argued that they are not) then they too must be thought capable of existing separately.

Buridan's revisions of Aristotelian ontology do not require any changes in his picture of predications of concrete accidental terms. "Socrates is white" can be given its usual truth conditions. But predication involving abstract accidental terms are more complex. On the view Buridan ascribes to Aristotle for example "Socrates is white" is equivalent to "Socrates' whitness exists" and that in turn to "A whiteness inheres in Socrates". But on Buridan's own view the second of these can be true though both the others are false. Moreover, "A whiteness inheres in Socrates" contains a term which stands, on Buridan's view, for whitenesses, whereas whitenesses are only connoted by "Socrates is white" so the truth conditions of the first of this pair cannot be simply reduced to those for the second. Sentences like "(A) Whiteness inheres in Socrates" appear to tell us not only that there is a whiteness and that there is Socrates but that these are related. What then is the ontological status of these 'modes' of relation?

Both Buridan and Ockham believe that things and whitenesses could exist without the things being white. For the things to be white it is required in addition that the whitenesses inhere in the appropriate things. Ockham never tells us just what this inherence is but Buridan is less coy: he devotes q. 8 of Bk V of his *Questiones in metaphysicam aristotelis* to the problem.

The ostensible subject of Bk V q. 8 is "Whether the causings (*causalitates*) or dependings (*dependentiae*) of things mutually connected (*adinvicem*) are dispositions added to those causes and those caused things" so it treats a special case of the problem of relations. Hence it should come as no surprise that the major problem faced by those who posit such 'dispositions' is what is now known as Bradley's Regress. As Buridan presents the problem:

Let us hold that a stone which is white is the cause of that whiteness and that the whiteness is dependent on the stone. However I prove that it is not necessary on account of this to posit an added disposition besides the whiteness and the stone because either this would be because the whiteness inheres in the stone or it would be because the whiteness depends on the stone. But neither way can be maintained. I first prove that it is not because it inheres because then that added disposition would also inhere in something and either this inheres without some third disposition or it inheres through some third disposition. If we say that it of itself [inheres] this will be able to be said of the whiteness itself and so it will not be necessary to posit the added disposition. If it be said that that disposition may inhere in its subject through another third disposition then we may ask of that as before and so to infinity . . . .[14]

On the other hand *some* metaphysical glue is required to connect the whitenesses to their subjects else we couldn't make sense of the notion that God might separate them. Buridan finds the glue in a special kind of added disposition:

To the other argument concerning the whiteness and the stone I say that it is necessary that there be an added disposition so that the whiteness may inhere in the stone or also that it may inhere in the stone because it is possible that it may remain not inhering and not depending – on the stone that is. But you say further that that disposition would inhere in a subject and depend on it. I concede it but this is inseparable – whence God is not able to make it that there be an inherence-of-the-whiteness-in-the stone and there not be a whiteness because it would imply a contradiction for such modes of relating this to that cannot be unless this or that exists. Therefore since they inhere and depend inseparably it should be said that they inhere [and] depend through themselves without another further disposition. Therefore there is not a process to infinity. And when it is asked in what subject such a disposition inheres I believe that it should be said that the disposition required for a man to be white besides the man and the whiteness is the inherence of the whiteness in the man and it is in it. Thus that whiteness is formally inhering in that man through that inherence and subjectively through itself.[15]

This doctrine seems clear enough. There is a mode which is the inhering of a quality in a substance and it is an "added disposition". Destroying the inherence is destroying this desposition. There is something unusual about these qualities – their inseparability – but there is nothing unusual about what they are. As Buridan adds "These dispositions truly are accidents which are thus inseparably related to their subjects" [*Ibid.*].

While Buridan seems clear enough about *what* these dispositions are, it is less clear whether they can be said to exist in the same sense in which substances and separable accidents can. His classifying them as accidents suggests that they can but his repeated insistence in Bk. V. q. 8 that these are just the way accidents would be on Aristotle's theory gives that suggestion the lie. It seems then that these added dispositions cannot be properly said to be or to be something (*aliquid*) but must be said to be *secundum quid* or to be somehow (*aliquale*). And so Buridan has a two tiered ontology after all, a tier of things and a tier of modes of things, modes which are nevertheless not different from other accidents in what they are but only in *how* they are.[16]

Although I have not been able to find explicit discussions of the other modes of appellation which parallel that of inherence it seems to me that we have here a model for the other modes. Some, like the mode "*in quantum*" can be treated as inherence of a magnitude. Others, like a wealthy man's possession of his goods, may best be treated as mental qualities of the man and his fellows. On this picture Buridan is a realist about modes but his modes are not new kinds of entities, rather they are unusual entities of an old kind – qualities.

It may seem at first as though Buridan has moved far from the austere nominalism of an Ockham or a Sellars but he has – at the price of some ontological extravagance but no abstract entities – opened a route which might solve a puzzling problem for nominalists, namely the explanation of change. Pictures of the sort Buridan and Ockham offer have no problem accounting for changes in the truth-values of sentences which arise from the generation or corruption of something because change of that sort will change the supposition of some term and it is no marvel that sentences have different truth values when the suppositions of their terms are different. But holders of the picture of language which Ockham and Buridan share do have a puzzle on their hands when sentences change their truth-value without any change in the supposition of their terms. How is the change of truth value then to be explained? The problem can be postponed by appeal to the appellation or connotation of terms, but only postponed, for there may be cases in which two sentences have different truth values though their terms stand for the same things and, apparently, appellate the same things.

One way this problem might be dealt with is precisely by giving ontological status to the ways in which terms appellate things; if two terms stand for the same things and appellate the same things in the same ways they are synonyms, and substitution of one synonym for another will preserve truth-value. Being a realist about modes in this way is a way of grounding every change of truth value in a generation or corruption – if only of a mode. And if modes are qualities this is not even especially puzzling.

This seems to me to be strongly suggested by some of his remarks in q. 3 Bk II of his ontological *Questions* on the physics. He is discussing whether shape (*figura*) is a thing distinct from the shaped object and in the course of the discussion has occasion to consider the ways in which a thing without being destroyed may come to have contradictory predicates applied to it. He seems to think that there are just three:

> The first way is if that predicate may be connotative of something extrinsic for then it is possible for that to happen on account of the existence or non-existence or on account of another change of the extrinsic thing ....
>
> The second way is if that predicate connotes the situation of the parts of that thing with respect to one another and then for it to be related differently there is required a local motion of those parts or of some of them and nothing more besides those things which were before. Therefore the motion, which was not permanent, having ceased, there is nothing afterwards which was not before and also nothing was before which is not afterwards and thus it is to sit or stand or be straight or spherically curved or cubical and so for other shapes ....
>
> The third way is if the predicate neither connotes something extrinsic nor connotes a relation of the parts of that thing to each other and then no cause appears natural to me unless it is because something added to that thing is generated or corrupted for thus is a man white and not white or white and black before and after and thus a distinction of forms and of accidents from their subjects can be known and argued for.[17]

The first and third of these cases are fairly unproblematic for the view I am suggesting. Even the "other change" mentioned in the first case poses no problem once it is realized that it too must fall into one of the three categories sketched. The difficult case is the second, for there Buridan explicitly claims that in such changes (changes of shape) nothing exists before the change which doesn't exist after and vice versa. Does this not undermine the project?

Perhaps not. I have suggested that Buridan accepts *possibilia* and accepts motions. Now each shape differs from every other in the motions which would be required to transform it into any given third. So, perhaps, we might distinguish shapes by the possible motions shape terms appellate. If shape words did appellate the (possible) motions which (might) create or destroy those shapes, then every change of truth-value would involve a generation or corruption (of a motion perhaps) and every difference in truth-value would involve a difference in the supposition or the appellation of the terms of the sentences involved. And that reduction of the problem of change to a form tractable by a Buridanian semantics would surely be worth its ontological price.

NOTES

1 For example Buridan writes

> The second conclusion is that significant spoken words signify passions (*passiones*) or concepts of the mind and do not signify other things except through the mediate signification of the concept. This is clear, first, because as letters relate to sounds, so analogously do sounds relate to intentions (*intentiones*) of the mind.

Buridan (1977) (*Sophismata*) cap. 1, IIa *Conclusio* p. 25, trans. In Buridan (1966), pp. 70–71.

2 Cf. Ockham (1974) (*Summa Logicae*) Pars I, cap. 1, lines 26–34.

3 For a discussion of some of the issues involved cf. Spade (1975).

4 Cf. Ockham (1974) (*Summa Logicae*) Pars I, cap. 33.

5 For the distinction between categorematic and syncategorematic terms cf. Ockham (1974) (*Summa Logicae*) Pars I, cap. 4.

6 Buridan (1977) (*Sophismata*) cap. 3, p. 50.

7 Buridan (1977) (*Sophismata*) cap. 3, p. 51.

8 Indeed, according to Albert of Saxony this theory is required to make sense of the rules for conversion:

> However it is true that some say that a categorical proposition only has two principal parts namely subject and predicate. Hence they call what precedes the verb the subject and if the verb has nothing following it as for example in '(a) man is' they call the verb the predicate. But if something follows the verb they say that it together with the verb is the predicate as in '(a) man is an animal' for they say that in it the whole [phrase] 'is an animal' is the predicate. But briefly this will not do (*non valet*) for then this would not be a case of conversion '(a) man is an animal therefore an animal is a man' because the subject (of the one) would not become the predicate (of the other) nor the predicate (of the one) the subject (of the other). For on their account in this '(a) man is an animal' this whole [phrase] 'is an animal' is predicate and 'man' is subject and in this 'an animal is a man' 'animal' is the subject and this whole phrase 'is a man' is predicate and therefore in the second the subject contains less than is in the predicate of the first and the predicate more than is in the subject of of the first. Therefore it is better to say that a categorical proposition is one which has, as principal parts, a subject, a predicate, and a copula. [Albert of Saxony (1522) (*Perutilis Logica*) *Tractatus* III, c. 1, f17rb]

Among those who defended a contrary view Peter Abelard and Vincent Ferrer deserve special mention. For Abelard's view cf. Tweedale (1976). For Ferrer's cf. Ferrer (1977).

9 Buridan summarizes the truth conditions for categorical propositions in Buridan (1977) (*Sophismata*) cap. 2 XIV and *Conclusio* p. 44. (Cf. Buridan (1966) p. 93.)

10 This picture of Ockham's view was first proposed in Trentman (1970).

11 For a discussion of the view that particular parcels of matter have independent existence cf. Weisheipl (1963).

12 In his *Questiones in metaphysicam aristotelis* Bk. IV q. 6 Buridan argues that it is precisely the capacity to stand on its own without inhering in anything which marks a subject.
13 Cf. Professor Adams' contribution to this volume.
14 Buridan, QM V q. 8 fol. 31rb–va

> Item ponamus quod lapis qui est albus sit causa albedinis illius et quod albedo sit dependens a lapide; tamen probo quod propter hoc non oportet ponere dispositionem additam praeter lapidem et albedinem: quia vel hoc esset ex eo quod albedo inhaeret lapidi vel hoc esset ex eo quod albedo dependet a lapide. Sed neutro modo potest dici. Probo primo non quia inhaeret: quia tunc etiam illa dispositio addita alicui inhaeret, et tunc vel ipsa inhaeret sine aliqua dispositione tertia vel inhaeret per aliquam dispositionem tertiam. Si dicamus quod se ipsa ita poterit dici de ipsa albedine, et sic non oportebit ponere illam dispositionem additam. Si vero dicatur quod illa dispositio inhaereat subiecto suo per aliam dispositionem tertiam, tunc quaeremus de illa ut prius, et sic in infinitum: quod est inconveniens dicere.

15 Buridan, QM V q. 8 fol. 33ra

> Ad aliam de albedine, et de lapide dico quod oportet quod sit dispositio addita ad hoc quod albedo inhaereat lapide, vel etiam ad hoc quod dependeat a lapide propter hoc quod possibile est quod ipsa maneat non inhaerens et non dependens, scilicet a lapide. Sed ultra: tu dices illa dispositio inhaeret subiecto et dependet ab eo. Concedo, sed hoc est inseparabiliter [FOR inseparabiter], sicut Aristoteles credidit quod albedo inhaeret lapide; unde Deus non possit facere quod esset inhaerentia albedinis ad lapidem et non esset albedo, quia implicaret contradictionem – tales enim modos se habendi hoc ad illud non est possibile esse, quin hoc sit vel illud. Ideo cum inseparabiliter [FOR separabiliter] inhaereant et dependeant, dicendum est quod se ipsis inhaerent <et> dependent sine alia dispositione ulteriori; ideo non proceditur in infinitum. Quando etiam quaerebatur cui subiecto inhaeret tales dispositio, credo quod dicendum est quod illa dispositio requisita ad hoc quod homo sit albus, praeter hominem et albedinem, est inhaerentia albedinis ad hominem, et est subiective per ipsum; ita illa albedo est formaliter inhaerens illi homini per illam inhaerentiam, et subiective per se ipsam.

16 The view that one can separate the ontological status of something from the question of what it is may appear strange but it is in fact required by medieval theology. The doctrine of the Incarnation, at least as understood in the fourteenth century, entailed that God could incarnate in anyone he chose. Were he to choose you then the human being which is you would cease to be a subject (or *suppositum*) and become something assumed by God. This assumption is sometimes compared with inherence and raises the possibility that something only contingently has the mode of existence it has.

17 Buridan, QSP 11 q. 3 fol. 31ra–rb

Primus modus est si illud praedicatum sit connotativum alcuius extrinseci, tunc enim propter existentiam vel non-existentiam (aut propter aliam mutationem) illius extrinseci possibile est illud contingere . . .

Secundus modus est si illud praedicatum connotat situm partium illius rei ad invicem, et tunc ad sic se habere aliter et aliter, requiritur motus locales illarum partium vel alicuius earum, et nihil plus praeter ea quae ante erant. Ideo cessante motu, qui non est permanens, nihil aliud est posterius quod non esset prius, et nihil etiam erat prius quod non sit posterius; et sic est de sedere et stare, de esse rectum aut curvum, sphaericum aut cubicum, et sic de aliis figuris.

Tertius modus est si illud praedicatum nec connotat aliquid extrinsecum nec connotat habitudinem partium illius rei ad invicem, et tunc nulla causa apparet mihi naturalis nisi quia illi rei est aliqua res addita generata vel corrupta; sic enim est homo albus et non albus, vel albus et niger, prius et posterius, et sic potest scire et argui distinctio formarum et accidentium a substantiis suis.

JAMES E. MCGUIRE

PHENOMENALISM, RELATIONS, AND MONADIC REPRESENTATION: LEIBNIZ ON PREDICATE LEVELS

A vexing problem in Leibniz's philosophy is whether his thesis that bodies are phenomena is compatible with the doctrine that they are aggregates of individual substances. Even in his mature philosophy, Leibniz holds that monads have an organic body in association with them that they dominate.[1] Moreover, the organic body is supposed, in some sense, to be an aggregate of monads, each of which is the dominant monad of another organic body. However the aggregation theory is interpreted, it demands that monads be associated in some fashion with physical bodies – the phenomena represented in inter-monadic representations.[2] But phenomena are perceived according to a relational network. And each monad can be assigned the position of its respective organic body. However, it must be recognized that there is no well-founded answer to the question where in the relational network a monad is located. For Leibniz is adamant that monads cannot have position as such.[3] It is clear that monadic representation, relations, and phenomena are intimately connected in Leibniz's metaphysics. For example, in his letter to Arnauld of October 9, 1687. Leibniz says "that the soul naturally expresses the whole universe in a particular sense and according to the relation that other bodies have to its own . . . " He goes on to define the term "expresses":

> . . . One thing expresses another (in my terminology) when there is a constant and fixed relation between what can be said of the one and of the other. It is thus that a perspectival projection expresses its ground-plan (G. 2. 112).

Thus, monads represent their respective organic bodies as standing in certain relationships, where 'expressing' is a representation of the monadic states that are these bodies. Leibniz is clearly stating that some determinate correspondence or mapping obtains between monadic representations of the diverse monadic states that pertain to different monads. But how can a correspondence of this sort obtain if there are no true relations among monads? The main task of the present study is to suggest an answer to this question. It will be argued that monads do not bear any non-formal relations to one another, that is relations such as causality, or space understood in a realist sense. But certain second-order relations do obtain, the sort that obtain in monadic representation as relations among organic bodies. Moreover, it will be argued

James Bogen and James E. McGuire (eds.), How Things Are, 205–233.

that it is the notion of a second-order relation (when properly explicated) that underscores Leibniz's claim that composite things are "... beings by aggregation and therefore phenomena" (G. 2. 252/L. 531).

In the course of the argument, the following issues will be addressed: (1) what are the theoretical inter-connections among Leibniz's views on monadology, relations, and his doctrine of representation of which monadic perception is a species? and (2) does Leibniz hold that relational accidents are, in some sense, reducible to non-relational ones? If so, what are the consequences for (a) Leibniz's understanding of the relational properties of every-day individuals like David and Solomon and such composite things as cannons and tanks, and (b) his claim that monads "mirror" the universe from their individual points of view. The latter part of the paper will consider whether Leibniz's conception of a phenomenalist reduction of the 'physical world' involves an explicit doctrine stating that there are different sorts of predicate levels. And if so, what are the criteria that distinguish them? On this score, I shall have something to say about Leibniz's attack on the labyrinth of the continuum and its connection with his theory of representation.

I. RELATIONS: THE BARE SCAFFOLDING

It is a commonplace that the basic doctrine of Leibniz's metaphysics is that "There is nothing in things [*in rebus*] except simple substances, and in them perception and appetite" (G. 2. 270/L. 537). Thus, reality is ultimately characterized by monads alone, together with their perceptions and appetitions. Accordingly, "a monad, like a soul, is a world by itself, having no intercourse of a dependent nature except with God" (G. 2. 436/L. 600). From this, it follows that monadic representation indicates a capacity that is intrinsic to each monadic reality conceived as a separate and discernible thing. It is this fundamental doctrine that the principles of the dissimilarity of the diverse and the identity of indiscernibles express.[4] It should be apparent that the doctrine of monadic autonomy is essential in any consideration of the status of relations in Leibniz's metaphysics. The doctrine of relations has been widely misunderstood, and its grounding in the monadology has not been properly appreciated.[5] In fact, too much attention has been focused – rather misleadingly – on the issue of the reducibility of relations. Consider the following two, well-known passages on relations:

My judgment about relations is that paternity in David is one thing, sonship in Solomon is another; but that the relationship common to both is merely a mental thing whose basis is the modifications of the individuals (To Des Bosses, 21 April 1714, G. 2. 486/ L. 609).

In his fifth letter to Clarke, he observes that the ratio between two lines *L* and *M* can be conceived in three ways: (1) as the ratio of the greater to the lesser; (2) the ratio of the lesser to the greater; and (3) as the ratio between *L* and *M* abstracted from both lines without considering which is subject and which is object:

> . . . in the first way of considering them, *L* the greater, in the second, *M* the lesser, is the subject of that accident which philosophers call 'relation'. But which of them will be the subject in the third way of considering them? It cannot be said that both of them, *L* and *M* together, are the subject of such an accident; for, it so, we would have an accident in two subjects, with one leg in the one and the other in the other, which is contrary to the notion of accidents (G. 7. 401/L. 704).

These passages combine two aspects of Leibniz's thought: his adherence to the view that relations are *entia rationis*, and his subject-predicate logic. On the first score, Leibniz denies that relational predicates express objective universals; on the second, he denies that relations are particularized accidents simultaneously instantiated in two or more subjects. There is nothing here about the reducibility of relations since Leibniz will clearly deny that objective relations exist on the level of monads. Nevertheless, he holds that certain accidents of phenomenal individuals are characteristically relational – the line *L*'s accidents include that of being longer than *M*, and David's accidents include that of being the father of Solomon. These are phenomenal relations, characteristic of things, not as they are on the level of monads, but as they appear to us.

This raises an important question. If "being longer than", "being the father of", etc. do not signify relational features of monadic reality, does Leibniz think that relational accidents are reducible to non-relational features? On this issue of phenomenal individuals, philosophers have failed to separate two distinct levels in Leibniz's thought. On one level, it can be asked: are relational properties of individuals like David, Solomon, cannons, and tanks, etc. – reducible to the non-relational properties of such individuals? Leibniz would probably answer no. But with regard to another level, it can be asked: are the relational properties of simple substances or monads – the ultimate realities – reducible to their non-relational properties? Here, at this level, the primary thrust of Leibniz's philosophy comes into play. Leibniz would respond that the question is ill-formed on the ground that monads do not have among their properties any first-order relational properties. To a large extent, the first part of the paper will be a defense of this claim and an exploration of its implications. In pursuing this task, I shall borrow some insights from the work of C. D. Broad and Wilfrid Sellars on Leibniz's treatment of relations.[6]

The first serious question that arises for my claim is this: if physical objects are founded in monadic existents, how is this to be reconciled with the fact that the first are conceived as characteristically relational, whereas, the second are said to be intrinsically non-relational. This brings me back immediately to the question whether Leibniz's claim that bodies are phenomena is compatible with the view that they 'result' from monadic aggregation.

Let me begin by borrowing from Broad a way of making the distinction between 'formal' and 'non-formal' relations as it pertains to Leibniz's thought. By formal relations, I mean such relations as similarity, identity, otherness, the subject-predicate relation, etc.; by non-formal relations I refer to spatial, temporal, and causal relations, and so on. Thus, for example, on this interpretation, there is no basis in any given monadic existent for inter-monadic causality of a transeunt sort; that is of causality that can be understood as a real feature *in* monads *and* one which also grounds an obtaining and mutual connection between them. If non-formal relations are not constitutive of monadic natures in the sense that they designate features of their individual actuality, formal relations obtain only insofar as they can be expressed *through* the language appropriate to inter-monadic aggregations.[7] Given this distinction, we can say the following: Leibniz holds that there are no non-formal relational properties of monads; and derivatively he holds that there can be none among a given set of monads. Call this claim *P*. What lies behind the claim is Leibniz's often-stated belief (a), that monads and their perceptions and appetitions alone constitute reality; and his belief (b), that monadic perception is a feature internal to each individual monad. In what follows, I shall neglect appetition, as it does not bear on those aspects of Leibniz's doctrine of representation that concern me. It will not do to reject (b). For example, if one takes the mirror analogy seriously, it is tempting to suppose that monads in perception can stand in a certain relation of connection among one another.[8] On one way of construing the connection, if monad m_1 perceives another m_2 this would entail $R(m_1, m_2)$, which contradicts *P*. That is, *P* is contradicted if we give a non-formal interpretation of the relational notion of connection. In other words, Leibniz cannot and does not claim that there exists an active and transitive relationship of causation between monadic perceivers. But what he does claim is that a *sine qua non* causal relationship can obtain between a set of monads: for example, if it were not for the fact that monad m_1 is in such and such a state, m_2 would not be in the state that it is now in. On this construal of causality, neither monad actively imposes any change of condition on the other. By taking clues from Broad and Sellars, the relation of connection can be clarified further by analyzing "m_1 perceives m_2" as follows: m_1 is in a certain inherent state S, which is characterized entirely by its non-relational properties; S_1 and S_2 represent one another isomorphically. Accordingly, to claim that m_1 perceives m_2 does entail $R^*(m_1, m_2)$; but now R^* is not a first-order relation, since $R^*(m_1, m_2)$ means that there exists certain first-order, non-relational features of m_1, and certain first-order, non-relational features of m_2, and further that there is a second-order relation between these two sets of first-order features. Thus, to claim in P that there are no non-formal relational properties of monads, implies that individual monads do not have any first-order relational properties, and also that non-formal first-order relations among sets of monads do not exist. The view just outlined partly explicates Leibniz's thesis that there are "no purely extrinsic denominations". To be sure, this notion is based on Leibniz's conceptual containment theory of truth, and his adherence to a subject-predicate logic. In fact, his claim that there are no purely extrinsic denominations follows from the doctrine that all true propositions are of subject-predicate form, as well as the view that in such propositions the features that predicates stand for inhere in the monads themselves. So, given these commitments, it is clear what is at stake: there can be no talk of formal second-order relations among monads; namely, of "extrinsic denominations" that are independent of reference to the first-order non-relational features that inhere in each individual monad. And for Leibniz, the status of second-order properties and relations is to be understood in terms of his theory of monadic representation. In other words, to speak of a character as an extrinsic denomination is to say that it is a supervenient or consequential character, one predicable of monads in virtue of their intrinsic properties and their capacity to represent, via their organic bodies, the states of one another either directly or indirectly.

II. RELATIONS AND REPRESENTATION

Although he nowhere uses the terminology of 'objective being' in contrast to 'formal being', it is clear that Leibniz thinks of monadic natures in terms of this distinction. It has its home in an act/content conception of thought and its objects. In virtue of its formal nature, a mind can perform an act which exhibits its power of representing, the content of which is what the representation is about. Thus, the content of any mental act is that something of which it is about; whether it is an act of imagining or understanding, in each case there 'exists' an intentional object *in* the act. It is that mode of being an item has when it is thought of *qua* thought; in Descartes' language, the "objective being" of the idea.[9] Both representings and representings *qua* what they represent (i.e. their contents) are modifications of mind. If a representative content of an act is *true*, it can be said to exist not only *in* the act, but also to possess a ground that exists independently in the world. In other words, everything that is represented by a *true* representation as existing in the world *is* in the world.[10]

This is essentially the Cartesian picture of the distinction, and one that Leibniz no doubt knew. In an important respect, Leibniz was forced to modify it. The ground of a true representation can never be in a world that exists independently of monadic representation. Ultimate reality comprises only the formal natures of individual monads together with their representative contents. For Leibniz, bodies are phenomena, that is things that appear. But they appear to monadic minds; as such, they are objects of awareness, that is intentional objects whose being is solely in the awareness that monadic perceivers have of them. Furthermore, phenomena "have their unity only in our minds, which bases itself upon the relations or modes of real substances" (G. 2. 97, to Arnauld, April 30, 1687). Thus, the unity of phenomena is merely an 'external denomination,' a feature that results from the fact that monadic representations express relationships among phenomena as objects of awareness. And, given that Leibniz holds that 'being' and 'one' are equivalent, if phenomenal unity is in the mind, so, too, is its mode of being.[11] In other words, phenomena cannot be said to possess being, apart from their being the objects of monadic representation. As such, phenomena are the expression of some nature or essence according to the representative capacity of each monadic perceiver.

With this background in mind, I shall now expand an interpretation of Leibniz's doctrine of relations that has been put forward by Sellars.[12] It helps to illuminate the connection between Leibniz's view of representation and the doctrine that monads do not have any first-order non-formal relational properties. Given that Leibniz mobilizes the distinction between 'representative' being and 'formal' being, S_1 is R to S_2 can be interpreted monadically, where S_1 and S_2 designate states inherent in individual monads. If "S_1 is R to S_2" is true for Leibniz, there must be an element R-to-S_2 inherent in S_1 such that S_2 is *in* S_1. But given Leibniz's theory of monadic representation, he can say that R-to-S_2 inherent in S_1 is a representing of

S_2 being inherent in S_1, such that S_2 is an element of the R-to-S_2 inherent in S_1, in virtue of its having objective or representative being in the representative content which is the R-to-S_2. According to this analysis, the truth of statements of the form S_1 is R-to-S_2 – where we have in mind formal relations of the first-order – is based on facts of the form S_1 represents in a determinate and specific manner S_2, such that talk of first-order relational facts is based on *phenomena bene fundata*; and phenomena are conceived as having intentional existence in the objective content of monadic representations. Of course, the mere fact that S_1 represents another set of monadic states appropriately, does not of itself make the corresponding relational statement true. For this to be so, the monad represented must have formal as well as objective being. As Sellars puts the point, "– a true representation is one whose subject matter is a representable which, in addition to having 'objective' being in the representation, has 'formal' being in the world."[13] As we shall see, this claim will have to be qualified somewhat in the Leibnizian frame of reference. But there is little doubt that Leibniz is committed to the view that the objective being of a representation must be *expressed* by the formal being of the perception insofar as it is an intrinsic modification of a monadic reality. This is, however, only a necessary condition for the objective being of a representation. It is not sufficient, in that Leibniz holds that "there is a constant and regulated relation" between elements as expressed by one monad and the same elements expressed by another. Thus, there is more than the formal being of a given monadic existent involved in the problem of what grounds the unity and harmony of inter-monadic representation. I shall return to this problem in a moment. But it is well to notice at this point that the above analysis of relations shows how they are intimately connected with the 'objective' being of monadic representation, grounded as it is in the formal being of monadic existence. Thus, when Leibniz says that monads "correspond to each other through their own phenomena and not by any other intercourse or connection" (G. 2. 444/L. 600), he understands the terms "intercourse" and "connection" as designating second-order relations expressed in the 'objective' being of monadic representation.

We can certainly understand Arnauld's difficulty with Leibniz's claim that monadic "expression" is "a constant and regulated relation between what can be said of the one and the other".[14] The explanation is so abstract as to accommodate any interpretation. Nor are the many analogies and metaphors that Leibniz uses to illustrate his theory of great help. To liken perception to a mirror, or to explain monadic "point-of-view" on analogy to the notion of spatial perspective are, at best, misleading and of use only to one who has already grasped Leibniz's metaphysical position. Always when Leibniz speaks of phenomena in relation to monadic representation, he takes it for granted that their analysis presupposes the conception of extended-objects-as-well-founded appearances to monadic perceivers.[15] The problem of their well-foundedness is the problem of the harmony of inter-monadic representation. Thus, if Leibniz's basic notion of perception denies that a monad perceives another monad directly, an hypothesis of harmony is necessary to insure that monads will correspond to each other through their phenomena, especially if they are said to represent the same state-of-affairs from their individual points-of-view and, as a consequence, mutually represent each other. On this showing, monadic perception can be seen as an artifact of harmony. The point is stated clearly in a passage from the *Theodicy*: "– there must always be an exact relation between the representation and the thing, and consequently between the different representations of one and the same thing." But even though each monad represents according to its nature, "a perfect harmony always subsists therein" (G. 6., par. 357, p. 327). Once again, it is clear that to claim that the objective reality of a monadic perception must be *expressed* by the monad's formal reality is to claim no more than a necessary condition for the having of that representative content. For one monad to represent another, Leibniz demands that there be a determinate mapping of their respective elements. But this is to place a particular burden on the doctrine of concomitance or pre-established harmony.

III. REPRESENTATION, MONADIC AGGREGATION, AND PRE-ESTABLISHED HARMONY.

At this point, it is necessary to give a more detailed account of the doctrine of monadic representation. In the *New System of Nature*, Leibniz tells us that:

> ... we must say that God has from the first created the soul or any other real unity (substance) in such a way that everything arises in it from its own internal nature through a perfect spontaneity relatively to itself and yet with a perfect conformity to external things (G. 4. 484).

Now there are two major claims here: (1) that all forms of monadic activity (including the capacity to represent) follow from the natures of individual monads according to a pre-determined life plan; (2) that inter-monadic activity is 'orchestrated' according to a pre-established harmony. With respect to the second claim, Leibniz adamantly rejects the view that the harmony between monads results from either a mutual causal influence, or from a Divine adjustment to keep them in agreement. Rather, there is only "the way of Harmony". From the beginning, each monad has such a nature that by "its own laws, given to it with its being, [it] still agrees with the other, just as though there were a mutual influence –" (G. 4. 499). But what is the relationship between the nature of monadic beings, and their power to represent one another? As we have seen, they are simple substances that consist in perceptions and volitions alone. Thus, we can say that the formal being of a monad is a set of mental acts that includes representings and willings, and with soul monads, apperception and higher order acts such as noticing and comparing.[16] Consequently, the monad is the formal being of a representation whose content is the objective being of a representable; and the transition from one representational state to another is regulated by the monad's internal appetition, which expresses its life plan. Accordingly, the objective being of a particular monad's act of representing is the expression of the internal states of another monad (either directly or indirectly), but always in accordance with the former's intrinsic nature.

The ontology is simple, but it has rich implications for the doctrine of inter-monadic representation. The second-level relations generated among monads through the objective contents of their representations are infinitely recursive. Consider the permutations possible with three monadic neighbors, where the term "neighbor" is understood non-spatially. *A* can represent *C* directly, or it can represent *C* indirectly by representing *B*'s representation of *C*, and similarly for *B* and *C*. But each monad can also represent (according to its nature) its own representation as it is represented by other monads. Since the Leibnizian world is a plenum, each monad has infinitely many neighbors. This means that each monad represents (albeit imperfectly) the infinity of monadic states of all other monads either directly or indirectly. But also, each monad can represent the individual representations that all other monads have of it, such that its representation of itself involves a multiplicity of embedded representations that are infinitely recursive. Given the multiplicity of infinity, it is obvious that no one monad can clearly and distinctively perceive the states of an infinitude of monads. Moreover, a monadic representation will embody some degree of confusion, in that many of the details involved in the representation are not consciously attended to. In the rational soul, Leibniz says that there are "numberless perceptions – of which we are not conscious, because the impressions are either too minute, or in too great a number, or too even, so that they have nothing sufficient to distinguish them one from another; but joined to others, they do not fail to produce their effect and to make themselves felt at least confusedly in the mass" (G. V. 46–7). Thus, many intrinsic characteristics of a representational state will be unconscious, because of the minuteness, number, and likeness to each other, of its multiple constituents. And since the parts of a perception that are conscious can alone be perceived distinctly and clearly, the minute and numberless perceptions that are always present unconsciously can be perceived only confusedly and obscurely.

Leibniz states these doctrines metaphorically in the *Discourse on Metaphysics.*

> ... That is why our confused sensations are the result of a variety of perceptions. This variety is infinite. It is almost like the confused murmuring that is heard by those who approach the shore of the sea. It comes from the continual beatings of innumerable waves. If now, out of the many perceptions which do not at all fit together to make one, no particular one perception surpasses the other, and if they make impressions about equally strong or equally capable of holding the attention of the soul, they can be perceived only confusedly (G. 4. 459/L. 325).

Notice how closely Leibniz identifies the perception of numerous and similiar monadic states with the notion of confusion. Actually what he says here describes his theory of the monadic perception of extended surfaces. What a requisite set of monadic perceivers perceive is a collection of very numerous and similar monads. Owing to their similarity, each attracts 'attention' in roughly the same degree. And, because of their number and likeness, many elements in the inter-monadic representation of this collection of 'bare' monads will be unconscious. But this is what is involved in the representation of extended-surfaces-appearing-to a perceiver; it is to perceive a collection of 'bare' monads in just this confused way. Thus, to perceive something as having extension, shape, and motion is to perceive confusedly an infinitely numerous collection of monads whose inter-monadic representations have similar recursive depth:

> ... Matter and motion – are not so much substance or things as they are the phenomena of percipient beings, whose reality is located in the harmony of the percipient with himself (at different times) and with other percipient beings (G. 2. 270/L. 537).

A similar view is expressed in a letter to Arnauld of October 9, 1687. Size, shape, and motion "are phenomena like colours and sounds, and although they involve a more distinct knowledge, they cannot stand up to the finest analysis" (G. 2. 119). Though Leibniz recognizes *a* distinction between the primary and secondary qualities of 'things,' all qualities alike are phenomenalistically 'reduced' to the theory of inter-monadic representation.

But so, too, are relations. The objective contents of monadic representations are grounded in the formal being of the monads; they are non-relational in that they do not involve any non-formal, first-order relations among the monads. Nevertheless, they express, inter-monadically, extended objects or phenomena as 'standing' in various relationships, including spatio-temporal ones. If extended objects themselves are the well-founded phenomena of monadic representation, the relations that obtain among them can also be said to be reduced to the intrinsic properties of the monads. But this sense of reduction does not favor the non-relational over the relational properties perceived in phenomena: both alike are reduced in the same way to the contents of monadic representation. This is precisely what is to be expected from a theory that explains the status of physical phenomena (together with their relations) in terms of collections of monads mutually representing the objective contents of one another's representations in a fashion that is infinitely recursive. In this setting, the 'body-monads' are the 'nearest neighbors' of each mind or soul monad, so that the universe is perceived through them. This is what Leibniz means by saying that "the soul naturally expresses the whole universe . . . according to the relationship which other bodies have to its own; consequently it expresses most directly that which belongs to the parts of its own body . . . ." (G. 2. 111).

I shall turn now to Leibniz's second claim, that inter-monadic activity is regulated by the principle of pre-established harmony. This principle is, of course, pertinent to the claim that extended things are both phenomena and aggregates of substances; but also to the question, "what insures that monads inter-monadically represent the same phenomena from their individual 'points-of-view?'" These two claims are closely related. And they involve a larger claim that is the main subject of the present study. This is that monads represent the same state of affairs from their individual points-of-view, as well as each other's objective contents as a result. This claim does not necessarily involve Leibniz in holding that monads mutually represent one another's apperceptions or individual awareness. I shall begin the discussion of the principle of harmony by showing that its foundation is in Leibniz's theory of creation, and shall consider as well how it is related to the conceptual containment theory of truth.

We can begin with the principle of sufficient reason (taken in its widest sense) as it relates to the doctrines of creation and compossibility. Possible substances only qualify for actualization as independent particulars when their possible and individual attributes are intrinsic to them and are also compatible with other possible particulars. In the *Discourse* and *On the*

Radical Origination of Things, Leibniz makes a clear distinction between the role of divine understanding and will (G. 4. 438, *Discourse*, section 13; G. 7. 302–8). Eternal truths exist in the divine mind and are in no way dependent on divine will. But concerning contingent, individual particulars, the case is different. For their existence, they depend on the fully actualized, free decrees of the divine will. Within the possibility of an individual substance is expressed its possible relations with other possible individual substances. And this also involves the possible purposes of God. To think of a possible substance in conjunction with other possibles, is to think of possible divine decrees, just as to think of an actual substance in conjunction with actual circumstances is to think of God's actual decrees.

But what determines that a system of possible individual substances will exist? At this point, the distinction between the divine will and understanding is crucial. All attributes that relate to the free exercise of God's will are contingent, as, for example, his choice in creating the world that actually exists (G. 7. 307). It is not a necessary truth in the mind of God that this present and actual world exist. Of all possible worlds that might exist, there must be a sufficient reason for the one that is in actual existence. There is a sufficient reason for God's creating this world, in that it uniquely expresses the principle of the best. But only those possible individual substances that can form a compatible or compossible world are candidates for existence. For "not all possibles are compossible" (G. 3. 573). The distinction is this: a possible individual is that which does not imply a contradictory nature, whereas a world of compossible individuals is such that there is the possibility of joint coexistence. Thus, a collection of possible individuals does not form a cosmos because the individuals do not necessarily form mutual connections among themselves. Now the compossible world that actually goes into existence can *only* be that which satisfies the principle of the best. This means that a world can come into existence if each of its individual substances harmoniously manifests its perfections, or quantity of existence, to a maximum degree. Such a world contains the greatest variety of jointly possible and compossible individuals. God, thus, actualizes the possible world in which perfection is at a maximum among all compossible substances. The principle of perfection, then, provides God with a means by which to decide among the infinite and mutually exclusive systems of compossibles: there is thus a sufficient reason why the actual world exists. Accordingly, possible individuals have a claim to exist only because God has chosen to act in selecting a compossible world according to the archetypes of compossibility and perfection.

I cannot, in the present study, pursue the difficulties in Leibniz's theory of necessity whereby he attempts to show that God acts freely in selecting the present world by means of the principle of perfection that follows from the necessity of divine nature alone.[17] But given that he can tell a consistent story at the level of *possibilia* (as they relate to those conditions that Leibniz thinks relevant for an intelligible account of creation), some important connections between his conceptual containment theory of truth and pre-established harmony can be established. To know the *haecceitas* of a substance is to know that it is "at the same time the foundation of, and the reason for, all the predicates that can truly be stated of it" (G. 9. 433; *Discourse*, Section 8). That the *haecceitas* of a substance is a "foundation" for all its predicates is explained by the thesis that a substance contains all its attributes. The latin term *subjectum*, "that which lies beneath" has for Leibniz, as it had for Medieval thought, the connotation both of a propositional subject and of a substance (G. 2. 486 to Des Bosses, April 1714). Accordingly, the term can refer to both an individual substance and a logical subject, and has ontological and logical import. Leibniz, of course, keeps those two senses separate. In his analysis of subject-predicate propositions, he seems to interpret the Greek term *huparchei*, as in "*x* belongs to every *y*" in an ontological sense, and seems to think of it through its Latin equivalent *in-esse*, 'to be in'. Hence, the Aristotelian doctrine that to say truly that *x* is *y* is to say that *x* belongs to *y* is construed by Leibniz as sufficient only when *x is contained in y*. It would be wrong to suppose, however, that the formal 'predicate-in-notion' model is alone responsible for the ontological theory that substances 'contain' their attributes. The *predicatum inest subjecto* thesis is at bottom a piece of metaphysics. Possible substances qualify for actualization as complete individuals if their possible individuating attributes are intrinsically inherent and compatible with other possible individuals; and, equally important, if they satisfy the principle of the best. Thus, the containment thesis with regard to actual substances has as its basis the principles of perfection and compossibility – precisely the same principles that ground the pre-established harmony of inter-monadic activity.

It is the principles of intelligibility that God employs in choosing the best possible world, which 'well-found' the doctrine of pre-established harmony and the conceptual containment theory of truth. Once the actual world is created, these doctrines ensure that the solipsistic worlds of the individual monads are regulated in are integrated fashion. If, in Leibniz's ontology, only mental acts exist formally, it is these that 'well-found' the truth of the contents of monadic representations according to the principle of harmony.

Anything less fails to enact the best possible world as prescribed by the principles of creation. In other words, according to Leibniz's theory of creation, a possible world is one that God could have actualized were it not the case that the actual world is the best. The fact that God chooses the world that He has actualized, in contrast to other possible worlds, is contained in God's defining concept. And it is precisely because God determines that the actual world is the best, in comparison to other possible worlds, that His determination follows from His wisdom and perfection. Thus, the actuality of created things follows from their individual concepts (together with God's concept and His comparisons of all alternative worlds that He could have created) according to the principle of maximal perfection that can obtain in a series of created things.

Against this background, it is clear that inter-monadic representations are global in their conception. To say that a world of monads is aware of an extended world is to say that the extended world obtains relative to the monadic collection iff each monad represents the same world. This global conception makes it clear that monadic representation involves the whole of the extended world as 'mirrored' from each monad's individual point of view. And it illustrates that the conception of harmony needed to 'well-found' the extended world is infinitely complex – so complex and global that only God can be responsible for it. But the ultimate basis for the principle of harmony that regulates inter-monadic representation is God's determination that the world which obtains is the best possible. This not only involves that each monad is self-consistent in itself, but also the determination that the monadic world that obtains is the most compossible of all possible worlds that are candidates for existence. This means that the actual history contained in the individual concept of each monad is its best possible history, and also ensures that these individual histories jointly constitute the best of all possible worlds.

It is important to see what my analysis thus far implies about the notions of false belief and illusion as they arise in a world of inter-monadic perceivers. If, as I suggest, veridical and false representations are both grounded in the constitutive natures of monadic existents, Leibniz cannot distinguish the latter from the former by denying that they correspond to the formal being of monadic reality. Were he to make such an appeal, the ontological basis of his epistemology would be threatened with incoherence. His recourse is the one that he, in fact, makes – to the doctrine of confusion. For example, if a monad has a false belief about what is occurring in another monad, this means that the first monad's representation of that state-of-affairs is confused concerning certain of the intrinsic features that occurrently obtain in that monad.

IV. INDIVIDUALS, THE CONTAINMENT THESIS, AND MONADIC ONTOLOGY

I can now assess some further aspects of Leibniz's doctrine that individual substances are the 'foundation and reason' of everything that can be said of them. Individual substances contain all their attributes, and these attributes express completely the individual *haecceitas* of a given substance. Leibniz develops this position in the thesis that diversity implies dissimilarity; namely, that two things cannot have the same nature (G. 2. 249, to De Volder, 20 June 1703). From this, two things are clear: that the doctrine of diversity of content is essential for Leibniz's general conception of substance, and that the principle of indiscernibles, "that there are not in nature two indiscernible real absolute beings" is logically subsequent (G. 7. 393). If it is impossible to have diversity without dissimilarity, the nature of a substance expresses completely what the substance is. A complete expression of what a substance is cannot be true of another substance. If there are really two numerically distinct substances, their content must be diverse in at least the minimal sense that the one possesses an attribute that the other does not. Hence the *haecceitas* of an individual substance is such that it 'contains,' independently of all else in existence, a diversity of content. It is the 'foundation' therefore of all the attributes that inhere in it.

Suppose someone claims that Leibniz's notion of *haecceitas* commits him to the view that substances can possess a diversity that is independent of dissimilarity. In other words, it could be claimed that a substance's individual *haecceitas* shows it to have an individuality *independent* of its attributes, since judgments using ostensive terms (such as "this is near that") have a clear use. It could then be said that two substances are numerically diverse without being diverse in content. If so, is there not sufficient ground for rejecting the doctrine of containment and the dissimilarity of diverse? But the 'thisness' of a substance is itself a quality. As such, it supposes the existence of 'something' of which it is an attribute. And the fact that a substance can be individuated independently of its attributes by means of judgments like 'this is here' does not prove that its individuality is independent of its attributes. Therefore, Leibniz is not committed to the view that a substance can be diverse without being dissimilar.

In what sense does *haecceitas* give a 'reason' for the attributes contained in an individual substance? What Leibniz seems to mean is that each substance has a constitutive basis that provides a mediating link between the concepts expressed by its subject and its predicates. Hence what appears in perception as a merely *de facto* association of related qualities is anchored in the internal reality of a substance's individual nature. In a polemic against the Cartesians and Occasionalists, Leibniz elaborated his position in *On Nature Itself* (1678). Without 'intermediaries' or mediating links, through which "what is remote in time and space can operate here and now – anything can be said to follow from anything else." Accordingly, there must be "some connection operating here and now between cause and effect." This mediating link is "a certain efficacy residing in things, a form or force such as we usually designate by the name of nature" (G. 4. 507). This provides a ground of truth that consists in the link that binds the predicates with the subject. Thus, the conceptual containment theory of truth is but an expression of the ontological doctrine that the link constitutive of a substance's *haecceitas* supports all change in its successive states. There is, thus, a sufficient reason in terms of a substance's nature by which the manifestation of all its attributes over time is determined. An individual nature is such that, were it known by the finite mind, all separate episodes in the substance's history would be seen as grounded in its essential nature. Under this description, the episodes would not be conceived merely as being hypothetically connected with the antecedent conditions related to the substance's behavior.

It is clear that Leibniz conceives the individual nature of monadic substances in terms of "a certain efficacy" or "a form or force"; in other words, in causal terms. It is this mediating causal link that provides the reason why different states at different times can be said to belong to one and the same substance: it explains identity through time. Leibniz also introduces the notion of *law* in this context:

> . . . And because from the time when I began to be, it could truly be said of me that this or that would happen to me, it must be acknowledged that these predicates were laws included in the subject or in my complete notion, which makes what is called me, which is the foundation of the connection of all my different states and which God knew perfectly from all eternity (G. 2. 43).

How are we to construe the notion of law here? Notice that Leibniz associates lawfulness with "the foundation of the connection of all my different states . . . " This is causal talk. For Leibniz, it expresses the fact that monadic individuals have at every time an ineluctable drive to realize the program that is written into their nature from the beginning of existence. Thus, all episodes and states that pertain to a monadic history are intrinsically connected. And the fact that the monad will become x and y at a certain time if other of its states have become a and b, results from a teleological plan generated by the internal efficacy of its individual nature. Leibniz sometimes puts this point in the language of representation. "Perception is the operation proper to the soul, and the nexus of perceptions, according to which the subsequent ones are derived from the preceding ones, makes the unity of the perceiver" (G. 2. 372/L. 599, to Des Bosses, 1709). But more often, the unity of monadic individuality over time is expressed in causal terms. To De Volder, he says that "there is a certain persisting law, which involves the future states of that which we conceive as the same, that is what I say constitutes the same substance" (G. 2. 264/L. 353, 1704).

This brings me to Leibniz's doctrine of monadic activity. It provides the fullest picture of Leibniz's views on trans-temporal identity and the grounds on which he would deny the contemporary doctrine of trans-world identity. I shall have little to say about the latter doctrine, as such, since it is fully discussed in Robert Adam's paper.[18] My purpose here is to throw some light on the manner in which Leibniz relates the doctrine of containment to the notions of individuality and activity. This will bring me back again to the doctrines of representation and relations.

As early as 1668, Leibniz linked the idea of substance with the notion that a substance is an entity that has an intrinsic principle of action. Anything that subsists in itself is a *suppositum* and is "denominated by action; hence, the rule that actions belong to *supposita*." This rule he attributed to Medieval thought. What acts is thus a *suppositum* or individual substance. The *actiones esse suppositorum* doctrine – that actions belong to substances – is discussed later in the *Discourse* (1686) and in *On Nature Itself* (1698) (G. 4. 432, *Discourse*, sec. 8; G. 4. 509). There, as in his earlier writings, Leibniz uses the term *suppositum*, rather than the more neutral *subjectum*, to link action with the nature of substance. For Leibniz, change is merely the transition from one state to that of another. Activity, however, is a teleological notion, the completed result of some predetermined action. It is Leibniz's aim to incorporate in one and the same theory of substance, an account of change, activity, and final causes.

Let us consider change first. At the most abstract level, change is accounted for in Leibniz's doctrine that a substance's predicates are contained in its individual concept. But, if each substance generates its own history, independently of any other substance, change must be founded in its individual nature. This means that its predicates get exfoliated over time and express conceptually different states of the substance at successive times. But, if a substance has all of its attributes from its beginning, such that each of its predicates now expresses some determined moment in its total history, can there be change? Leibniz, of course, distinguished between the state of a substance at a time and the fact that it possesses an attribute to be in that state at that time. The former is a temporary episode occuring to the substance and expressed by the appropriate predicate; but the latter constitutes a permanent feature of the substance, programmed to express itself according to the monad's pre-determined lifespan. Change is thus accounted for because each substance is the ground of different states that occur at successive times. And if each substance contains all its attributes and generates its own history, there is no change that is not at the same time an intrinsic modification of the substance itself. So considered the persistent nature of each substance provides a sufficient reason for each mode of change that occurs in its history. As Leibniz says to Arnauld. "The concept of an individual substance involves all its changes and all its denominations" (G. 2. 56, 14 July 1686). Here Leibniz intends to imply more than the doctrine that a complete definition of an individual concept of a substance contains all the predicates that can be truly ascribed to that subject. He refers as well to change, a notion that ultimately finds its basis in the ongoing activity of the substance itself.

But a theory of change does not explain activity. And Leibniz is committed to holding "that not only whatever acts is an individual substance,but also that every individual substance acts without intermission" (G. 4. 509, *On Nature Itself*). In the same work, Leibniz makes an important distinction between the source of activity, and activity itself. ". . . there can be no action without a power of acting, and conversely, a power which can never be exercised is meaningless. Yet, activity and power are different things, the former a matter of succession and the latter permanent" (G. 4. 509). It is clear that a substance can act because it possesses an inherent power or capacity to do so: this is an essential part of its defining nature. Moreover, a substance exercises its power or efficacy so that fully actualized change, as manifested through time in successive stages, is the direct consequence. There is always present, therefore, a perpetual and pre-existent tendency to change such that every state of each substance ineluctably passes to the next. Since activity results from an inherent tendency to change which always strives to actualize itself, substances are intrinsically directed teleologically. And since Leibniz invokes both the source and end of action in his theory of substance, there is a ground for each episodic change and for the end to which it strives, a completed action. Thus, the internal constitution of each monad contains a power or law which is the ground of its trans-temporal identity. And viewed from the perspective of Leibniz's theory of creation, all monadic individuals mutually satisfy the principle of the best, the end for which they are created.

In his *The Concept of Substance* (1694), Leibniz makes it quite clear that his doctrine of active power is not that of a mere potentiality. Rather, to have an active capacity is a permanent attribute of a substance: "Active power – contains a certain act or entelechy and is thus midway between the power of acting and the act itself and involves a conatus. It is thus carried into action by itself and needs no help but only the removal of any impediment" (G. 4. 469).

I said earlier that a monad is a set of mental acts: it is the formal being of a representation whose content is the objective being of a representable. In the light of the present discussion, it is apparent that Leibniz is not content to characterize monadic ontology in purely mental terms. To say that a monad has an internal appetition that regulates its transition from one representational state to another is to refer ultimately to its intrinsic efficacy, an efficacy which drives the monad's activity transtemporally according to the individual law of its nature. Thus, to say that a monad is a set of mental acts is to say that there is a ground or efficacy in its nature that provides a permanent causal nexus requisite to produce objective representations that follow the laws of its intrinsic individuality. This is a strong notion of individuality. It embodies the thesis that the complete concept of the individual contains the predetermination of all possible changes and circumstances intrinsic to its individual history, together with the compossibility of that history with the actualization of the best possible world. For Leibniz, the nature of monadic substances precludes the possibility that there are purely extrinsic denominations, with the consequence that every substance represents the entire world from its individual 'point-of-view'. Thus, the relational aspects of the world are expressed through inter-monadic representations that are grounded in the intrinsic nature of each individual substance; just as each substance's intra-monadic history is generated by the primitive force derivative from each individual nature. If my analysis is along the correct lines, the inter-monadic representations of monadic aggregations are such that the representations of each monad are logically inseparable from the totality of the best possible world that goes into existence. This is a presumption that is evident in Leibniz's theory of creation and in the doctrine of pre-established harmony. This framework precludes the possibility that the structure of the world could contain individuals that each possess a separate internal history, such that they merely relate conjunctively to form that world. A world comprised of such individuals violates Leibniz's dictum that there are no purely external denominations; for if the internal history of each individual is independent logically, their relational properties will not follow from their internal natures alone, but rather from the conjunctive set of all the individuals considered together. In this case, relations will fall entirely into the category of external denominations. But I have argued that relations are second-order formal properties that are grounded in the intrinsic nature of each monadic reality. So, if monads are only 'related' through their representations, relational properties are a function of monadic perception; but this involves the Leibnizian conception that the intrinsic properties of all substances (together with the notion of inter-monadic existence) are united by the compossibility of the best of possible worlds.

It is clear that Leibniz would deny trans-world identity. If the internal history of one monadic substance were different in some respect from what it in fact is, the possible world that contains that substance would differ qualitatively from the actual world and would hence be discernibly non-identical. In other words, though there is no conceptual inconsistency in a world in which Arnauld marries, in Leibniz's view, God would not actualize it since it is an inferior world.[19] The notion of abiding celibacy is contained in Arnauld's individual concept because it is the only possible history for him that is logically tied to the best of all possible worlds. And it is only thus that the world is a candidate for existence, a world in which Arnauld never marries. It is this train of reasoning that lies behind Leibniz's insistence that "the Existent would be defined as what is more compatible with things than anything else is incompatible with it" (C. 390/p. 51).

Similarly, since Leibniz takes seriously the doctrine that there is a ground or efficacy in each monadic nature according to which it is identically the same transtemporally, he denies that a monad's history can diverge along an alternative path from the one it will in fact pursue and, yet, remain the same individual. If the law which programs a monad's history is constitutive of its nature, it predetermines from the beginning all changes and circumstances that pertain to it. Thus, the laws that program an individual's history are internal to it alone and remain unchanging throughout its goal-directed life. And, since the inherent power of a substance is at all times exercised, there is always present a pre-existent tendency to change – a *telos* which strives to enact the plan of the substance's predetermined nature through time. This teleological conception of what it is to be an individual substance (i.e. that which contains from the beginning what it will become) rules out identical counterparts in other possible worlds. That is, Leibniz would disallow the possibility that an individual in another possible world could be identical with an individual in the actual world up to time t, after which time its history begins to diverge. Given Leibniz's strong criteria of individuality, the counterpart would be qualitatively different from the very beginning of its history. It again appears that Leibniz's conception of individuals that go into existence in a compossible world rules out trans-world identity. One can remain skeptical, however, whether the containment theory of truth provides compelling grounds for denying trans-world identity. In accessing this view, Adams notes correctly Leibniz's unstated opinion that identity across worlds needs explanation more than its denial.

V. DIVISIO ENTIS AND PREDICATE LEVELS

It is clear that Leibniz's ontology involves the demarcation of reality onto the actual, the phenomenal, and the ideal, for example, space and time conceived as possible orders of existence. I shall have something to say about this shortly. But first, if my analysis of his theory of representation and pre-established harmony is correct, Leibniz has no problem in reconciling the thesis that extended things are phenomena with the doctrine that they result from an aggregation of individual substances. As I have urged, this involves an understanding of the implications in the view that relations are formal second-order properties expressed in the theory of inter-monadic representation. The question still remains, of course, whether the notion of what it is for a physical world to obtain is satisfactorily captured by Leibniz's notion that principles of phenomenal ordering can be established relative to monadic perception. Leibniz says little about this reductive question. And what he does say is of little help:

> . . . We mean nothing else when we say that Socrates is sitting down than what we understand by 'Socrates' and by 'sitting down' is appearing to us and to others concerned . . . . (G. 2. 451–52/L. 605).

Here, Leibniz is merely stipulative: "We mean nothing else." This scarcely constitutes a justification of a translation of statements about physical things in terms of statements about perceptions. In fact, it is no more than a restatement of his phenomenalism: "– these beings by aggregation have no other complete unity but what is mental; and consequently their entity also is in some way mental or phenomenal, like that of the rainbow" (*NE*, Bk. 2, ch. 13, p. 149, n. 6).

I shall not pursue this question here, nor shall I have anything to say about the semantics of Leibniz's theory of truth in connection with the predicate-in-notion principle. This has been dealt with at great length in the literature and is freshly discussed by Robert Adams in the present volume. Instead, I shall devote the remainder of this essay to a discussion of the various sorts of attributes that are involved in Leibniz's ontology and to the question of predicate levels. I shall draw on the foregoing analysis, as well as incorporate some new material.

I have suggested that Leibniz demarcates reality into the actual, the phenomenal, and the ideal. An analysis of this claim will show that Leibniz's theory of predication involves attributions that fall into these three categories. I shall begin with phenomenal predication by considering Leibniz's treatment of extension. This notion has a peculiar status in Leibniz's thought. It is clear from the above analysis that extended objects are well-founded appearances to monadic perceivers, so that they "are not so much substances or things as they are the phenomena of percipient beings" (G. 2. 270); for "perception is involved in extension" (G. 2. 183). Thus "the aggregates themselves are nothing but phenomena, since besides the monads that enter into them, the rest is added by perception alone, by the very fact that they are perceived together" (G. 2. 517). Accordingly, the unity of phenomena is the result of inter-monadic representation; but equally, the fact that the extended appears to be continuous is a result of the continuity of monadic perception.

But the category of the extended also falls under an incomplete notion:

> ... The notion of extension is relative, or extension is the extension of something, as we say that multitude or duration is the multitude or duration of something. But the nature that is presupposed as diffused, repeated, continued, is what constitutes the physical body (*corpus physicum*) (G. 2. 269).

The "nature which is presupposed" is the capacity that monads have for inter-monadic representation, together with the claim that a correspondence obtains between the successive representations of all monadic natures considered together. Thus at once the theory of simple substances and the doctrine of representational activity are linked with the notion that extended things are perceived phenomena. And the extended continuum is not a complete notion since it depends on something that can act, as in the case of numbers, where aggregate parts are discerned "something is neccessary to be numbered, repeated and continued" (G. 2. 164). Thus, the very existence of extension is based on a principle of perceptual activity involving the simultaneous projection of a non-geometrical nature; namely, those second-order characteristics that are expressed through inter-monadic representations.

But Leibniz goes further. He tells us that extension "may be analyzed into plurality, which it has in common with number; into continuity, which it has in common with time; and into coexistence in common even with things that are not extended. I should not have thought that plurality is to be denied in the extended thing, especially if we admit that it has actual parts; we should, in that case, also have to deny plurality to a herd or an army; in other words, everywhere" (G. 2. 183, to De Volder, 23 June 1699). This is an important passage. In the first place, Leibniz maintains that only extension is resolvable into plurality, continuity, and "the coexistence of parts at one and the same time" (G. 2. 169). Time and space have continuity alone. Time like duration is successive, and so "perishes continuously" (G. 7. 408, fourth letter to Clarke, 1715). But extended things have a plurality of coexisting "actual parts" that they have "in common even with things that are not extended." In other words, Leibniz is giving an alternative expression to his thesis that extended things are phenomena that result from the aggregation of individual substances. The claim involves two things. First, that the "actual" parts of extended things are non-extended substances. At this level, the level of what is ultimately real, extended things are simply the representations of a plurality of coexisting and active substance. But, second, since the non-extended cannot be a part of the extended in any spatial sense, extended things result from the perceptions of a plurality of substances and so *appear* to be continuous and extended. Put otherwise, the phenomenal comprises second-order non-formal properties which are expressed through the objective content of monadic representations. Thus, the perception of extended things involves nothing more than the reality of individual substances, taken one at a time, together with their first-order properties. It is the unity of being of simple substances with their diverse contents that, through continuous activity, well-found the appearance of continuity in the perception of phenomenal entities and their changes. And it is thus that extended phenomena are perceived as unities by the mind.

It is clear that Leibniz considers ultimate reality to comprise a plurality of discrete and individual substances, each complete and separate from every other. Consequently, the monadic world is a radically discontinuous collection of entities, each one entire unto itself. But this is precisely what extended things are not. To say that something is extended is not to say anything about the actual. To be actual is to be a complete individual determined to all episodes and circumstances in the development of a life-plan; it is to be an *individuum ens positivum*. But to say that something is extended is to bring it under an abstract and incomplete concept, one that applies equally to a number of individuals. Just as from the abstract concept of kingship, nothing particular about Alexander can be inferred, e.g. that he is the son of Philip and the conqueror of Darius, so from the abstract concept "extended sphere" no inference to the diameter of an actual sphere is possible. But, if a concept is complete if, and only if, it cannot be contained in a concept more complex than itself, it is clear that only individual substances can fall under the criteria of a complete concept. But equally, it is clear that, whatever the notion of an incomplete concept picks out, it has its foundation in Leibniz's theory of representation. We have seen how phenomenal extension presupposes the plurality, continuity, and coexistence of monadic perceivers. Accordingly, that which an incomplete concept picks out involves indirect reference to first-order properties that are expressed through inter-monadic representation. In other words, phenomenal predication has the same monadic basis as relational predication. In contrast to this, complete concepts pertain to the attributes of individual monads together with their histories. Thus, they pertain to the realm of the actual, that is, to simple substances together with their representations and volitions. Against this background, Leibniz's doctrine that there are "no purely extrinsic denominations" takes on a deeper meaning. He tells us that it follows from his containment theory of truth, and the thesis that all true statements are of subject-predicate form, "that there are no purely extrinsic denominations which have no basis at all in the denominated thing itself. For the concept of the denominated subject necessarily involves the concept of the predicate" (*First Truths*, L. 268). Since relations and phenomenal properties are "extrinsic denominations", they supervene on the intrinsic and non-relational properties of the monadic world; as a result, the latter are necessarily presupposed by the former.

Leibniz's picture of the world is this. Analysis of perceptual phenomena shows them to be extended aggregates; extension is essentially plurality; plurality is infinite since there is no limit to the divisibility of the extended. Since all extension is composite, if there is to be ultimate unity in reality, the plurality of the extended must be a result of entities that are indivisible and non-extended. So what gives rise to the appearance of matter is not material if what is material appears to be extended. If the extended results from the coexistence of monadic perceptions, it arises from the perceptual activity of individuals that are intrinsically active. This immediately rules out atoms; they are divisible (at least conceptually) and accordingly lack the right sort of unity. Moreover, they lack the capacity to represent. Mathematical points are also ruled out. They are mere modalities, e.g. the limits of a process of subdivision. As such, though simple, they are not real since only an individual with the power to act, like a soul, can be real and simple and can account for the perception of the extended. Monadic substances alone satisfy these criteria. Against this background, Leibniz views space and time (with motion conceived as a function of these) as a conceptual grid that the mind imposes upon phenomenal change. Everything phenomenal is perceived as having a determinate extension and duration, but spatio-temporal changes that occur to phenomena are conceived as possible orders or *entia rationis*. Space is a possible order of coexistence; time, a possible order pertaining to change and succession. Space is founded in the perceived ordering of simultaneous phenomena and time in the perceived ordering of successive events. It is clear that Leibniz divides the central categories in terms of which reality is to be understood into three distinct groups; space, time, and motion are *entia rationis*; extension, changes, and relations in general are well-founded phenomena; and substances together with their attributes are actual existents (G. 7. 552–3 to Sophia, 12 June 1700).

Thus far, I have established that Leibniz thinks that entities exist in different senses. The primary sense is the sense in which individual substances, the ultimate subjects of all reference, exist. The sense in which *entia rationis* and phenomenal entities exist is variously derived from this primary sense. In every case, the existence of these non-substances presupposes the existence of intrinsically monadic properties, an ontology that underlies the sense in which Leibniz holds that there cannot be any "purely extrinsic denominations". A related point is this: predicate-expressions are thought by Leibniz to introduce various sorts of attributes that function at different levels in his ontology. In this connection, I do not intend a distinction between *per se* and accidental predication (though a distinction in predicate levels in compatible with such a distinction), but rather the more benign point that the attributes that various predicates pick out are not all constitutive of what is actual, that is, of the realm of monadic individuals: Leibniz can distinguish between the absolutely essential and the absolutely accidental in an individual nature.

Leibniz relates this view of reality, with its distinctions between ideal, phenomenal, and real entities, to his analysis of the structure of the continuum. Consider the following passage from a letter to Des Bosses in 1709:

> ... Space, like time, is a certain order – which embraces not only actuals, but possibles also. Hence, it is something indefinite like every continuum whose parts are not actual, but can be taken arbitrarily – Space is something continuous but ideal, mass [*massa*] is discrete, namely an actual multitude, or being by aggregation, but composed of an infinity of units. In actuals, simples are prior to aggregates, in ideals the whole is prior to the part. The neglect of this consideration has brought forth the labyrinth of the continuum (G. 2. 379).[20]

Notice that the pasage involves the familiar trichotomy of the ideal, the phenomenal, and the actual. The criterion that is stressed here is the one that distinguishes between wholes that are prior to parts and wholes to which simples are prior. In an earlier letter to Des Bosses, he argues the same point, saying that a "continuum – involves indeterminate parts, but, on the other hand, there is nothing indefinite in actuals – but we confuse ideal with real substances when we seek for actual parts in the order of possibles, and indeterminate parts in the aggregate of actuals – (G. 2. 282, 1706). But not only are wholes distinguished into those that involve indeterminate parts and those that are determined, but those "who have philosophized – have become confused for want of distinguishing between resolution into notions and division into parts" (G. 2. 583).

It is clear that Leibniz's related distinctions are based on a conception of the unity of being, in particular on the view that only monadic simples can satisfy the criterion that "that which is not truly *one* being is not truly a *being*" (G. 2. 297, to Arnauld, 1687). Consider space and time in accordance with Leibniz's distinctions. The notion of the spatial continuum involves, for Leibniz, the idea of a *possible order* of coexistence. As such, it does not involve actual distances that are extended and presuppose division into parts. Equally, it is not composed of parts. Given the nature of space, however, it is resolvable into ideal modalities, such as a point "which is nothing but a mode, namely an extremity" (G. 2. 347). Time is also an ideal principle of ordering. It is neither composed of, nor divisible into, instants since these are "not properly a part of time" (G. 3. 591). Being an ideal continuum, it contains instants potentially and is arbitrarily resolvable into them. It is clear that resolvability pertains to ideal notions whose differentiation is based on conceptual discrimination. In this connection, the notion of spatial distance is a phenomenal, but not an ideal notion. It is based on the contemporaneous perceptions of bare monads. Monad A can be said to have a perception of the distance that obtains between it and *C* on the basis of perceiving *C* indirectly through *B*'s direct perception of former's occurrent states. Since the phenomenal distance between *A* and *C* is a second-level relation it is not composed of actual parts; but it can be conceptually divided into parts according to a determinate rule of division.

Leibniz speaks of actuals that are fully determined and definite. The only entities that satisfy this criterion are individual substances. They fall under complete concepts and are determined to all possible circumstances, unlike abstract and indeterminate concepts such as kinship. Leibniz says that "in actuals, simples are prior to aggregates." In a basic sense, simples are incomposite because they lack actual parts: straightaway this precludes the possibility that they are divisible, composable, or resolvable in terms of parts. If they are partless in the spatial sense, the fact that monads possess an internal organization of great complexity is in no way affected. The various intrinsic attributes that constitute the defining nature of a monad are not 'parts' in any spatial sense; yet, they give the monad its reality and individuality. Simplicity, then, does not deny to a substance a unique diversity of content. If a monad is not resolvable into spatial parts, it is certainly resolvable into the determinate attributes that constitute its inner organization and of which it is the foundation and reason. After all, the multiplicity in unity that characterizes every monad is simply the representative and volitional states that its determinate history is programmed to manifest over time. In this connection, Leibniz defines perception as the representation of the many in the one (G. 2. 121). And the world of monads is for Leibniz a composite actual. For it is composed of simple individuals and is resolvable and divisible into a plurality of coexisting substances.

Given his distinctions, Leibniz can also demarcate the composite and the simple. The phenomenal continuum is resolvable into the inter-monadic representations from which it results. So considered, it is merely an ***unum per accidens***. Thus, phenominal entities have an ordered coherence in perception to the extent that they are "modes of genuine substances" (G. 2. 97). But in this context, to be a mode of substance is to have "no other complete unity but what is mental": thus the entity of phenomenal things "is in some way mental or phenomenal, like that of the rainbow" (*NE*, Bk. 2, ch. 13, p. 149, n. 6). Leibniz is suggesting that, if extended things are phenomena which result from the aggregation of actuals, they are resolvable into the representative contents of monadic perceivers that coexist at one and the same time. And since there is no determinate way by which this can be done, phenomenal entities lack the determinate multiplicity that characterizes the autonomy of monadic existences.

I have concentrated on the metaphysical foundations of Leibniz's philosophy of relations. This question is intimately related to his treatment of phenomena and their basis in inter-monadic perception. An analysis of these features leads into the question of the status of the various sorts of entities that Leibniz posits in his analysis of the representative appearances of which any philosophy must give an account. In this connection, I suggest that Leibniz's metaphysics involves predications that fall into the categories of the actual, the phenomenal, and the ideal. There is much in Leibniz's philosophy of perception that I have not dealt with, notably the doctrine of degrees of perception in the inter-monadic setting. It has recently been argued that inference is the primitive notion in Leibniz's theory of representation that anchors the link that binds a determined set of perceptual episodes to those which follow specifically from them.[21] I have argued that monads are more than a set of mental acts. Section 4 argues that monads are principles of efficacy that act to fulfill determined ends according to laws intrinsic to their defining natures. It could be argued that the laws intrinsic to monadic actuality comprise those features that are primitive to their natures. If this is so, the inferential patterns implicit in Leibniz's theory of representation are not primitive, but rather have their 'foundation' and 'reason' in the constitutive natures of monadic realities.

NOTES

[1] See C. I. Gerhardt (ed.), *Die Philosophischen Schriften von G. W. Leibniz*, 7 vols. (Berlin: N. p., 1875–90), Vol. 2, pp. 113–4 and 253. Hereafter designated G. In some cases, L. E. Loemker (trans.), *G. W. Leibniz: Philosophical Papers and Letters*, 2d. ed., 2 vols. (Dordrecht: D. Reidel, 1969) is cited. Hereafter, it is designed L. NE = *New Essary Concerning Human Understanding* (London: Macmillan, 1896). C = L. Courturat, *Opuscules et Fragments Inedits de Leibniz* (Hildescheim: Olms, 1961).

[2] G. 2. 444 "Unaquaeque [of the monads] est velut separatus quidam mundus, et hi per phaenomena sua consentiunt inter se, nullo alio per se commercio nexuque".

[3] G. 2. 253 and 444. For a discussion of this view, see C. D. Broad, *Leibniz: An Introduction* (Cambridge: Cambridge University Press, 1975). Chap. 4, pp. 103–4.

[4] For a discussion of these principles in the context of Leibniz's monadic metaphysics, see J. E. McGuire, '*Labyrinthus Continui*: Leibniz on Substance, Activity, and Matter', in *Motion and Time Space and Matter* (Columbus: Ohio State University Press, 1976), pp. 292–9.

[5] The literature is vast, and I have learned much from it. Representative studies are as follows: G. H. R. Parkinson, *Logic and Reality in Leibniz's Metaphysics* (Oxford: Clarendon Press, 1965), pp. 141–50; H. Ishiguro, 'Leibniz's Theory of the Ideality of Relations', in H. Frankfurt (ed.), *Leibniz* (Garden City: Anchor Books, 1972); N. Rescher, *The Philosophy of Leibniz* (Englewood Cliffs: Prentice-Hall, 1967), Ch. 6; C. D. Broad, *Leibniz: An Intorduction*; B. Mates, 'Leibniz on Possible Worlds', in Frankfurt (ed.), *Leibniz*.

[6] After first giving this paper as a talk and before producing the present version for publication, I encountered John Earman's perceptive and suggestive study. See 'Perceptions and Relations in the Monadology', in *Studia Leibnitiana*, Band IX/2 (1977), pp. 212–30. Earman states the same thesis as I have developed in Section I of my paper. In what follows, I am indebted to the manner in which he approaches Leibniz on the problem of relations, even though the substance of my position was established independently of his work and under the influence of Sellars, Broad, and Leibniz's Medieval background on the conception of relations. For Sellars see Note 12; for Broad, Note 5, Chap. 2, p. 39.

[7] For an interesting discussion of formal relations that has a bearing on the distinction I am drawing, see Peter Long, 'Formal Relations', *The Philosophical Quarterly* 32 (April 1982).

[8] See Ishiguro's 'Leibniz's Theory of the Ideality of Relations', *op. cit.*, Note 5, p. 211, for a discussion of this issue.
[9] See *Meditation Three* in *Oeuvres de Descartes*, 12 vols., (ed.), Charles Adam and Paul Tannery, Vol. VII, pp. 39–42.
[10] *Ibid.*, *Meditation Six*, AT. VII. pp. 78–80.
[11] G. 11.97. Here, of course, Leibniz has non-aggregational unity in mind, insofar as it is an act of mental awareness.
[12] See Wilfrid Sellars, '*Meditations Leibniziennes*', *American Philosophical Quarterly* 2 (1965), 1–14. Sellars's stimulating paper deserves to be better known among Leibnizian scholars.
[13] *Ibid.*, p. 4.
[14] Arnauld to Leibniz, 28 August, 1687. G. 2. 105–106.
[15] See Robert Adams, 'Leibniz's Phenomenalism' for an account of this interpretation to appear in Midwestern Studies in Philosophy.
[16] For a discussion of Leibniz's views on higher order mental acts see Robert B. Brandom, 'Leibniz and Degrees of Perception', *Journal of the Hist. of Phil.* **XIX** (1981), 447–479.
[17] See Robert Adams, 'Leibniz's Theories of Contingency', *Rice University Studies*, Vol. 63, No. 4, and Adams, in this volume.
[18] See Adams, in this volume.
[19] *Ibid.*, p. 000.
[20] See McGuire, *op. cit.*, Note 4, pp. 301–318 for an elaboration of these levels. Also see A. T. Winterbourne, 'On the Metaphysics of Leibnizian Space and Time', *Studies in History and Philosophy of Science*, Vol. 13 (1982), pp. 208–214 for a further development of these ideas as presented in my '*Labryinthus Continui . . .*'. See also the diagram in Loemker on p. 617 which is a supplement to Leibniz's letter to Des Bosses on August 19, 1715. It gives a useful overview of the interrelations in Leibniz's ontology that I am attempting to adumbrate in this section.
[21] Brandom, *op. cit.*, Note 16.

ROBERT M. ADAMS

PREDICATION, TRUTH AND TRANSWORLD IDENTITY IN LEIBNIZ

I. INTRODUCTION

Among the theses that Leibniz sent to Arnauld in February 1686, the one that aroused Arnauld's initial objection was the statement,

> Since the individual concept of each person contains once for all everything that will ever happen to him, one sees in it the proofs *a priori* or reasons for the truth of each event, or why one has occurred rather than another (G II, 12/L–A 5).[1]

All the predicates of an individual substance are contained in the concept of that individual, according to Leibniz. This thesis gives rise to many questions about the relation of individuals to their predicates. Why should the predicates be contained in the *concept* of the individual, and not just in the individual itself? Why does Leibniz infer from the conceptual containment thesis, as he does (DM 14), that all the states of an individual substance are *caused* by previous states of that individual alone? We will come to these questions, but the prevent investigation is organized around another issue. Today we might call it the issue of "trans-world identity"; Arnauld raised it in these words:

> Since it is impossible that I should not always have remained myself, whether I had married or lived in celibacy, the individual concept of myself contained neither of these two states; just as it is well to conclude: this square of marble is the same whether it be at rest or be moved; so neither rest nor motion is contained in its individual concept (G II, 30/L-A 30).

Arnauld affirms trans-world, or counterfactual identity as a reason for rejecting Leibniz's conceptual containment thesis. He denies that his actual predicate of life-long celibacy is contained in his individual concept on the ground that he is the same individual as one who would, under some possible circumstances, have been married. In his response, as in a number of other places in his writings, Leibniz made clear that he did not accept Arnauld's assumption of trans-world identity. He held that no individual creature exists in more than one possible world – that if Arnauld, for example, had married, he would not have been Arnauld (or more precisely, that anyone who got

James Bogen and James E. McGuire (eds.), How Things Are, 235–283.

married, in any possible world, would not have been Arnauld [cf. Gr 358]). The issue of trans-world identity being very much alive today, this aspect of Leibniz's philosophy has received a lot of attention from recent interpreters.[2] But I think there remains (and perhaps always will remain) enough obscurity about his reasons and motives to invite further exploration.

Right at the outset, we should note an important difference between Leibniz and present-day philosophers who are interested in trans-world identity. For the latter, talk about possible worlds serves to explicate the structure, if not the basis, of logical possibility and necessity. What is possible is what is true in (or at) some possible world; what is necessary is what is true in all possible worlds. From this point of view, the denial of trans-world identity seems to entail that no actual individual could possibly have had different properties from those it actually has. To say that Caesar, for example, could have turned back from the Rubicon is to say that there is a possible world in which he does; and that obviously must be a possible world in which he exists. But if there is no trans-world identity, then Caesar exists only in the actual world, in which he crosses the Rubicon. So it seems to follow that he could not have turned back. Leibniz emphatically rejects this conclusion, however. Much of Section 13 of his *Discourse on Metaphysics* is devoted to developing the thesis that, while the crossing of the Rubicon is contained in the individual concept of Julius Caesar, it follows "that it was reasonable and consequently assured that that would happen, but not that it is necessary in itself, nor that the contrary implies a contradiction."

David Lewis has developed his well-known "counterpart theory" as a way of rejecting trans-world identity without denying that we could have had somewhat different properties from those we actually have. Acording to Lewis no individual exists in more than one possible world, but there is a "counterpart relation" that obtains among sufficiently similar individuals in different possible worlds. Your counterparts are "people you might have been", so to speak. To say that it would have been possible for you to have done a certain thing that you did not do is to say that there is a possible world in which a counterpart of yours does it. Several recent interpreters (especially Mondadori, 1975, pp. 94–101) have pointed out that Leibniz has a use of proper names that suggests something like Lewis's counterpart theory. He speaks, for example, of "possible Adams" in the plural, meaning "possible persons, different from each other, who fit" a general description consisting of a part of the predicates of the first man in the actual world (G II, 41/L–A 45).

Margaret Wilson has argued convincingly, however, that it is a mistake to ascribe to Leibniz a counterpart-theoretical account of possibility and necessity *de re* in terms of alternative possible Adams, alternative possible Caesars, alternative possible Arnaulds, and so forth. Leibniz never (so far as I know) gives such an account. And as Wilson points out, it would be inconsistent with his conception of God's freedom. For Leibniz holds that there are infinitely many possible worlds that God could possibly have created. And he never speaks of alternative "possible Gods" who could have created the different worlds. He seems to be committed, not only by this silence, but also by the doctrine of God's necessary existence, to the view that it would have been *the same* God that created whatever possible world was created. In other words, Leibniz seems to have accepted trans-world identity for God while rejecting it for every other individual.

These apparent inconsistencies can be removed by a correct understanding of Leibniz's conception of possible worlds and its relation to his conception of necessity. I shall be somewhat dogmatic on this subject, having argued elsewhere (Adams, 1977) for my interpretation. According to Leibniz (from 1686 on), the difference between necessary and contingent truths is that necessary truths can be proved by analysis of their terms in a finite number of steps, whereas it could take an infinite analysis to establish a contingent truth. Even contingent truths, however, are seen by God to follow from the concepts of their terms, the predicate being contained in the subject, or the consequent in the antecedent. A contingent falsehood is possible, on Leibniz's view, in the sense that its denial cannot be proved by a finite analysis; but it remains *conceptually inconsistent* in the sense that the negation of its predicate is contained in the concept of its subject, although it would take an infinite analysis to render that containment explicit. (It will be useful for us to employ 'conceptually inconsistent' in this sense, although Leibniz does not.)

Leibniz's conception (at any rate, his principal conception) of possible worlds does not provide a semantics for this conception of necessity and possibility. In particular, he does not want to say that what is possible is true in some possible world. For it is possible to have a world as good as the actual world in other respects without the horrors of the Thirty Years' War; that is, such a state of affairs cannot be proved inconsistent by a finite analysis. For obvious reasons of theodicy, however, Leibniz will not admit that there is a possible world in which it is true that that much goodness occurs without the Thirty Years' War. For similar reasons, it will not do for Leibniz to say that a possible world is a world whose complete concept is possible in the sense that no finite analysis can prove it inconsistent. Possible worlds are worlds that God could have created, in a sense in which

He could not have created a world as good as the actual world without the Thirty Year's War.

On the other hand, it also will not do for Leibniz to say that a possible world is a world that God could have created without any conceptual inconsistency at all. For Leibniz is committed to the belief that it follows (though not by a finite analysis) from the concepts of God and of the actual world that He does create it, and from the concepts of God and the other possible worlds that He does not create them. I think that Leibniz's principal conception of possible worlds – the conception on which he depends in his theodicy – can be captured by saying that a possible world is a world that God could have created were it not for His goodness (or that He could have created were it not for the fact that some other world is better). For the way in which the actuality of actual things and the non-actuality of non-actual things follows from their concepts, according to Leibniz, is that it follows from their concepts (together with the concept of God and of all the alternatives He could have chosen) that the actual things belong to the best series of things He could have chosen, and the non-actual things do not; but this does not follow by a finite analysis. Suppose we say that the basic concept of a possible world is a concept that contains everything that would be true if that world were actual *except* that it contains nothing of the comparison that God would have made with other possible worlds in deciding to create that one. Then we can say that a possible world is one whose basic concept contains no conceptual inconsistency at all. It is a world which contains no conceptual inconsistency "in itself", though an inconsistency that cannot be finitely proved may arise when its internal features are put together with the idea of its having been created by a perfectly good God and the fact that some other possible world is better.

Now I take it that for Leibniz the individuality of Julius Caesar is defined by a (complete) concept of him that contains crossing the Rubicon at a fateful juncture in his career. This containment cannot be proved by a finite analysis; and, therefore, it is not necessary that he crossed the Rubicon then. (Cf. C 376f./P66, where exactly this is asserted about Peter's denying Jesus, which was the example Leibniz first intended to use in the place now occpied by Caesar's crossing the Rubicon in DM 13.) But a world in which Caesar turned back from the Rubicon at that juncture world contain a conceptual inconsistency in itself, and would therefore not be a possible world. Its basic concept would contain Caesar's not crossing the Rubicon at a time at which it follows from his definitive concept that he does cross it. Both his crossing and his non-crossing would thus be contained, at least implicitly, in the basic concept of that world. In saying this, I assume that the containment of Caesar's crossing the Rubicon in his complete concept does *not* depend on the fact that if God compared all possible worlds He would find that the best is one in which Caesar crosses the Rubicon. Doubts may be raised about ascribing this assumption to Leibniz, but without it, the denial of trans-world indentity cannot be justified. For the assumption could hardly be false unless God had a choice between a possible world in which Caesar crosses the Rubicon and another possible world in which the same individual Caesar exists but does not cross the Rubicon.

Leibniz is not committed to making a corresponding assumption about God. He is, indeed, committed to the view that God's choosing this world that He has actually created, and His not creating any other possible world, are contained in the concept of God. But it is precisely because this is the world that He would find to be best, in comparing all possible worlds, that these choices follow from God's essential wisdom and goodness. I think this enables us to understand why Leibniz allowed God to have trans-world identity. One of Leibniz's main concerns in his thinking about possible worlds is to maintain the reality of God's choice. And the crux of this matter, as Leibniz perceived it, is that God (one and the same God, presumably) should have had a plurality of *internally* consistent worlds to choose from. It is not enough that the alternative worlds be free of finitely provable inconsistency; they must not have any conceptual inconsistency "in themselves." It is *not* required of each possible world, however, that there be no conceptual inconsistency in God's choosing to create it instead of all other possible worlds. On the contrary, it is required by the principle of sufficient reason, as I think Leibniz understood it, that there be a conceptual inconsistency in God's creating any other possible world than the one He has created. The reason why Leibniz is prepared to say that the same individual God could have been the creator of different possible worlds is that the sense in which He "could" have is that were it not for His wisdom and goodness, or were it not for their inferiority, He could have – not that His creating them would be perfectly consistent with His nature. But this very same conception of the sense in which God could have created a different world requires that no *creature* occur in any possible world in which something occurs which is contrary to the nature of that creature (considered apart from the reasons for God's choice among worlds). For then it would not be because of its inferiority, and His wisdom and goodness, but because of an internal conceptual inconsistency in that world, that God could not create it.

There is much more to be said, however, about Leibniz's reasons for denying trans-world identity to creatures. As Margaret Wilson (1979, p. 729) has remarked, "we know what *use* Leibniz wants to make of" this doctrine, theologically. "He wants to use it as a basis for denying that God is responsible for a created individual's misfortunes or bad choices, since *that individual* couldn't have existed unless he made those choices and experiences whatever he in fact experiences." Leibniz says that God "did not make Sextus wicked .... All he did was grant him existence, which his wisdom could not refuse to the world in which he is included" (T 416). Leibniz thought that this entitled him to hold that evil is not caused by God, but by the limitations inherent in the concepts of the creatures that it was best, on the whole, for God to create (DM 30, T 20). He also thought it provides an answer to complaints that individuals might be tempted to make against God. "You will insist that you can complain, why did God not give you more strength. I reply: if he had done that, you would not be, for he would have produced not you but another creature" (GR 327: cf. Conf 128 f).[3]

These theological motives are important for understanding Leibniz's denial of trans-world identity; but as Wilson suggests, he undoubtedly thought he could give other *reasons* for it. Two of them will detain us only briefly. A particularly bad argument is found in a letter that Leibniz sent to Arnauld in July 1686, but is missing (whether as a result of a later correction, I do not know) from the copy retained by Leibniz at Hanover. Responding to the claim "I find clearly in the individual concept that I have of myself that I shall be myself whether or not I take the trip that I have planned", Leibniz says,

> If it is certain that *A* is *B*, what is not *B* is not *A* either. So if *A* signifies me, and *B* signifies the one who will take this trip, it can be concluded that the one who will not take this trip is not me; and this conclusion can be drawn from the certitude alone of my future trip, without having to impute it to the proposition in question (LW 39).

The last phrase ("without ... ") in this passage is completely obscure to me, but Leibniz seems to be arguing that a contradiction would be involved in trans-world identity, on the ground that I, who am going to take a certain trip, cannot also have the property of not taking that trip. But of course, the question is not whether I could *also* have the property of not taking the trip, but whether I could have had that property *instead of* my actual property of taking the trip.

Another argument is suggested by the statement, "It follows also that it would not have been our Adam, but another, if he had had other events, for nothing prevents us from saying that it would be another. Hence it is another" (G II 42/L–A 46). This looks like an appeal to the principle of sufficient reason: There is no compelling reason to say it would be the same individual if different events happened to him; therefore, it would not be the same individual. This is a weak argument. It leaps over vast stretches of disputed territory. Arnauld was at least initially inclined to say, for example, that our ordinary conception of a person keeps us from saying it would have been another Adam if he had never sinned. One wonders, moreover, whether the argument does not cut both ways. As Margaret Wilson (1979, p. 729) says, "The question presents itself insistently: what prevents us from *denying* it would be a different Adam if circumstances had been different . . . . ?" Would it be any easier to find a sufficient reason for denying trans-world identity than for affirming it? Leibniz seems to assume here that presumption favors non-identity – that identity needs to be explained in a way that distinctness does not, and hence that there is more need of reasons for affirming than for denying identity. Perhaps that is correct. I think it is not without plausibility in some cases. But not seeing how Leibniz would defend it, I will pass on to other arguments.

II. THE CONCEPTUAL CONTAINMENT THEORY OF TRUTH

It is natural to turn next to the argument on which Leibniz himself laid the greatest weight for the defense of his whole position regarding individual concepts, including his denial of trans-world identity.

> Finally [he wrote to Arnauld] I have given a decisive reason, which in my opinion ranks as a demonstration; it is that always, in every true affirmative proposition, necessary or contingent, universal or singular, the concept of the predicate is included in some way in that of the subject, *praedicatum inest subjecto* [the predicate inheres in the subject]; or else I do not know what truth is (G II, 56/L–A 63).

This is the argument by which Arnauld confessed himself especially impressed, when he gave up the debate with Leibniz about individual concepts (G II, 64/L–A 77). We shall consider what is the theory of truth to which Leibniz here appeals, what were his reasons for holding it, and whether it does entail his denial of trans-world identity.

II.a. *What is the theory?*

As usually stated by Leibniz, the conceptual containment theory of truth applies to categorical propositions – that is, to propositions of subject-predicate form. He sometimes applied it to conditional propositions, holding that they are true if and only if the consequent is contained in the antecedent (e.g. C 401); but that need not concern us here. The six traditional types of categorical proposition, with an example of each, are:

universal affirmative	All men are married.
universal negative	No man is married.
particular affirmative	Some man is married.
particular negative	Some man is not married.
singular affirmative	Arnauld is married.
singular negative	Arnauld is not married.

Only two of these types, the universal and singular affirmatives, are mentioned in the formulation of the theory that I have quoted (as also in C 16f., 519). This is no accident, for Leibniz applies the theory more straightforwardly to these types than to the other four. Nevertheless, it is clear that he intends the theory to determine an interpretation of all six types. And he sometimes expresses it in a more sweeping form, claiming that in *every* true proposition without restriction as to type, the concept of the predicate is contained in the concept of the subject (DM 8; C 388/P 77).

The conceptual containment theory is closely connected with Leibniz's preference for what is nowadays called an "intensional" as opposed to an "extensional" interpretation of the categorical propositions. (Perhaps in some sense it *is* that preference.) Roughly, an *extensional* interpretation is one that treats the truth or falsity of propositions as depending on relations among the extensions of their terms, where the extension of a term is the class of things that satisfy it, or which it characterizes.[4] And an *intensional* interpretation is one that treats the truth or falsity of propositions as depending on relations among the intensions – that is, the concepts – of their terms.

Since Leibniz's discussion partner, Arnauld, was co-author of a famous logic text, the *Port Royal Logic*, it is of interest to note that a similar distinction is made in that book between the comprehension (*compréhension*) and the extension (*étendue* or *extension*) of an idea. The comprehension of an idea consists of "the attributes which it includes in itself, and which cannot be taken away from it without destroying it." The extension of an idea consists of "the subjects with which that idea agrees", or which contain it (Arnauld and Nicole, 1965, I.6, II.17). The extension of a general term, however, may be viewed in the *Port Royal Logic* as constituted by the species that fall under it as well as the individuals that fall under it. Both relations of the comprehensions of terms and relations of the extensions of terms are used in the *Port Royal Logic* in justifying basic rules of traditional logic (Arnauld and Nicole, 1965, II.17–20).

Scholars have noted that Leibniz worked out quite a variety both of intensional and of extensional treatments of the logic of predicates, and that he preferred the intensional approach.[5] For the sake of brevity, I will abstract somewhat from this variety and discuss only the one intensional and the one extensional interpretation that are most closely related to the controversy between Leibniz and Arnauld. Leibniz did not have fixed terminology corresponding to "intensional" and "extensional"; so I will use these more modern terms. He distinguished the two principal types of treatment variously as in terms of ideas [*secundum ideas* (C 300) or *per ideas* (G VII, 215)] on the one hand, and in terms of instances [*per exampla subjecta* (G VII, 215)] or in terms of individuals belonging to the terms [*ex individuis terminorum* (C 300)] on the other hand (cf. C 53/P 20). In Leibniz, unlike the *Port Royal Logic*, the extensional treatment is almost always in terms of the individuals, not the species. that fall under the general terms.[6]

The contrast between the two interpretations is easiest in the case of *universal affirmative* propositions. As an example, take the proposition, 'All gold is metal'. Extensionally interpreted, it means that the extension of 'gold' is contained in the extension of 'metal': the class of gold things is contained or included in the class of metal things. Intensionally interpreted, it means that the concept of metal is contained in the concept of gold. Extensionally, the subject is contained in the predicate in this type of proposition, if the proposition is true. Intensionally, the predicate is contained in the subject. Leibniz noted this "inversion" of the containment relation, contrasting his own intensional interpretation with the extensional interpretation of "the Schools", who

> . . . say that metal is wider than gold, and if we wished to enumerate the individuals of gold on the one hand and the individuals of metal on the other, the latter would certainly be more than the former, which will therefore be contained in the latter as a part in the whole (C 53/P 20).

'Some metal is gold' is a *particular affirmative* proposition. Extensionally interpreted, it means that the extension of 'metal' intersects the extension of 'gold': the class of metal things and the class of gold things have at least one member in common. The intensional interpretation of particular affirmatives is less obvious. Leibniz wants to say that in these propositions too "the concept of the predicate is contained in some way in the concept of the subject" – but in what way? It does not seem to be true that the concept of gold is included in the concept of metal. For something can be metal without being gold.

Leibniz's solution to this problem is that "in a particular affirmative proposition it is enough that the thing should follow when something is added." He gives an example to show what sort of addition he means: "although metal does not by itself contain gold, nevertheless some metal, with an addition or specification (for example, that which makes up the greater part of a Hungarian ducat) is of such a nature as to involve the nature of gold" (C 51/P 19). This choice of an example is not a happy one. For it seems doubtful that the concept of gold is contained in the *concept* of metal that makes up the greater part of a Hungarian ducat.

Leibniz gives a better example elsewhere, suggesting that 'Some man is a laugher' could be understood as 'A laugher-man is a laugher'. Here the "addition" of the specification 'laugher' to the concept of man yields a subject concept (of laugher-man) which manifestly does contain the predicate of being a laugher. It might seem that this method of intensional interpretation has the unacceptable result that all particular affirmatives turn out to be true, since we can always "add" the predicate concept to the subject concept. But Leibniz denies that we can always make such an addition. The qualified subject concept resulting from the addition must be *consistent* in order to verify a particular affirmative proposition, on his interpretation. Thus, he says that "a laugher-stone would not be a laugher since a laugher-stone involves a contradiction" (G VII, 214/P 118). Intensionally interpreted, therefore, 'Some *A* is *B*' means that the concept of *A*, with some consistent addition, includes the concept of *B*.

The intensional interpretation of universal and particular *negative* propositions may be seen as following from that of the affirmatives. Leibniz says that "negative propositions merely contradict the affirmatives, and assert that they are false". A particular negative does nothing but deny a universal affirmative, and "a universal negative proposition merely contradicts a particular affirmative" (C 52/P 19f). This treatment of the negative propositions, as negations of the affirmatives, can, of course, be applied to the extensional as well as the intensional interpretation.

The two interpretations of the universal and particular propositions can be summarized as follows, using the traditional square of opposition. (In each case, I have numbered the extensional interpretation (1) and Leibniz's intensional interpretation (2).)

All A is B.
(1) The class of B contains the class of A.
(2) The concept of A contains the concept of B.

No A is B.
(1) The class of B does not intersect the class of A.
(2) The concept of A, with any consistent addition, does not contain the concept of B.

Some A is B.
(1) The class of B intersects the class of A.
(2) The concept of A, with some consistent addition, contains the concept of B.

Some A is not B.
(1) The class of B does not contain the class of A.
(2) The concept of A does not contain the concept of B.

With regard to universal and particular categorical propositions at least, the intensional interpretation is substantially equivalent to the conceptual containment theory of truth. The intensional interpretation of singular propositions, however, is both more problematic and more relevant to our present investigation. At this point, it will be helpful to take up some terminology from the Leibniz-Arnauld correspondence. Arnauld agreed with Leibniz that there is a *complete concept* which God forms of every individual and which contains all of the individual's predicates – everything that will ever happen to that individual. That much clearly follows from the doctrine of divine omniscience that was common property to the two thinkers. They agreed that complete concepts of individuals are known only to God, not to us; we have to think individuals by incomplete concepts. They also spoke of an *individual concept* of each individual, which is minimally characterized by Arnauld in the following words:

> It seems to me that I ought to regard as contained in the individual concept of *me* only what is such that I would no longer be *me* if it were not in *me*; and that everything that is on the contrary such that it could be or not be in *me*, without my ceasing to be *me*, cannot be considered as being contained in my individual concept (G II, 30f/L–A 30).

Leibniz maintained that the individual concept of every created individual is its complete concept. Arnauld disputed this, at least partly on the ground that it seemed to imply the denial of trans-world identity. For it follows from the definition of an individual concept that I could not be me if I lacked any predicate contained in my individual concept; so if my complete concept is my individual concept, it seems to follow that I could not be me if I lacked any of my actual predicates.

It seems clear that Leibniz interprets *singular affirmative* propositions as meaning that the concept of the predicate is contained in the individual concept of the subject. For instance, 'Arnauld is married' means that the concept of being married is contained in the individual concept of Arnauld. Thus, the singular affirmative is treated in the same way as the universal affirmative.[7]

The question naturally arises, however, why the singular affirmative should not rather be treated in the same way as the particular affirmative. Leibniz's intensional interpretation demands of particular affirmatives only that the concept of the subject *with some consistent addition* contain the concept of the predicate. Why should more be demanded of singular affirmatives? Why shouldn't 'Arnauld is married' mean that the individual concept of Arnauld, *with some consistent addition*, contains the concept of being married? And why wouldn't Arnauld be correct in saying that the concept of *married Arnauld* is his individual concept with a *consistent* addition?

These questions appear even more pressing when we consider the interpretation of *singular negative* propositions. The singular negative is surely supposed to be the simple negation, the contradictory, of the singular affirmative. Now, suppose the singular affirmative means that the individual concept of the subject contains the concept of the predicate. Then its negation must mean that the individual concept of the subject does not contain the concept of the predicate. Thus, 'Arnauld is not married' will mean that the individual concept of Arnauld does not contain the concept of being married. On this interpretation, the singular negative is treated in the same way as the particular negative.

But, suppose, on the other hand, that the singular affirmative means that the individual concept of the subject, with some consistent addition, contains the concept of the predicate. In that case, its negation must mean that the individual concept of the subject, with any consistent addition, does not contain the concept of the predicate. Or, in other words, that the individual concept of the subject excludes the concept of the predicate, so that the predicate cannot consistently be added to it.[8] 'Arnauld is not married' will, therefore, mean that the individual concept of Arnauld excludes the concept of being married. On this interpretation, which Leibniz seems to prefer, the singular negative is treated in the same way as the universal negative.

The intensional interpretation thus seems to generate a complete new square of opposition, with four types of proposition, out of the two original types of singular proposition:

x is F. The individual concept of *x* contains the concept of *F*.	*x is not F.* The individual concept of *x*, with any consistent addition, does not contain the concept of *F*.
x is F. The individual concept of *x*, with some consistent addition, contains the concept of *F*.	*x is not F.* The individual concept of *x* does not contain the concept of *F*.

In this square, as in the traditional square of universal and particular propositions, the diagonally opposite propositions are mutually contradictory. All the logical relationships of the square of opposition are reproduced.

This is a strange result. Leibniz certainly wants to have two intensional types of singular proposition instead of four. Yet, any choice of only two out of the four types in the square seems to run some risk of being arbitrary. Furthermore, Leibniz seems inclined to choose the top two types in the square, the two that correspond to the universal propositions; and there is a serious objection to that choice, in that those two types are not contradictories but contraries, which can both be false together. On the other hand, if he refused to choose, he would be left with four types of singular proposition, and would have to regard the two ordinary forms of singular proposition ('*x* is *F*' and '*x* is not *F*') as ambiguous – which also seems undesirable.

We do not have to guess what Leibniz's solution to this problem would be, for a considerable part of the problem occurred to him and was discussed by him.

Some logical difficulties worth solution have occurred to me. How is it that opposition works in the case of the singular propositions – e.g., 'The Apostle Peter is a soldier' and 'The Apostle Peter is not a soldier' – whereas elsewhere a universal affirmative and a particular negative are opposed? Should we say that a singular proposition is equivalent to a particular and to a universal proposition? That's right . . . . For some Apostle Peter and every Apostle Peter coincide, since the term is singular (G VII, 211/P 115).

Here Leibniz sees that there is a problem about treating the singular negative as the contradictory of the singular affirmative, and that the problem arises from the possibility of interpreting the singular propositions either as analogous to universal propositions or as analogous to particular propositions. His solution is to insist that the two interpretations of each singular proposition are equivalent, "coincide", and so collapse into one.

But how is that equivalence possible, within the framework of a Leibnizian intensional interpretation? Inspection of the apparent square of opposition for singular propositions will show that the reduction of the four types to the two desired types requires that the individual concept of x with any consistent addition be equivalent to the individual concept of x. This requirement will be satisfied if and only if there is no predicate not already contained in the individual concept of x which can consistently be added to it. What is required, in other words, is precisely that the individual concept of x be a complete concept.[9] For a complete concept, as we shall see in Section II.f, is one to which no new predicate can consistently be added.

II.b. *The treatment of Actuality*

Leibniz says that as opposed to an extensional calculus based on relations of classes of individuals, "I have preferred to consider universal concepts or ideas and their combinations, because they do not depend on the existence of individuals" (C 53/P 20). I think we can guess what the advantage is that Leibniz thus claims for the intensional interpretation. There are well-known difficulties in maintaining the validity of the traditional inferences in the square of opposition, under an extensional interpretation, if the subject or predicate concept (or one of their complements) has an empty extension, because of the way in which the truth of the propositions, extensionally interpreted, depends on the existence of individuals satisfying (or not satisfying) the concepts. Leibniz could justify all the traditional inferences, under his intensional interpretation, without assuming anything about the actual existence of individuals, but assuming only that the subject and predicate concepts (and their complements) are consistent (and hence *possibly* exemplified).

From another point of view, however, the fact that under an intensional interpretation, the truth of the categorical propositions does not depend on the existence of individuals can be seen as a serious *disadvantage*. Many of the propositions to be interpreted are normally understood to be about the universe of actually existing things in such a way that their truth or falsity *ought* to depend on the existence of individuals. From the standpoint of common sense, it may be implausible to say (as the usual extensional interpretation does) that 'All centaurs are made of metal' is true if no centaurs exist, but it is at least as implausible to say (in accordance with the intensional interpretation) that 'Some dogs are green' is true if the concept of a green dog is consistent.

Leibniz will not be speechless in the face of this problem. He had a method for dealing with it. If 'Some pious man is poor' means that the poverty of a pious man is not just possible but actual, Leibniz will state it as 'A poor pious man is existent', where 'existent' means existing in the actual world (C 270–273). Intensionally interpreted, this yields 'The concept of a poor pious man, with some consistent addition, contains the concept of existing in the actual world.' Here, we must think more about what is meant by 'consistent addition'. Does it mean an addition that could be made without engendering a *demonstrable* inconsistency – that is, an inconsistency that could be proved by a finite analysis? Or does it mean an addition that would not result in any conceptual inconsistency at all? I think the latter must be meant. For the interpretation is intended to apply to contingent propositions, which it would take an infinite analysis to prove or disprove (cf. C 272). Indeed, 'A poor pious man is existent' is such a proposition.

So interpreted, 'Some pious man is poor' will be true if and only if the concept of existing in the actual world can be added to the concept of a poor pious man without generating any conceptual inconsistency. And that will be true if and only if a poor pious man exists in the best of all possible worlds. For given the necessary existence of a God of perfect wisdom, power, and goodness, a conceptual inconsistency is involved in the actuality of any world but the best. This interpretation is borne out by Leibniz's attempts to define 'existence' and 'existent' in terms of entering into the most perfect series of things, or being compatible with more things than anything incompatible with it is, or being pleasing to God (C 9, 360, 376, 405/P 51, 65f; cf. Gr 325). I doubt that he was committed to any such definition, but I think it is clear that his intensional interpretation of propositions that are intended to be about the way things actually are will work only on the assumption that the concept of existing in the actual world and the concept of existing in the best of all possible worlds can be added, without any conceptual inconsistency, to exactly the same concepts.

Universal propositions can be treated in an analogous way. For example, if 'Every man sins' means, not that sinlessness is conceptually inconsistent with humanity, but only that sin is actually universal among men, it can be rendered as 'The concept of a man that does not sin contains the concept of not existing in the actual world' (cf. C 271). As for singular propositions, 'Arnauld is (actually) celibate' can presumably be rendered as 'The concept of celibate Arnauld contains the concept of existing in the actual world' – which will be true if and only if Arnauld exists, and is celibate, in the best of all possible worlds.

II.c. *An Anti-semantical Theory of Truth*

Leibniz held a very austere view of conceptual containment. He thought that all true propositions are either "identities" or reducible by (a perhaps infinite) analysis of terms to identities. Among categorical propositions, identities have such forms as '*A* is *A*', '*A* is not non-*A*', and '*AB* is *A*' (C 518, 369/L 267, P 59). And the reduction by analysis of terms proceeds by the use of definitions, or by substitution of (more complex) equivalent terms for the original terms of the proposition (see C 518, 68; G VII, 44/L 267). I take it from this that in truths that are not explicitly identical, conceptual containment is supposed to be based on a formal logical construction of the terms of the proposition out of simpler concepts. And this construction, at least insofar as it is relevant to the truth of the proposition, proceeds by the simple logical operations of conjunction and negation. Leibniz does speak of other "*simple particles* or primitive syncategorematic terms", such as 'in' (C 358/P 49), which play a part in the composition of concepts; but conceptual containment and exclusion seem always to turn on conjunction and negation. For if a predicate is contained in a subject concept, it is because they are identical or the subject concept is a conjunction one of whose conjuncts is identical with the predicate. And if a subject concept excludes a predicate, that is because it is the negation of the predicate (or *vice versa*), or because the subject or predicate or both are conjunctions of concepts and one of them is, or includes as a conjunct, a concept that is the negation of the other or of one of the other's conjuncts. I cannot see that Leibniz's theory of conceptual containment allows for any other relation of implication or inconsistency between concepts than these.

Leibniz may be contrasted with Descartes on this point. In his early *Rules for the Direction of the Mind*, Descartes held that we are acquainted with a number of "simple natures", which cannot be analyzed into anything simpler. In spite of their simplicity, he thought these natures may have a necessary connection with each other; one may be

> . . . so implied in the concept of another, by some confused reason, that we cannot conceive either distinctly, if we judge that they are separated from each other. Thus figure is conjoined with extension, motion with duration or time, and so on, because it is impossible to conceive of a figure that has no extension nor of a motion that has no duration (Descartes, 1913, X, 421/1955, I, 42).

Descartes could not explain these implications in the way proposed by Leibniz, for Descartes maintained that figure, extension, motion, and duration are simple, and explicitly denied that they can be analyzed. He could not, therefore, regard them as conjunctions of simpler properties. In other words, Descartes supposed that there are primitive necessary implications between concepts, which are not to be explained by structures of conjunction and negation within the concepts. He was less explicit about primitive inconsistencies between concepts, but he did say that the simple natures can be sorted into those that are ascribed to the intellect, those that are ascribed to bodies, and those that are common to both (Descartes, 1913, X, 419/1955, 41). Implicit in this division, I think, is the assumption that there is some sort of necessary incompatibility between the purely intellectual and the purely material natures, even though they are simple. Knowledge and doubt, for instance, cannot have a shape.

Descartes's position is plausible. If there are any simple concepts, I should think there probably are primitive implications and inconsistencies between them. Leibniz will have none of it, however. He maintains an account of conceptual implication and inconsistency which is an ancestor of the attempt to explain all logical necessity and all logical implication in terms of analyticity. One place in his philosophy where something hangs on this is in the proofs of the divine existence. Early and late he gives an argument for possibility of God's existence that presupposes that conceptual inconsistency cannot arise except by conjunction of concepts one of which is the negation of the other (A II, i, 271f/L 167f; Mon 45; cf. texts published in Janke).

The consequence of defining truth in terms of conceptual containment, so formally understood, is a radically anti-semantical theory of truth. That is, it is a theory in which the truth or falsity of a proposition does not depend at all on what objects – or even on what properties – are represented by the concepts making up the proposition. Truth or falsity depends only on the logical structure of the proposition and the internal logical structure of its concepts. In this sense, truth is not a semantical but a purely syntactical property of propositions, according to Leibniz's theory.

The point is underlined by two consequences that follow from the theory, although Leibniz may not have recognized them. (1) Suppose, as Leibniz did, that all concepts are ultimately composed of simple concepts. It follows from Leibniz's theory that God does not need to know what simple concepts are involved in a proposition *P*, or what properties are represented by the simple concepts in *P*, in order to know whether *P* is true or false. He has only to know the logical structure of *P* in order to know whether *P* is true. Suppose, for example, that God intuits a representation of a complete analysis of *P*, in which each simple concept is represented consistently and uniquely by a symbol. He could know whether the represented proposition is true or false without knowing what the symbols stand for. Of course, Leibniz would insist that God does know what simple concepts are involved in every proposition (and what properties they represent, if that is different); but his knowledge of the truth and falsity of propositions does not depend on that.

(2) The world is perfectly symmetrical with respect to simple concepts, if Leibniz's theory of truth is correct. Take any true proposition *P*. Suppose *A* is a complete analysis of *P*; perhaps *A* is infinitely complex. And let *F* and *G* be two simple concepts occurring in *A*. Now consider the proposition *Q* formed by interchanging *F* and *G* at all of their occurrences in *A*. It follows from Leibniz's theory that *Q* is true if *P* is, because they have exactly the same logical structure and therefore *Q*'s subject concept will contain *Q*'s predicate if and only if *P*'s subject concept contains *P*'s predicate.

This is not to say that there is no place in Leibniz's philosophy for a semantics – that is, for a theory of how thought and/or language represent things distinct from themselves. On the contrary, Leibniz believed that each of us has infinitely many thoughts that "express", and constitute "perceptions of", things outside our own minds; and he devoted a lot of attention to this relation, which is, in a broad sense, a semantical relation. It is just that in his philosophy the truth and falsity of propositions are determined independently of this semantical relation (or perhaps more accurately, prior to it). This makes it all the more difficult to answer the question to which we turn next.

II.d. *Why Did Leibniz Hold This Theory?*

Clearly, the conceptual containment theory of truth is an extraordinary doctrine. Why did Leibniz believe it? Leibniz saw some advantages for formal logic in an intensional interpretation of the logic of predicates; but those formal considerations are inconclusive, and Leibniz knew it. (See especially Kauppi, 1960, pp. 247–252.) Recently, Robert Sleigh (1982) has expressed confidence that the "explanation of why [Leibniz] thought that true propositions about actual substances were true in virtue of certain connections among concepts" is to be found in "Leibniz's story of creation". I am sure at least a large part of the explanation is there. Specifically, I do not believe Leibniz would have adopted the conceptual containment theory of truth had it not been for his prior adherence to a form of the Principle of Sufficient Reason that makes every truth follow, in some sense, from the necessary existence and nature of God. In the works of the 1680s and after, Leibniz sometimes proposes a derivation of the Principle of Sufficient Reason from the conceptual containment theory of truth (C 519/L 268; cf. G VI, 414). Couturat (1972, p. 20f) even held that Leibniz identified the Principle with the theory. But it is not plausible to suppose that Leibniz accepted the Principle of Sufficient Reason *because* he believed the conceptual containment theory of truth. The Principle was one of his first metaphysical commitments. In a letter to Magnus Wedderkopf (in 1671) and in *The Philosopher's Confession* (in 1673), we see the young Leibniz wrestling with the apparently deterministic implications of the Principle; but the conceptual containment theory of truth makes no appearance there (A II, i, 117f/L 146f; Conf). Leibniz would have believed in the Principle of Sufficient Reason even if he had never thought of the conceptual containment theory of truth. But the theory does provide an *explanation* of *how* there is a sufficient reason for every truth; and I think that is the point of the passages in which Leibniz seems to derive the Principle of Sufficient Reason from the theory. (Cf. Brody, 1977, p. 53f.)

Perhaps the theory was practically forced on him by his commitment to the Principle. He may have been unable to see a *sufficient* reason as having been given for any truth that has not been shown to follow from concepts. That is not the only way in which the Principle of Sufficient Reason can be understood, but it is one way. Historically, the most important alternative, I think, is to suppose that the decision of a free will can constitute a sufficient reason for a truth that does not follow from concepts; but Leibniz firmly rejected this view when it was proposed to him by Clarke (G VII, 359f, 363, 367, 371f/L–C 20f, 25, 30, 36; cf. Rowe, 1975, Ch. 2, and Adams, 1978).

Leibniz presents his conceptual containment theory to Arnauld, however, not as a consequence of the Principle of Sufficient Reason, but as an analysis of the notion of truth. Apart from the theory, he says, "I do not know what truth is" (G II, 56/L–A 63). So, we still want an explanation of why he thought the theory is required for an understanding of the notion of *truth*. And the first question that will occur to most of us in this connection is, 'Why didn't Leibniz define the truth of propositions in terms of their *correspondence with facts*?' After all, as I noted at the end of Section II.c, Leibniz did believe that our concepts and true beliefs express a world of things distinct from us.

Some remarks of Michael Dummett (1981, p. 442) about Frege may shed some light on this aspect of Leibniz's thought:

> Frege, although a realist, did not believe in the correspondence theory of truth .... The truth of a (complete) sentence or of the thought which it expresses is not relational: there is no question of our having first to discover the state of affairs which the sentence is intended to describe, and then to compare the sentence with it to see whether or not it

corresponds; the sentence is simply true or false without qualifications. Facts, in Frege's ontology, are not further constituents of reality, . . . alongside objects, truth-values, concepts, relations, and functions. They are, rather, to be identified with true thoughts.

Much of what Dummett says here about Frege could also be said about Leibniz. Facts have no more independent a place in Leibniz's ontology than in Frege's. This world, for Leibniz, is the totality of (finite) things, not of facts. It is virtually defined as *aggregatum rerum finitarum*, "the aggregate of finite things" (G VII, 302/L 486). If reality is composed of things rather than of facts, it falls primarily to concepts rather than propositions to correspond with reality; for it is concepts that express (or fail to express) things. That may be why it was natural for Leibniz to think of a proposition as a complex of concepts or terms – a conception of the proposition for which there was also precedent in medieval logic. And given that what one is doing in framing or asserting a proposition is connecting concepts or terms in a certain way, the realist's demand that every truth should have "some foundation in the nature of things" (DM 8) naturally comes to expression as a demand that there should be "some real connection between" the terms of the proposition that is true (G VII, 300/L 226; cf. C 518f, G II, 56/L 267, L–A 63).

I think that what has just been said is, indeed, important for an appreciation of the conceptual framework in which Leibniz saw the issue, but the objection that was raised in terms of a correspondence theory of truth can still be raised in these new terms. At least for propositions about actual things, we may ask, why can't the real connection between the terms of the proposition be a containment of the predicate in the concrete thing to which the subject concept corresponds? Isn't that, indeed, the normal understanding of the Aristotelian formula "*praedicatum inest subjecto*" ("the predicate inheres in the subject") to which Leibniz so often appeals? Why must the predicate be contained in the *concept* of the subject?

The *Port Royal Logic*, which was largely Arnauld's work, seems to hold this view that the predicate of a true proposition is contained in the *concrete* subject, without implying that it is contained in the *concept* of the subject. It states, for example, that "it is true that lions are all animals, that is to say, that each one of the lions includes the idea of animal" (Arnauld and Nicole, 1965, II, 17). This statement presents an easy target for Leibniz. For how does an individual, concrete lion include the *idea* of animal? Presumably, lions do not conceive or understand the idea of animal. And we who do have the idea in the sense of conceiving or understanding it are not animals *because* we conceive or understand it; we also understand the idea of lion, but are not therefore lions. If lions are animals because the *idea* of animal is contained in something, the idea or concept of lion seems to be the likeliest container. And, indeed, the concept of lion does in the relevant sense include the concept of animal.

But, if lions are animals by virtue of *containing* something other than the *idea* of animal, what would it be? Two candidates come to mind. (1) Could it be the universal property, animality? Like other early modern philosophers, Leibniz was no Platonist about universals. The only universals he recognized were concepts and "possibilities in resemblances" (NE III, vi, 32). And since he thought that relations between substances exist only in the mind, he would have to say that a lion's resemblance to other animals is contained, strictly speaking, in the concept that a mind forms of lions by comparing them with other animals, rather than in the lion itself – though he would also say that it must be "expressed" by something in the lion itself. (2) There are, of course, concrete particular animals; and if identity can be regarded as a (degenerate) case of containment, we could say that each of the lions contains an animal – namely, itself. But Leibniz might well have thought this *too* degenerate a case of containment to satisfy his belief that predication should be explained in terms of containment. These considerations may also be at least part of the explanation of Arnauld's ready approval of Leibniz's conceptual containment theory of truth.

These are technicalities, however; and we ought not to lay too much weight on them. If Leibniz had wanted to avoid the conceptual containment theory, he could have said that (at least in some cases) the real connection between the terms of a proposition on which its truth depends is that its predicate corresponds with, or expresses, the concrete individual things that its subject concept corresponds with or expresses (or some of them, if the proposition is particular). Besides which, we shall see that the containment of predicates in concrete individual substances was not for Leibniz a rejected alternative to the conceptual containment theory of truth; there is a sense in which he was inclined to affirm it as a concomitant of the conceptual containment theory. In fact, I believe it was largely because of his views about the structure of concrete individual substances that he found it plausible to include all their predicates in their concepts.

II.e. *Conceptual Containment and Trans-World Identity*

Before taking up these considerations about the structure of individual substances, however, I want to point out an important flaw in the argument against trans-world identity from the conceptual containment theory of truth. Leibniz and Arnauld appear to have assumed that the denial of trans-world identity follows from Leibniz's theory of truth. This is a natural assumption to make, but it is incorrect. The two doctrines are logically independent: neither entails the other.

Let us note first that the denial of trans-world identity does not entail the full conceptual containment theory of truth. If each possible individual exists in only one possible world, I think it does follow that the concept of each possible individual contains (in some relevant sense) its whole history in the one possible world in which it occurs. But it does not follow that the concept of each possible individual contains an answer to the question whether that individual (actually) exists, or whether its possible world is actual. For there would be no inconsistency in denying trans-world identity (in agreement with Leibniz) while holding a voluntaristic doctrine of creation (in sharp disagreement with Leibniz). In such a voluntaristic doctrine it would be held that God could create an inferior world – indeed, any world at all – without doing anything conceptually inconsistent, and hence that no answer to the question which possible world is actual, or which possible creatures exist, is contained in any concepts. And, so far as I can see, that is perfectly compatible with the claim that each possible individual creature exists in exactly one possible world; but it is not consistent with Leibniz's conceptual containment theory of truth, which is explicitly applied to *all* propositions (including existential propositions).[10]

More important, for our present purpose, is the fact that the denial of trans-world identity does not follow from the conceptual containment theory of truth. I shall try to establish this fact by sketching a view in which both the conceptual containment theory and the trans-world identity of created individuals are affirmed. This view starts with Leibniz's thesis that individual concepts are complete concepts. Completeness can be understood here as a purely logical property of concepts. "An individual or complete term" may be defined as one to which it is "superfluous" to add any other term whatsoever (C 375/P 65; cf. Kauppi, 1960, pp. 168, 231). A complete concept, in other words, is one to which no predicate not already contained in it can be added (without conceptual inconsistency). Or what comes to the same thing, it is a concept which contains one member of every pair of mutually contradictory predicates. Such a concept obviously will contain all the predicates of anything that satisfies it.

Completeness is a very interesting property of concepts from a logical point of view. The part of logic that most interested Leibniz is the logic of predicates. He thought of all concepts as generated, principally by conjunction and negation, from predicates which are absolutely simple. In this generation, there are two types of concepts that are distinguished from all others as the extreme cases of simplicity and complexity. Both of the distinguished types of concepts are associated with distinguished features of Leibniz's metaphysics. At one extreme, the absolutely simple concepts are associated with God, who may be defined as the being that satisfies all the absolutely simple concepts (see A II, i 271f/L 167f). At the other extreme are the complete concepts, which are reached by conjoining more and more predicates until no new predicate can be added without some conceptual inconsistency; and these concepts are associated with individuals. Indeed, Leibniz uses the idea of completeness to provide a purely logical characterization of individual substance; he virtually *defines* an individual substance as a thing whose definitive concept is complete (DM 8; cf. C 375/P 65).

What are the predicates with respect to which individual concepts are complete? That is, what are the predicates with respect to which an individual concept, according to Leibniz, contains one number of every mutually contradictory pair? Does the individual concept of Leibniz, for example, contain eactly one of the predicates 'is more than one meter tall' and 'is not more than a meter tall'? Presumably not; for Leibniz was less than one meter tall in 1647 and more than one meter tall in 1670. For this reason, the predicates with respect to which Leibnizian individual concepts are complete will in general be indexed to times, as Benson Mates (1972, 108f) has pointed out. The individual concept of Leibniz will contain the predicates 'is not more than one meter tall in 1647' and 'is more than one meter tall in 1670' (which are perfectly consistent with each other).

As Mates also points out,

> But, if Leibniz considered [an individual] concept as containing a time-parameter, it is hard to see why he did not also build in a parameter relating to the different possible worlds. Thus the complete individual concept of Adam would not only indicate what attributes he had at what time, but what attributes at what time (or other time-like relation) *in what world*.

In this way, Mates suggests, Leibniz could accept trans-world identity "and yet could retain the principle so important to him that every individual concept involves all that would ever happen to a corresponding individual and reflects all the other individual concepts that are compossible with it." (Mates, 1970, p. 109). In other words, the individual concept of an individual that exists in more than one possible world could be complete – complete with respect to world-indexed predicates.

Completeness with respect to world-indexed predicates is not in itself enough for Leibniz's conceptual containment theory of truth. For, if the individual concept of Arnauld contains the predicates 'is married at some time in some possible world' and 'is always celibate in some possible world', the conceptual containment theory will require that the individual concept of Arnauld also contain an answer to the question whether the *actual* world is one in which Arnauld is married or celibate. And the world-indexed predicates do not directly tell us that. The solution to this problem would be easy for Leibniz, however. It is essentially the same as his solution to the problem of accounting for actual existence and actual non-existence in terms of conceptual containment. On the view suggested by Mates there would be no *internal* conceptual inconsistency in a possible world in which Arnauld marries, but Leibniz could still say that there would be an external conceptual inconsistency in God's actualizing it, because it is an inferior world. The predicate of *actually* marrying is, therefore, not contained in Arnauld's individual concept, which contains rather the predicate of *actual* perpetual celibacy because it contains a history of perpetual celibacy for Arnauld, not as the only possible history for him, but as the history that is indexed to the world that is, in fact, the best of all possible worlds.[11] Thus, actual celibacy and actual existence would be contained in Arnauld's individual concept in exactly the same way. Since an account of trans-world identity in terms of world-indexed concepts can satisfy the conceptual containment theory of truth in this way, the theory does not require a denial of trans-world identity.

III CAUSATION AND PERCEPTION AS INTERNAL TO SUBSTANCES

The richly deserved prestige of Russell's and Couturat's work on Leibniz set a fashion of trying to see as much of his philosophy as possible as derived from his logic and philosophy of logic. And much in Leibniz's own work, including his anti-semantical theory of truth as well as such papers as 'First Truths (C 518–523/L 267–270) seems to invite this approach. But the deductive order in which Leibniz presents his doctrines, especially in his writings of the mid-1680s, is misleading. An adequate foundation of the system – or even a fully adequate explanation of his having held it – cannot be found in his philosophy of logic. We have to turn to his nonformal views, not only about God and creation, but also about causation and perception, in order to understand the motivation for his philosophy of logic, which was as much shaped by them as they by it (Cf. Mondadori, 1977.) In particular, it is clear that he found in his views about causation and perception reasons for denying trans-world identity.

Trying to respond to the common-sense appeal of Arnauld's affirmation of trans-world identity, he wrote,

> It seems to us indeed that this square of marble brought from Genoa would have been exactly the same if it had been left there, because our senses let us judge only superficially, but at bottom because of the connection of things the whole universe with all its parts would be entirely different, and would have been another universe from the beginning, if the least thing in it happened otherwise than it does (G II, 42/L–A 46).

This passage does not present a fully articulated argument. But I take it Leibniz means that, if we combined our ordinary conceptions of individuality and individual identity with a correct understanding of the connections of things, we would not make the "superficial" judgments of counterfactual identity that common sense, in fact, makes, but would conclude that no actual individual would exist in a world that differed from the actual in respect of "the least thing".

In particular, Leibniz probably supposes, plausibly, that Arnauld's belief that he could have married and would still have been himself is based on the assumption,

(A) In some possible world, *w*, there exists a man whose characteristics and history are exactly like those of Antoine Arnauld in the actual world until some time at which, in *w*, the man marries.

Leibniz seems to be saying that this assumption is mistaken. Indeed, it appears that he is not only denying (A) but also committing himself to

(B) There are no two possible worlds that are qualitatively identical before a certain time but different thereafter,

and

(C) No individual substances in (qualitatively) different possible worlds are exactly alike, qualitatively, during an initial portion of their histories.

Leibniz seems still to be committed to (C) twenty-four years later in the *Theodicy*, where he says that among those possible worlds that "differ from the actual world only in one particular thing and its consequences" there are worlds

> . . . in which will be found, not exactly the same Sextus that you have seen [in the actual world] (that is impossible; he always carries with him what he will be), but similar Sextuses, who will have everything that you already know of the real Sextus, but *not everything that is already in him*, without being noticed, nor consequently everything that will yet happen to him (T 414, my italics).

It follows from (C) that an Arnauld who married, in another possible world, must have had there, from the very beginning of his existence, different characteristics from those of the actual Arnauld. And Leibniz evidently thinks those differences would be sufficient grounds for denying trans-world identity. This is not the place to discuss at length whether trans-world identity must, indeed, be denied if (C) is accepted and (A) is rejected. I think Leibniz's inference at this point is at least plausible. Our strongest intuitions of trans-world identity are of individual identity in alternative possible continuations of the history of a world – that is, in alternative possible continuations of exactly the same history of the world and of the individual until the time at which the continuations diverge. A convincing argument against the possibility of such alternative continuations of the same history would greatly weaken the intuitive support for trans-world identity, in my opinion.

But does Leibniz have a convincing argument against that possibility? He rejects it "because of the connection of things". But the "connection" referred to is presumably a feature of the *actual* world – the laws or order in accordance with which every actual event has always been prefigured in its causes and will always be recorded in its effects, as Leibniz believes. Why couldn't some other *logically* possible world, by virtue of a difference in its laws, or by a miracle, diverge qualitatively form the actual world only after a certain time? And why wouldn't that permit trans-world identity? Our next task is to investigate these questions, considering how individual substance is related, first, to laws of nature, and second, to the possibility of miracles, according to Leibniz. Then, in the third place, we shall consider the extent to which the "connection of things" referred to here is to be understood in terms of perception.

III.a. *Substance and Law*

To the question, why there couldn't be a possible world that is just like the actual world up to some time but diverges qualitatively thereafter, due to some difference in the *laws* of that possible world, I think Leibniz would reply that any difference in the laws of the universe would imply a qualitative difference between the possible worlds from the very beginning – indeed, that it would imply a qualitative difference in every individual substance from the very beginning. This is connected with Leibniz's views about trans-temporal identity. Almost immediately after the passage I have quoted about the marble from Genoa, Leibniz goes on to say,

Since then we suppose . . . that it is I who exist during the time AB and who am then in Paris, and that it is still I who exist during the time BC and who am then in Germany, there must necessarily be a reason that makes it true to say that we endure – that is to say, that I, who have been in Paris, am now in Germany. For if there is no [reason], one would have as much right to say it is someone else. It is true that my inner experience has convinced me *a posteriori* of this identicalness, but there must also be a reason *a priori* (G II, 43/L–A 46).

The unsupported assertion that there must be a reason *a priori* may strike us at first as merely a question-begging assertion that the reason must be conceptual containment; but I think that is an anachronistic reading of the text. Leibniz does not usually use '*a priori*' in the epistemological sense to which we have become accustomed since Kant, but in its older, original sense, in which *a reason a priori* is an argument from the causes rather than the effects of the fact to be proved. For Leibniz a proof *a priori* explains the fact proved; a proof *a posteriori* does not; that is the crucial difference between them.[12] Here Leibniz is saying that there must be a reason that explains why different states at different times should be said to belong to the same substance. Or, in other words, he is saying that there must be a reason that explains his own identity through time; he does not think that that identity can be an inexplicable primitive.

Its explanation is, indeed, to be sought in a conceptual containment, according to Leibniz.

Now it is impossible to find another reason [for this identicalness] except that my attributes of the preceding time and state are predicates of one and the same subject; *insunt eidem subjecto* [they inhere in the same subject]. Now what is it to say that the predicate is included in the subject, except that the concept of the predicate is included in some way in the concept of the subject? And because, from the time I began to be, it could truly be said about me that this or that would happen to me, it must be acknowledged that these predicates were laws included in the subject or in my complete concept, which makes what is called me, which is the foundation of the connection of all my different states and which God knew perfectly from all eternity (G II, 43/L–A 47).

We may wonder why Leibniz thought that the inclusion of earlier and later states in a single *concept* is peculiarly apt to explain a trans-temporal identity. Surely he did not suppose that a concept arbitrarily framed to include his own youth and the old age of Sir Isaac Newton could ground the trans-temporal identity of an individual as well as his own identity through time is grounded by his individual concept. What is it, then, about individual concepts that fits them to explain trans-temporal identity?

The answer to this question is to be found in the claim that "these predicates were *laws* included in the subject or in my complete concept, . . . which is the foundation of the connection of all my different states". I think the laws here are conceived of as *causal* laws, and as founding a connection of successive states not only in thought but also in reality. The greatest threat to our understanding of this part of Leibniz's philosophy is a prejudice that may have its historical roots in Locke and Hume – an assumption that conceptual connections cannot also be causal connections, that they impose necessity only on our thought, and cannot explain why anything occurs in reality. Leibniz certainly did not make this assumption, as is clear from his conceptual containment theory of truth and his use of it to explain the Principle of Sufficient Reason. Leibniz thought that conceptual connections are precisely what do ultimately explain the existence of all real things and the occurrence of all real events. He adhered to the Aristotelian notion of formal cause [cf. NE III, vi, 13 (G V, 288)], according to which a good definition is not just a structure of thought but expresses an essence which is a causal power in anything that satisfies the definition, making it act in accordance with the definition.

This view of the matter is more fully developed in Leibniz's later writings, with the aid of his metaphysical conception of "force". There are several texts in which Leibniz seems to explain the trans-temporal identity of a substance in terms of "the perpetual law which makes the sequence of perceptions that are allotted to it" and in which "its individuality consists" [T 291 (1710)]. "That there is a certain persisting law, which involves the future states of that which we conceive as the same: that is what I say constitutes the same substance" [G II, 264/L 535 (1704)].[13] Successive momentary states belong to the same individual substance, according to Leibniz, if and only if they are produced by the same persisting individual *law*. Now, if we think of causal laws as abstract objects, we might think that the same law could be responsible for the sequence of states of several different individual substances. But Leibniz evidently has something else in mind. As Louis Loeb (1981, pp. 317–319) has pointed out, Leibniz's criterion of trans-temporal identity depends on his denial of causal interaction between created substances. The succession of states of each individual substance or "monad" forms an isolated causal series, so that two states occurring at different times belong to the same individual substance if and only if the earlier is a cause of the later. "Perception is the operation proper to the soul, and the nexus of perceptions, according to which the subsequent ones are derived from the preceding ones, makes the unity of the perceiver" [G II, 372/L 599 (1709)].

The "law" or "nexus" that constitutes the unity through time of an individual substance is for Leibniz not merely a formula that describes the series of states. It is concretely realized in the substance – indeed in every momentary state of the substance – and can be called a "force" that produces all its subsequent states. The concept of force is given a central metaphysical role in Leibniz's later writings. "The very substance of things consists in the force to act and be acted on", and therefore a denial of "lasting force" to created things, reserving it to God alone, comes to the same thing as Spinoza's "doctrine of most evil repute", "that God is the very nature and substance of all things" (G IV, 508f/L 502; cf. RML 421, G II, 133/L–A 167, G IV, 594). The force that constitutes the very substance of things is the "primitive force" that Leibniz discusses in his correspondence with de Volder. Primitive forces are dispositional properties; they are "internal tendencies of simple substances, by which according to a certain law of their nature they pass from perception to perception" (G II, 275). But they can also be identified with laws; "the primitive force is as it were the law of the series" of successive states of an enduring thing (G II, 262/L 533). This "law of the series" is surely also the "persisting law" which constitutes the trans-temporal identity of an individual substance and which is included in its individual concept. Thus a part at least of the individual concept is more or less identified with a concrete force or tendency in the concrete substance.

We began this section of our investigation with the question, why there couldn't be a possible world that is just like the actual world up to some time but diverges qualitatively thereafter, due to some difference in the laws of that possible world. I think we are now in a position to see that Leibniz rejects an assumption that gives rise to that question. He rejects the assumption that causal laws and relations are imposed from the outside on individuals that are causally neutral in their individual nature (G IV, 507, 584/L 500; cf. Nobis). Causal powers, dispositions, or forces are primitive features of reality as Leibniz conceives of it; they are not to be analyzed in terms of causally neutral states plus extrinsic laws. Leibniz thinks that the laws are *internal* to individual substances, and that they are permanent and unchanging throughout the substance's history. In a possible world that had different laws from the actual world, therefore, every individual substance affected by the difference must always have been constituted by a "primitive force" that is qualitatively different from that which constitutes the corresponding individual in the actual world. In that respect, the individuals of that other possible world differ qualitatively from their actual counterparts from the very beginning of their existence. I think Leibniz could plausibly claim that that difference, in something so basic as the primitive force that constitutes them, is enough to rule out trans-world identity and make them distinct individuals.

We are also in a better position now to understand Leibniz's assumptions about the relation between concepts and reality. If the internal constitution of an individual substance contains a causal force or law which is the foundation of the trans-temporal identity and trans-world non-identity of that individual, surely the individual or definitive concept of that substance should be or include a representation of its internal causal structure. "It will be insisted that [Julius Caesar's] nature or form corresponds to" his complete concept – and Leibniz does not reject this assumption, though he combats an objection that is based on it (DM 13). He took for granted that the definitive concept (God's concept) of a substance would represent its internal causal structure, as his Aristotelian ideal of "real" and "causal" definition (DM 24, G IV 424f./L 293) and conception of formal causation would lead one to expect. That, I think, is why he saw no need to justify his inference from the claim that all the states of a substance follow from its individual concept to the claim that the substance by itself is a sufficient cause of all its states.[14] In the same way, we can see that the containment of Adam's sin in Adam's individual concept is not an *alternative* to the containment of the sin in Adam himself. For Adam's individual concept represents Adam's permanent internal causal structure, so that whatever sin follows from the one follows from the other also.

III.b. *Substance and Miracle*

What I have been suggesting is that Leibniz may have assumed that two possible worlds cannot be qualitatively identical before a certain time but then diverge by virtue of a difference in their laws, on the ground that, in his opinion, a difference in the laws implies a qualitative difference in the individual substances of the two worlds from the very beginning of their existence – a qualitative difference in the primitive forces of the substances, as he would have said in 1704. It remains to consider why he should not have thought that two possible worlds, with the same laws, could differ qualitatively only after a certain time in one of the worlds. Shouldn't an omnipotent God be able to suspend the operation of any created forces and work a miracle for which there was no preparation in the previous qualitative history of the world?

Admitting that God could, by a miracle, prevent Antoine Arnauld's primitive force from having its natural effect of lifelong celibacy would undermine the case we have been developing for denying Arnauld's trans-world identity with a possibly married man. It is easy enough to see why Leibniz would think that a man whose primitive force was qualitatively different from the beginning of his existence would not be identical with Arnauld; but why would a miracle preclude trans-world identity? Suppose there is a possible world, *w*, whose history is exactly like that of the actual world up to 1632. In *w*, of course, there is a man, *a*, born in 1612, who is qualitatively identical with the actual Arnauld for the first twenty years of his life. In particular, the primitive force that *a* has in *w* is qualitatively identical with the primitive force that Arnauld has in the actual world, and would have lifelong celibacy as its natural effect. But we are to suppose that in *w, a* marries in 1632 as a result of a miracle that prevents *a*'s primitive force from having its natural effect. The occurrence of one such miracle does not seem to me a plausible reason for denying that *a* could be, in *w*, numerically the same person as the actual Antoine Arnauld. Leibniz could support his rejection of trans-world identity more easily if he denied that there is any such possible world as *w*. But would his commitment to divine omnipotence permit him to deny it?

The tradition of Christian Aristotelianism, to which broadly speaking Leibniz's views on causality belong, had limited the dependence of causality on God's will more than many moderns would. Many medieval philosophers held, for example, that while God could choose to create or not to create fire, He could not have withheld the (causal) power of heating from fire if He did create it. These predecessors of Leibniz would generally have said, however, that once God has created a substance with certain causal powers, He could miraculously *suspend* those powers, or prevent them from operating, and could miraculously annihilate the creature, and could equally refrain from doing any such miracle. God's miraculous powers are so important to orthodox Christianity that it would have been difficult for Leibniz to deny them. We are not surprised to find him saying, in the *Preliminary Discourse to the Theodicy* (Section 3), that "the laws of nature are subject to the dispensation of the Legislator", and that "God can dispense creatures from the laws that he has prescribed for them, and produce in them what their nature does not hold, in doing a *Miracle*". We shall have to see in what sense he can have accepted this.

The possibility of miracle was a sensitive issue for Leibniz. It engages his attention in several sections of the *Discourse on Metaphysics*, where he explicitly asks "how it is possible for God sometimes to influence men or other substances by an extraordinary or miraculous occurrence, since it seems that nothing extraordinary or supernatural can happen to them, given that all their events are only consequences of their nature" (DM 16). His reply depends on a distinction between two types of law. Miracles "are always in conformity with the universal law of the general order" (DM 16), but not with certain "subordinate maxims, or laws of nature" (DM 17: cf. DM 6–7). The same view, which we must now examine more closely, is stated briefly when Leibniz introduces the subject of miracles in his correspondence with Arnauld (G II, 51/L–A 57).

The universal law of the general order is a law that governs the mutual relations of absolutely all substances and events in the universe. By definition, it is without exception; any law to which there is an exception is not universal in the relevant sense. Leibniz thought we can see that every possible world has an order of this sort. He appealed to geometrical examples. No matter what sort of line is drawn,

> . . . it is possible to find a concept or rule, or equation common to all the points of that line, in virtue of which [its] changes [of direction] must occur . . . . Thus it can be said that in whatever way God had created the world, it would always have been regular and in a certain general order. But God has chosen the one that is the most perfect, that is to say, that is the simplest in hypotheses and the richest in phenomena (DM 6, G VII, 312)

Leibniz regarded it as a trivially necessary truth, therefore, that the universe is ordered by a universal law to which there are no exceptions – not even miraculous exceptions.

The general order of the actual world, however, cannot be comprehended by any created mind (DM 16). (How that is consistent with the preeminent simplicity of the actual general order, Leibniz does not explain, so far as I am aware.) On the other hand, "everything that is called natural depends on less general maxims that creatures can comprehend" (DM 16). These less general maxims include the "laws of nature" that human science can in principle discover. And there can be exceptions to them; indeed, Leibniz believes that miracles are such exceptions.

For our present purpose, the most important question here is how these two types of law are related to the primitive force that constitutes a created substance. For if the universal law of the general order is incorporated in the primitive force of a particular substance to which a miracle is not going to happen, that substance would have to have had a qualitatively different primitive force if a miracle were going to happen to it, and two possible worlds could no more differ with respect to miracles than with respect to laws without differing qualitatively from the beginning. But if it is only the laws of nature that are incorporated in the primitive forces of individual substances, then two possible histories of a substance could be qualitatively the same up to a certain time, proceeding from the same primitive force, and could then diverge by a miracle, which would require an interference with the operation of the primitive force in one of the histories.

So far as I know, Leibniz does not directly answer this question about primitive forces. The conception of primitive force is not even fully developed (nor the term 'primitive force' employed) in the *Discourse on Metaphysics*. But he does propose there a distinction between the "essence" and the "nature" of an individual substance, corresponding to the distinction between the two types of law.

> That could be called our essence or idea which includes all that we express, and as it expresses our union with God himself, it has no limits and nothing goes beyond it. But that which is limited in us can be called our nature or our power [*puissance*], and in this regard what goes beyond the natures of all created substances is supernatural (DM 16).

Two interpretations of this passage are possible. (1) It is the "nature or power" that here corresponds to the "primitive force" of Leibniz's later writings. It is a causal force in the concrete substance, but the "essence" is only an idea in the mind of God. It is only the subordinate laws of nature that are incorporated into concrete substances. (2) It is the "essence" that here corresponds to the "primitive force" of the later writings. It is not only an idea in the mind of God, but also a causal force in the concrete substance, which thus incorporates the universal law of the general order. I believe the second interpretation is to be preferred, at least as an understanding of the predominant tendency of Leibniz's thought; but there is a case to be made for the first interpretation, and we will begin with it.

One might try to argue for the first interpretation on the ground that, for Leibniz, later states of a free creature (Caesar's crossing the Rubicon, for example) are only contingently connected with their earlier (DM 13; cf. Gr 387). But this is a weak argument. For Leibniz also holds that actions of a free creature are only contingently connected with the creature's complete concept, but at the same time denies that there is a possible world in which a creature satisfying Caesar's complete concept turns back from the Rubicon.

So, why couldn't he also deny that there is a possible world in which the contingent connection between Caesar's previous states and his action fails? The contingency in both cases can be understood in terms of the non-existence of a finite proof, which is not enough to establish a possible world (as discussed in Section I, above).

The strongest evidence I have found for the first interpretation is in a letter of April 1687 to Arnauld, a letter which surely must have left Arnauld thinking that Leibniz believed in miracles and helps of grace ("extraordinary cooperations of God") that are not produced by anything in creatures. I am inclined, in the end, to think that Leibniz meant to be (misleadingly) non-committal, asserting nothing about extraordinary cooperation, when he wrote in this letter, "Setting aside extraordinary cooperation, [God's] ordinary cooperation consists only in conserving the substance itself in conformity with its previous state and the changes that [that state] holds" (G II, 91f./L–A 115). But a reader might be pardoned for taking Leibniz to mean that in metaphysical rigor, God does something more than conserve a substance when He makes it an object of His extraordinary cooperation. And a similar implication might easily be drawn when Leibniz, in the same letter, asserting that "the actions of minds change nothing at all in the nature of bodies, nor bodies in that of minds", adds that "God changes nothing there [i.e. in their nature] on their occasion, except when he does a miracle" (G II, 93/L–A 117). But what is especially striking about the treatment of miracle in the letter is that Leibniz offers an explicit criterion of the miraculous:

> I admit that the authors of occasional causes will be able to give another definition of the term, but it seems that according to usage a miracle differs internally and by the substance of the act from an ordinary action, and not by an external accident of frequent repetition; and that strictly speaking God works a miracle when he does a thing that surpasses the forces that he has given to creatures and conserves in them (G II, 93/L–A 116; cf. G VII, 366/L–C 30).

In reading this definition, we should bear in mind that Leibniz and Arnauld attacked occasionalism from quite different directions. Leibniz repeatedly accused the occasionalists of implying that all events in nature are miraculous (e.g. in G II, 57f./L–A 65). Occasionalists would not admit this; Malebranche held that most events in nature are not miraculous because God produces them not by particular but by general volitions (for instance, not by willing that I see red on this occasion, but by willing that all human minds shall see red when their bodies are affected in a certain way, in which man happens now to be affected.) Leibniz here suggests that this leaves the occasionalists with too superficial a conception of miracle. His comments accurately reflect a fundamental difference between him and Malebranche, which is that Leibniz recognizes created substances as true causes, whereas Malebranche accepts only God as a true cause. And Leibniz proposes a criterion of the miraculous that does leave Malebranche with too many miracles.

It threatens to leave Leibniz himself with too few miracles, however. This is a serious matter in his correspondence with Arnauld, who already feared that Malebranche would end up, not with too many miracles, but with none, or with miracles inadequately distinguished from God's ordinary activity (RML 219–222). Leibniz will end up with no miracles, by his own criterion, unless he thinks that God sometimes does to creatures things that "surpass the forces that he has given to" them. This letter clearly implies that Leibniz did think that, and it was at best disingenuous if he really thought that there is in each created substance a causal structure that produces *all* its states, miraculous as well as ordinary.

Yet, I think there is a good case for supposing that Leibniz was disingenuous in just this way, and that by 'forces' in this letter, and 'nature or power' in DM 16, he means only a part of the primitive force or total causal structure of a created substance. This is strongly suggested by DM 16, where he says,

> If we include in our nature everything that it expresses, nothing is supernatural to it, for it extends to everything, since an effect always expresses its cause and God is the true cause of substances. But as that which our nature expresses more perfectly belongs to it in a special way, since it is in that that its power consists, and since [its power] is limited, as I have just explained, there are plenty of things that surpass the forces of our nature, and even those of all limited natures. Consequently, in order to speak more clearly, I say that miracles and extraordinary cooperations of God have this peculiarity, and that they could not be foreseen by the reasoning of any created mind, however enlightened it might be, because the distinct understanding of the general order surpasses them all.

There are three points to be noted in this passage. (1) The "effect" of which Leibniz says here that it expresses God as its cause is surely not just an idea in God's mind, but a concrete structure in the substances of which "God is the true cause". This agrees with the fact that Leibniz is prepared to say, not merely of an individual *concept*, but of "each individual substance", that in expressing the whole universe "it expresses also the aforesaid miracles" (G II, 40f/L–A 44). The nature construed as including everything expressed by the substance is clearly the same as "our essence ... which includes all that we express", and can be identified with what Leibniz later calls "primitive force", if this talk about substances expressing even miracles is to be believed.

(2) The "power" of our nature consists only in "that which our nature

expresses more perfectly". What Leibniz "has just explained" (in DM 15) about the limited power of finite substances is that "practical" talk about such substances "acting" on each other, and thus "exercising power", can be understood in terms of differences in how well each substance "expresses" what is going on. The forms in which this recurrent Leibnizian theme is developed in DM 15, however, seem to me to shed less light on DM 16 than does Leibniz's later statement that a substance is "*active* insofar as what is distinctly known in it serves to give a reason for what happens in another, and *passive* insofar as the reason for what happens in it is found in what is distinctly known in another" (Mon 52). Now if something happens in a substance miraculously, in a way that does not agree with the laws of nature that finite minds can understand, the reason for that event will presumably be vastly less distinctly "known in" the substance (that is, it can much less easily be read off the previous states of the substance) than if it were produced in accordance with those laws of nature. And this will be a reason for saying that in the miraculous event the substance is not active, nor exercising any power, and that the event exceeds anything it has the "power" to produce – even though, in metaphysical rigor, the miraculous states of the substance are produced by the substance itself in accordance with the universal law of the general order that is included in the "essence" that God has given the substance. This interpretation of DM 16 is supported by the fact that in the summary of DM 16 that Leibniz sent to Arnauld in February 1686, " our nature", whose forces are surpassed by "the extraordinary cooperation of God", is equated with "our distinct expression, which is finite and follows certain subordinate maxims" (G II, 13/L–A 5f). Similarly, when Leibniz implies to Arnauld that God sometimes "does a thing that surpasses the forces that he has given to creatures", he may have meant only that God sometimes does a thing that is not distinctly expressed by creatures because it does not agree with the laws of nature that finite minds can grasp – though he can hardly have expected Arnauld to understand him in this sense. And, when Leibniz's letter carries a strong suggestion that in miracles God changes something in the "nature" of created substances, he may be thinking only of changes in (or exceptions to) the subordinate laws that are intelligible to us.

(3) The inability of created minds to foresee miracles and extraordinary cooperations of God, or to understand the order on which they depend, plays a central role in this interpretation. It is a point that is often emphasized in Leibniz's discussions of miracle (RML 203, C 508, G III, 353). The reason why we cannot foresee miracles is that the order on which they depend is too complex for us to grasp. Accordingly, Leibniz sometimes indicates that

miracles are distinguished from non-miraculous events by their *not* having *simple* explanations. In the copy he kept of his April 1687 letter to Arnauld, immediately after the definition of miracle that I have quoted, Leibniz says that if God decreed that any body swung in a circle in a sling should *always* continue in circular motion when released from the sling, "without being pushed or restrained" by any other body, this would be a miracle even though it happened regularly, "since this motion cannot be explained by something simpler" (G II, 93/L–A 116).[15] The simplicity of explanations plays a particularly decisive part in the discussion of miracles in a draft Leibniz wrote in 1702 for a reply to François Lamy, an occasionalist critic of the system of pre-established harmony. Here Leibniz repeats his objection that, if God is the sole cause of natural events and produces natural events whenever He acts according to general laws, then natural events will differ from miracles only by an external accident of frequent repetition; "the natural and the miraculous would not differ in themselves, but only by an extrinsic denomination taken from the antecedents and consequences. For what is preceded and followed by [events] like it would be natural, and what is not such would be a miracle". But in going on to explain that "there is an essential difference between the natural and the miraculous," what Leibniz emphasizes is that "not every sort of rule or law is fit to make a law of nature". For example, if God established a rule that a planet should always move in an elliptical orbit, without any mechanical action of other bodies to keep its path elliptical,

> God would have established a perpetual miracle, and . . . it could not be said that the Planet went that way in virtue of its nature or following natural laws, since it is not possible . . . to give a reason for such a phenomenon . . . because that [elliptical] motion is composite, hence the reason must come from simpler motions.

Whereas, if the composite motion is caused by simpler motions of other bodies, "then it will be in these simpler reasons that the nature of the thing will consist" (G IV, 587f). In short, the essential, internal distinction of the miraculous from the natural seems here to depend on the *simplicity* of the explanations, or the system of laws, in accordance with which natural events are produced.

I would not claim that this conception of the difference between natural and miraculous events was fully or adequately thought through by Leibniz, nor that it is perfectly consistent with all his utterances. One problem that he does not notice (so far as I am aware) is that the free actions of creatures appear to be miraculous by the criteria of DM 16. For Leibniz certainly thinks that free actions "could not be foreseen by the reasoning of any created mind" because of the complexity of the order on which they depend.

Despite such problems, I think the tenor of Leibniz's philosophy in general favors the view that the whole universal law of the general order is incorporated in the primitive force or concrete causal structure of each created substance. One further argument for this interpretation is the following. According to Leibniz, we – not only our concepts but also we concretely – express our whole universe, including all the miracles that occur in it. How do we express it? By virtue of our primitive force, from which always proceed perceptions of the whole universe. If our primitive force incorporated only the laws of nature and not the whole law of the general order of the universe, then the divine "clock-maker" would have to be adjusting the pre-established harmony at every instant in order to keep us perceiving the miracles that occur in the universe – a consequence that would certainly be unpalatable to Leibniz.

Leibniz himself points out a similar problem, in a curious section of the *Theodicy* where he suggests that the miracle of Cana (John 2: 1–11) may have been worked "by the ministry of some invisible substances, such as the Angels, as the Rev. Father de Malebranche also holds",[16] rather than by a direct divine intervention in nature.

> If the changing of water into wine at Cana were a miracle of the first rank, God would have changed thereby the whole course of the universe, because of the connection of bodies; or else he would have been obliged to obstruct that connection, miraculously again, and to make the bodies not involved in the miracle to act as if no miracle had happened; and after the miracle was over, he would have had to restore everything, even in the bodies that were involved, to the state it would have come to without the miracle, after which everything would be returned to its original channel. Thus this miracle demanded more than it seemed to (T 249).

This is one of a group of texts from the last years of Leibniz's life, in which he speaks somewhat differently about miracles but draws the line between divine and creaturely action exactly where (in my opinion) his fundamental principles require it to be drawn. It is only "miracles of the first rank" (cf. G VII, 418/L–C 93) that require direct divine action. "The changing of water into wine would be a miracle of that [inferior] species" that angels might be naturally able to do. "But creation, incarnation, and some other actions of God surpass all force of creatures, and are truly miracles, or even mysteries" (T 249). Let us consider the principal actions that are reserved to God on this view.

One is creation (which presumably included conserving creatures in existence). Leibniz nowhere suggests that any creature or finite being can cause the existence of another creature. God is the sole cause of the actual existence of things. But given the existence of a creature, produced by God, the creature itself produces *all* its own *states*; nothing more than the existence of the creature has to be supplied by God. That is what Leibniz generally says when he is not thinking about miracles, and it would be good if his conception of miracles could be made to agree with it.

Incarnation is a special case. Obviously, nothing would be an incarnation of God unless God participated directly in it. But incarnation is not a matter of causing states of a created substance; rather, it is a union of the divine nature with a human nature. There is another text in which Leibniz, proposing to account for most apparent miracles as produced through the ministry of angels, says,

> Thus there would be no absolute miracles in these things, except the creation of things and the union of God with the first creature [that is, with the human nature of Jesus Christ]; the rest would happen by the laws of nature, but by laws that would apply to the first creature and the angels. Through the union, moreover, God would not change the laws of the first creature, since he would have no need of this change This union would therefore change nothing in the phenomena, even though the state of union differs internally from non-union (RML 413).

A third action that is reserved to God alone is annihilation. In his correspondence with Clarke, Leibniz introduces his distinction between "miracles of an inferior sort, which an angel can work", and "miracles which none but God can work . . . , of which kind are creating and annihilating" (G VII, 377/L–C 43).[17] Leibniz may have thought that God never, in fact, annihilates a created substance. In the text quoted in the previous paragraph, he says, "I prefer to say that God has not acted miraculously except in the Creation and the incarnation" (RML 413). But Leibniz certainly believed that God *can* annihilate created substances; that is the reverse of the coin of his frequently repeated orthodox belief that created substances are continually conserved in existence by God (DM 14, 28, 30; Gr 381). "*God is the conserver of all things*. That is, things are not only produced by God when they begin to exist, but they also would not continue to exist unless a certain continuous action of God terminated in them, on the cessation of which they themselves would also cease" (Gr 307). This does not mean that created substances are not the causes of all their own states. "I grant in some way . . . that God continually produces all that is real in creatures. But I hold that in doing it he also continuously produces or conserves in us that energy or activity which according to me constitutes the nature of substance and the source of its modifications. And so I do not grant that God alone acts in substances, or alone causes their changes" (G IV, 588f).

What particularly concerns us here is *how* God can annihilate created substances, according to Leibniz. There are two possibilities. (1) No matter what individual substances, and the world, have been like hitherto, God can at any time withdraw His conserving action from any one created substance, or from all, thereby annihilating them, causing them to lapse into non-existence. (2) God does not act in time, although He causes the existence of things in time. The existence of created substances at any time (and at all times) is caused by a single timeless act of God, by which He actualizes the possible world that He has chosen. He can annihilate a created substance, in that He can, in His single creative act, cause the existence of a substance to have an end in time; but if He does so, that end must, by the very nature of substance, have been expressed and perceived in the substance from the very beginning of its existence, so that a substance that will not be annihilated must always have been qualitatively different from what any substance that God annihilated could have been.

Of these two interpretations, the second is consistent with propositions (B) and (C), and the first is not. The second is therefore more favorable to the rigorous exclusion of trans-world identity, insofar as that exclusion is supported by (B) and (C). Part of the second interpretation (God's acting timelessly to cause the existence of things in time) is certainly in agreement with Leibniz's views. I do not know that he ever addressed the issue of whether a creature annihilated must always have been qualitatively different from any possible creature that would not be annihilated. But I think an affirmative answer (as in the second interpretation) would agree best with the general structure of his philosophy, although it would be the more daring answer, in that it would impose an additional (presumably conceptual) limitation on the scope of God's omnipotence. Thus, while there is certainly a tension in this part of Leibniz's thought, we have not been compelled to conclude that Leibniz was committed to a treatment of the miraculous that is inconsistent with (B) and (C).

III.c. *Perception and Relations*

There is a similar tension in Leibniz's thought about relations, and here I think we will have to conclude that he was less successful in tying an individual substance's predicates as tightly to its internal structure as to its concept, and hence less successful in grounding his denial of transworld identity. Issues about the possible existence of an actual substance in a world in which some of its actual relations would fail are not prominent in Leibniz's discussions of trans-world identity, which are much more concerned with whether there are possible worlds in which actual creatures act or suffer otherwise than they do in the actual world. A focus on the latter question is easily explained by Leibniz's (and Arnauld's) religious and ethical interests. Nonetheless, we can ask whether according to Leibniz there is a possible world in which Adolf Hitler, for example, does not exist, or is less malicious than in the actual world, but in which there is a man both numerically and qualitatively identical with the actual Antoine Arnauld (except, of course, for the merely "extrinsic denominations" that Arnauld actually has by virtue of his [distant] relations to Hitler).

This question can be divided into two parts: (1) Are there possible worlds in which there is a man *qualitatively* identical with the actual Antoine Arnauld, but no one qualitatively identical with the actual Adolf Hitler, except for purely relational properties of the individuals? (2) If there are such possible worlds, does at least one of them contain someone *numerically* identical with the actual Antoine Arnauld? Leibniz has little to say about the second of these questions, whether differences in purely relational properties would always preclude trans-world identity. His general denials of trans-world identity seem to commit him to answer the second question negatively if he answers the first affirmatively; but it might be difficult to make a negative answer to the second question plausible. In what follows, we will focus on the first question. Leibniz has a lot to say that bears on it – but perhaps not consistently.

A negative answer to it seems to be implied in the passage about the block of marble from Genoa. For Leibniz says there not merely that the *stone* would have been qualitatively different from the beginning if *it* were going to stay in Genoa, but that "because of the connection of things the whole universe *with all its parts* would be entirely different, and would have been another universe from the beginning, if *the least thing in it* happened otherwise than it does" (G II, 42/L–A 46, my italics). So, if Leibniz is denying that there are possible worlds in which the block of marble, or its counterpart, is qualitatively the same as in the actual world until the time of its actual transportation, but then stays in Genoa, he is equally denying that there are possible worlds in which one substance is internally different from its actual counterpart but other substances start out only relationally different from their actual counterparts. What is "the connection of things" which precludes such possible worlds? We have seen that for Leibniz the primitive force of each individual substance provides the connection that precludes (if anything does) pairs of possible individual histories that differ internally after but not before a certain time. Is there something in Leibniz's conception of things that might play an analogous role in precluding pairs of possible individual histories that differ relationally but not internally?

The obvious candidate for this role is *perception*. Leibniz held that each substance always perceives everything that ever occurs in the whole universe. Throughout his life, Antoine Arnauld, for instance, perceived the existence of Adolf Hitler and every detail of Hitler's career, though of course Arnauld was not conscious of these perceptions. It seems to follow that if a possible world in which Hitler does not exist contained a man who is internally just like the actual Arnauld, that man would, in that world, "perceive" some things that do not exist. Perhaps this inference could be resisted by holding that any sufficiently complex internal state of a perceiving substance could be placed in one-to-one correspondence with *any* sufficiently complex possible world, with or without Hitler, and would therefore constitute an accurate perception of whatever world the substance existed in. Leibniz does often explain perception in terms of one-to-one correspondence, but I doubt that he would accept this trivialization of his concept of perception.

If he does not accept it, his alternatives are to admit that there are possible worlds in which some substance fails to perceive the whole world just as it is (fails to perceive it as not containing Hitler, for instance), or to deny that there are possible worlds that differ qualitatively solely with respect to the existence or internal properties of some but not all of their substances. There is much to incline Leibniz to the second alternative. He thought that it follows from the nature of substance (and, indeed, from the nature of truth) that there are no purely extrinsic denominations, and hence that every substance perceives the whole universe (C 520f/L 268f). Perhaps he thought that this precludes possible worlds in which any substance fails to perceive the whole world just as it is. (Of course, he recognized that there are false beliefs, but he thought that the beings that have them also, unconsciously, perceive the world exactly as it is.) On this view, the whole general order of the universe would be incorporated in the concrete substance as it is included in the substance's complete individual concept. The relational laws of the general order would be incorporated in the substance's perceptions as the laws of intrasubstantial causation are incorporated in its primitive force. And all of the substance's states and relations would follow from the force and the perceptions it has at any given time, so that there would be no possibility of a predicate's being contained in the individual concept and not being contained in the concrete substance.

This has the very important advantage for Leibniz of providing a non-trivial sense in which possible individuals can fail to be *com*possible with each other. In thinking about the problem of evil in more or less Leibnizian terms, it is natural to ask, 'Why couldn't God have improved the world by replacing Hitler with a morally better individual, while leaving the excellences of more excellent individuals (such as Arnauld) unchanged?' One answer that Leibniz seems to want to give to this sort of question is that God could not do that because Arnauld (with his actual excellences) is not compossible with any such Hitler-substitute. But how are we to understand this incompossibility? Suppose that Hitler is replaced by a more virtuous substitute in some possible worlds that contain a man whose internal, non-relational qualities and states are exactly the same as Arnauld has. Of course if Arnauld's individual concept includes the whole history of the actual world, Arnauld himself will be absent, by definition, from any possible world containing such a substitute in place of Hitler. In that case, there will be a trivial sense in which Arnauld is incompossible with a Hitler-substitute. But this sort of incompossibility does not answer the question, why God could not have made a better world by replacing particularly vicious or unhappy creatures with better and happier ones while replacing his more excellent creatures with counterparts[18] that would have exactly the same excellences internally and would differ only in their extrinsic denominations. To this question, I suppose, Leibniz could reply that in the proposed replacement world the creatures that would be internally just like the excellent actual creatures, but in a world that is not just like the actual world, would thereby lack an important (relational) excellence of all actual creatures, in that they would not perceive their whole world just as it is. But in this reply it is not incompossibility but harmony that is called in to solve the fundamental problem.

Perhaps indeed the appeal to incompossibility just is a covert appeal to considerations of harmony. But it would be something more if Leibniz thought that the internal states of possible substances are perceptions that are conceptually connected with the existence and states of other substances, so that no individual in any possible world has exactly the same history of internal states as any individual in any other possible world. Then the internal histories of substances in one world would be incompossible with the internal histories of substances in any other world; and this incompossibility would put a constraint on God's options that would be quite independent of the constraints of harmony.

There is another, more speculative line of argument that might have influenced Leibniz to think that the same internal history of a substance cannot occur in more than one possible world. For suppose that is not the case. Then possible internal histories of substances will be independent building blocks of possible worlds, the same history occurring in different worlds. And the structure of each possible world will be of the form, 'There exists a substance with internal history H_1 *and* there exists a substance with internal history H_2 *and* so forth'. The coexistence of the substances, and their other relational properties, will then be conceptual consequences of this conjunction of existences; for Leibniz certainly did not think there could be relations between substances that are not conceptually contained in the internal properties of those substances. There are no primitive spatial relations, for example, between substances in Leibnizian possible worlds. If the internal histories of substances are logically independent, however, the relational properties of any one substance will not follow logically from *its* internal properties (its perceptions) alone, but rather from a conjunction asserting the existence and internal properties of all the substances involved in the relations. In this way, existence will be conceptually prior to coexistence.

There are indications, however, that Leibniz seriously entertained the idea of analyzing existence in terms of coexistence. "The Existent could be defined", he wrote, "as what is compatible with more things than anything else that is incompatible with it" (C 360/P 51), and "Nothing else is explicable in existence than entering into the most perfect series of things" (C 9). Now these are clearly suggested as analyses of the notion of *actual* existence. Strictly speaking, they could presuppose the notion of *existing in a possible world* as a primitive in terms of which the notions of being compatible with things and entering into a series of things are to be analyzed in their turn. But Leibniz would have a more elegant and radical analysis of existence if he did not presuppose the notion of existing in a world in this way; and I suspect he did not mean to presuppose it. He can avoid presupposing it if he does not think of possible worlds as composed of logically independent internal histories. For then he could say that to perceive a certain possible world *is*, by definition, to enter into that world; and he could define actual existence as entering into (i.e. perceiving) the best possible world. This complete reduction of the notion of existence would be achieved by making the perceptions (internal states) of each substance logically inseparable from its whole world.

Despite these reasons for ascribing to Leibniz the view that God could not have actualized an individual substance without actualizing the whole possible world perceived by that substance, there is at least one text in which Leibniz explicitly contradicts it; and indeed, it is hardly to be reconciled with his conception of the pre-established harmony. Every student of Leibniz is familiar with his statement that the series of future perceptions that follows from his present nature "would not fail, and would happen to me just the same, if everything that is outside of me were destroyed, provided that there remained only God and I" (DM 14; cf. G I, 382, IV, 484/L 457). This strongly suggests that Leibniz thought that without creating any other substance, God could create a substance with qualitatively the same primitive force and series of perceptions as Leibniz actually has. But this is not the strongest evidence, for Leibniz might conceivably be using a counterfactual conditional with an impossible antecedent, as a rhetorical device to express the mutual causal independence of created substances.

More significant is the fact that in explaining the pre-established harmony Leibniz repeatedly says that only the action of God can cause created substances to correspond to each other's perceptions. "There is nothing but God . . . to be the cause of this correspondence of their phenomena, and to make what is particular to one to be public to all" (DM 14). The "ideal influence of one Monad on another . . . cannot have its effect except by the intervention of God . . . For since one created Monad cannot have a physical influence on the interior of another, it is only by this means that one can have any dependence on another" (Mon 51). Leibniz presents the difficulty of achieving this correspondence as a proof of the infinite knowledge and power of God.

> For since each of these Souls expresses in its way what goes on outside, and cannot have this from any influence of other particular Beings, or rather must draw this expression from the resources of its own nature, each one must necessarily have received this nature (or this internal reason of the expression of what is outside) from a universal cause on which these Beings all depend and which makes one agree and correspond perfectly with another; which is not possible without an infinite knowledge and power, and by a great artifice, especially with respect to the spontaneous agreement of the machine with the actions of the reasonable soul, [so that Bayle] doubted, as it were, whether it did not surpass all possible wisdom, saying that the wisdom of God did not seem to him too great for such an effect, and recognized at least that the feeble conceptions that we can have of the divine perfection had never been put in such high relief (NE IV, x, 10; cf. G IV, 578, VII, 411/L–C 83f).

But the harmony or correspondence of the perceptions of created substances with each other could hardly be such a powerful proof of God's infinite wisdom and power if no creator could have created a world that lacked such correspondence. The use of the pre-established harmony to prove the greatness of the creator seems therefore to presuppose that, among the possible worlds that God could have actualized, were some in which created substances fail to correspond with each other's perceptions.

Particularly interesting is a statement that Leibniz made in 1698, that "God was able to give each substance its phenomena independent of those of others, but in that way he would have made, so to speak, as many worlds without connection as there are substances" (G IV, 519). The first thing that is noteworthy here is that Leibniz says explicitly that God could have created substances that would not have corresponded with each other's perceptions. And I believe that this was his consciously held opinion, at least from 1686 to the end of his life. The evidence to the contrary is outweighed by the statements, scattered throughout his mature writings, describing the pre-established harmony as explainable only by the action of God.

The other thing that is noteworthy in this text is the suggestion that a collection of substances that did not correspond with each other's perceptions would not be sufficiently connected to constitute a single "world". Perhaps Leibniz did not think that every state of affairs or set of creatures that God could have actualized should be counted as a "possible world" (or part of a possible world); perhaps only complete sets of perfectly harmonious created substances are to count as possible worlds. Certainly Leibniz was so convinced of the supreme worth of harmony that he would not have expected God seriously to consider actualizing any state of affairs that did not satisfy this constraint. And I do not know of any passage in which Leibniz himself speaks of "possible worlds" that do not satisfy it. If Leibniz assumed that the notion of a "possible world" is subject to this requirement of harmony, he would have a clear reason to hold that substances with different relational properties in different possible worlds must have had different internal properties (different perceptions) from the very beginning; and this would help to explain and support his denial of trans-world identity. But it would also move him still farther from the conception of possible worlds in terms of which the question of trans-world identity has been raised in our time.

III.d. *Conclusion*

In pursuit of a rationale for Leibniz's denial of trans-world identity, I have sketched many ways in which Leibniz might have thought that all the states and relations of every possible created substance are contained in the primitive force and the perceptions of substance at all times just as rigorously as they are contained in the complete individual concept of the substance. I do not claim to have presented a complete explanation of the rigor that would be involved. The sketch works out well for Leibniz's views about causality within the monad, and I think it is consistent with his treatment of miracles. But it runs into problems in his treatment of inter-monadic relations, with respect to which I cannot now see how to reconcile all of his intentions.

Even on his own terms, I think Leibniz is left without a completely satisfying justification for his denial of trans-world identity. He seems to have thought that it was solidly grounded in his conceptual containment theory of truth. But I have argued that it does not follow from that theory even if the theory is correct.[19]

NOTES

[1] See the Bibliography of this volume for the abbreviations by which works of Leibniz are cited. Entries separated by a slash refer to the original and a published English translation of the same passage; but I take responsibility for the translation of all quotations, except as noted. All works are cited by page number, except as noted in the Bibliography.

[2] Notably from Mates (1972), Mondadori (1975) and (1977), and Wilson (1979). Here, as elsewhere in both text and notes, authors other than Leibniz are cited by surname, an index number (if needed to distinguish two or more publications of the same author), and page numbers (where relevant, with exceptions noted in the Bibliography). Consult the Bibliography of this volume for other details.

[3] I have discussed this reply at some length in Adams (1979).

[4] See Kauppi (1960, pp. 254–256) for a comparison of Leibnizian and modern conceptions of classes.

[5] The interpreter to whom I owe the most on this subject is Kauppi. The whole of her excellent book is a study on the intensional and extensional approaches in Leibniz's logic. She seems to me correct in her reply (pp. 210, 251f.) to the charge of Couturat (1901, pp. 19–32) that Leibniz ought to have preferred the extensional interpretation.

[6] G VII, 244/P 141 seems to present an exception, as noticed by Kauppi (1960, p. 43).

[7] Cf. Arnauld and Nicole (1965, II. 3), where it is held that singular propositions are more like universal than particular propositions in their logical properties.

[8] For Leibniz's treatment of universal negatives in terms of the predicate being excluded by the subject, see G VII, 208/P 112.

[9] The connection of this solution with the doctrine that individual concepts are complete is noted by Kauppi (1960, p. 213).

[10] Some scholars have thought that Leibniz meant existential propositions to be an exception to the theory; but I believe that is an untenable interpretation, as I have argued in Adams (1977, 30f).

[11] This solution does require a revision of Arnauld's definition (or partial definition) of 'individual concept' (G II, 30f/L–A 30, quoted above). For in this solution the predicate of perpetual celibacy (i.e. *actual* perpetual celibacy) is supposed to be contained in Arnauld's individual concept although there are possible worlds in which Arnauld would not be perpetually celibate and would still be himself. What remains true in this solution is that nothing is contained in Arnauld's individual concept, *independently of the question which world would be chosen by God as the best possible*, except predicates such that Arnauld would not be himself if they were not true of him.

[12] The old sense of '*a priori*' and '*a posteriori*' is still found in the *Port Royal Logic*, which speaks of matters in which our mind may be "capable of finding and comprehending the truth, either in proving the effects by the causes, which is called demonstrating *a priori*, or inversely in demonstrating the causes by the effects, which is called proving *a posteriori*" (Arnauld and Nicole, 1965, IV, 1). The conception is clear, although the authors go on immediately to remark that "it is necessary to extend these terms a little in order to reduce all sorts of demonstrations to them". There are many indications in Leibniz's writings that he understood '*a priori*' in the older sense that I have indicated. He says, "Proof *a priori* or Apodeixis is explanation of the truth" (C 408). He equates knowledge *a priori* with knowledge through causes (C 272). Reason "would make known the reality [of a definition] *a priori* in presenting the cause or possible generation of the thing defined" (NE III, iii, 15). Most telling, perhaps is the fact that Leibniz treated indirect proof (*reductio ad absurdum*) as a case of proof *a posteriori* – presumably because it does not explain *why* the proposition proved is true (C 154, which may have been written in 1677). Nonetheless, there are passages in Leibniz which could easily be read in the newer sense in which '*a priori*' means simply 'non-empirical'. He regularly connects the *a posteriori* with experience, and can contrast "truths *a posteriori*, or of fact" with "truths *a priori*, or of Reason" (NE IV, ix, 2). I suspect, indeed, that Leibniz played a crucial role in the transformation of the meaning of '*a priori*'. I believe that 'proof *a priori*' always *meant* for him 'proof that explains the reason for the fact that is proved.' But it follows from his epistemology and his conceptual containment theory of truth that *a priori* knowledge in this sense coincides with knowledge that is independent of experience and with knowledge by analysis of concepts. It was therefore easy for him to use the term '*a priori*' when the idea that was foremost in his mind was not that of knowledge by causes but one of these other ideas that were, for him, coextensive with it. By the same token, it was easy for a Leibnizian like Alexander Gottlieb Baumgarten to *define* the *a priori* as the non-empirical (Risse, II, 649). Christian August Crusius, on the other hand, who was more resistant to Leibnizian views about sufficient reason, was still using '*a priori*' and '*a posteriori*' in the old sense in the middle of the eighteenth century: "A demonstration *a posteriori* is one from which is known only that a thing is – for example by experience, or by reduction to absurdity, or by comparison with some other thing. A demonstration *a priori* is one from which is known why a thing is – for example where we deduce the attributes of things from definitions, or draw out the effect from the causes and determining reasons" (Risse, 1970, II, 690; a similar conception is found in Crusius, 1753, Section 35).

[13] These words are part of a passage that Gerhardt prints in brackets, meaning (I presume) that there is some indication in the copy retained by Leibniz at Hanover that they may have been omitted from the copy sent to de Volder. But there are enough similar texts to remove any doubt that Leibniz did, indeed, hold the view expressed here. For other passages, and a good discussion, see Loeb (1981, pp. 317–319).

14 For development of the question to which I am suggesting an answer here, see Broad (1975, 24f), and Loeb (1981, pp. 279–291).
15 This statement was omitted from the copy sent to Arnauld (Le Roy, 1957, p. 302); I don't know why. L-A, which is usually a rather reliable translation, renders 'simpler' ('*plus simple*') here by 'similar' (perhaps by a typographical error).
16 See Robinet (1965, p. 112), on the development of this theme in the thought of Malebranche.
17 I follow Clarke's translation here, except for punctuation. Leibniz goes on to say that mutual attraction of bodies at a distance, and continued circular motion of a body, would be "supernatural" unless there were a mechanical explanation. He could be taken here to be assigning these to the first rank of miracles, but I think it likelier that his thought about these cases, which were very relevant to his controversy with Clarke, had more to do with the idea, discussed above, that "laws of nature" must provide explanations of a certain simplicity.
18 I am *not* using 'counterpart' here in a technical sense like that of David Lewis.
19 I am indebted to students in several classes on Leibniz in which I have been working out my ideas on these subjects for more than a decade. Tyler Burge read an ancestor of this paper, and gave most helpful comments. I am also particularly grateful to Marilyn McCord Adams, David Blumenfeld, John Earman, and J. E. McGuire for valuable discussion of issues treated here.

WILFRID SELLARS

TOWARDS A THEORY OF PREDICATION

I

1. Consider the statement

a is red,

where, since we are beginning *in medias res* (where else?), *a* is a sensum and the statement predicates sensible redness of *a*.

2. Entering the dialogue at a point of our choice, let us assume that the primary role of the 'is' is to indicate tense, and to relate the when of *a*'s being red to the past, present, and future. Thus, let us assume that the predicative force of the statement can be represented by the PMese form

red ***a***,

i.e. by concatenating the predicate 'red' with the referring expression '*a*'.

3. And, assuming that our problem has already achieved a certain measure of understanding, let us provisionally take the stance of what can, I believe, legitimately be called the 'standard' (indeed, the 'perennial') theory. This theory is built around the following diagram.

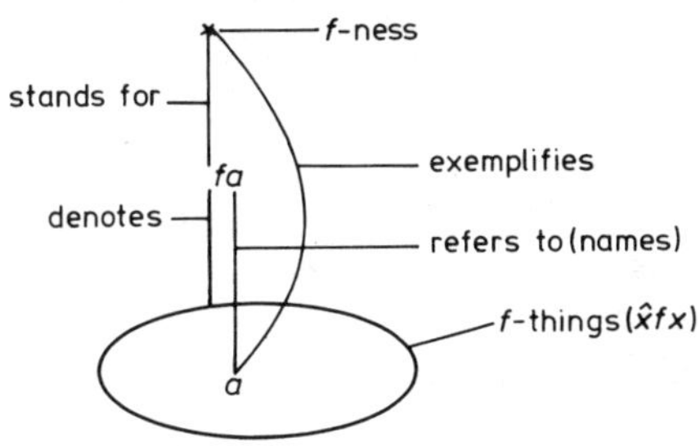

4. The following points give the essentials of the theory:

James Bogen and James E. McGuire (eds.), How Things Are, 285–322.

a. '*f*' and '*a*' are both construed as names: '*a*' of a particular, '*f*' of a universal.
b. Why is there both the predicate '*f*' and the name '*f*-ness'? Because '*f*' is semantically related both to *f*-things and to *f*-ness, the property '*f*-things' have in common.
c. Hence

   fa

   is equivalent to

   f-ness *a*

   and the fact that *a* exemplifies *f*-ness is expressed by concatenating the names '*a*' and '*f*-ness' as above. In other words, the relation of a particular to the universal it exemplifies is expressed by a relation between tokens of the names of these entities.

   That an '*f*-ness' is R_1 to an '*a*' expresses that *a* exemplifies *f*-ness.

d. This obviously requires a distinction between *ordinary* relations such as those expressed by two place open sentences such as

   --- adjoins . . .

   and those, call them 'ties' or *nexus*, expressed by

   - - exemplifies - - .

   For (ordinary) relational predication, according to the theory, requires sentences of the form, for example,

   being adjoined to x, y,

   where 'being adjoined to' stands for (names) the relation of *adjoining* as 'redness' stands for the property of *being red*. This preserves the parallel between the case of relational predication and the non-relational case with which we began.
e. Thus, in the present case the principle in (c) becomes

   That tokens of '*a*' and '*b*' are jointly R_2 to a token of 'being adjoined to' expresses that *a* and *b* (jointly) exemplify *adjoining* where R_2, like R_1 in (c) above, is a mode of concatenation.

II

5. It is no news that a theory of predication is no sooner formulated than it generates puzzles. But before we look at our perennial theory with this in mind, it needs to be placed in a larger context.

6. The example of predication with which we began was one of *adjectival* predication:

a is red.

We have also saluted *relational* predication, and must be prepared, of course, to take into account such other forms as *sortal* predication, e.g.

Leo is a lion

not to mention *verb* predication, thus:

Socrates runs.

7. Thus, although I am ostensibly dealing with a single topic, many of the central issues of metaphysics are lurking in the underbrush. After all, the problem of predication is but one form of the many puzzles which traditionally fall under the heading of the One and the Many. If redness is a one which can be truly predicated of many red things, it is equally true that red things are ones which have many features, of which redness is but one.

8. And, indeed, to get a better grip on the current stage of the argument, it will be useful to consider an example in which a *linguistic one* is predicated of a *linguistic many*.

9. Thus it has become customary to contrast words as *types* (e.g. the word 'and') with words as *tokens* (thus the many cases of the word 'and'). Is not the word (types) 'and' as much a universal as redness? Is it not *exemplified* by its tokens? Consider the following token – let us call it '*t*' – –

and

10. If we say of our favorite sensum *a* that

a is red

should we not also say

t is an 'and'

and if *a* is a red-thing, is not *t* an 'and'-thing? Indeed, if 'red' is the name of redness, is not "and" the name of 'and'-ness, the character of *being an 'and'*?

11. Or, to come right to the point, is not the relation of type to token simply a special case of exemplification? Are not linguistic *ones* as types universals, and the corresponding *manys* the particulars which exemplify them?

12. A closer look, however, generates a sense of disquiet. We are easily pulled in another direction. It is not so much that we are pulled *away* from the idea that tokens of 'and' have the common character of being 'and's – as red things have the common character of being red, it is rather that when we reflect on the fact that we can make predications of linguistic *types*, a whole new line of thought opens up.

13. Thus we can say not only, as in (10)

t is an 'and'

but also

t is a conjunction

and add

because *t* is an 'and' and 'and's properly so-called are conjunctions.

14. In other words we are confronted by two intimately related readings of

'and' is a conjunction

15. According to the first, "and" is interpreted along the lines of (10) as a *name*, the name of an object, a linguistic character, comparable, ontologically, to the non-linguistic character redness. In this case, the statement becomes,

'and'-ness is a conjunction

which predicates *being a conjunction* of the object in question, and by doing so *entails* that the items which exemplify this character (its tokens) are conjunctions.

16. According to the second, "and" is interpreted *not* as the name of an object ('and'-ness) in accordance with the schema

a is $f \rightarrow a$ exemplifies f-ness

so that we would have

t is an 'and' $\rightarrow t$ exemplifies 'and'-ness

but rather as a *distributive singular term*[1] in accordance with the schema

(Standard K's are $\phi \rightarrow$ The (or a) K is ϕ,

e.g.

(Standard lions are tawny → The lion is tawny.

The latter does not predicate being tawny of lionhood. It predicates it of *the lion*, which is, indeed, an *ens rationis*, but one which is only under penalty of confusion to be identified with the character of being a lion.

17. Thus, in the case with which we are concerned, the relevant grammatical transformation is

(Standard) 'and's are conjunctions → The (an) 'and' is a conjunction.

III

18. Now the point of entering the dialogue on predication with a diagram and supporting statements which present the gist of the perennial, or as I have also called it, the *standard*[2] theory of predication is to express my conviction that here, as in the case of so many classical problems, the platonic tradition has the essential structure of the truth, but, to use Emily's invaluable metaphor, tells it 'slant'. In other words, I hope to show, in part by constructing a plausible lexicon, that the natives of our jungle came far closer than is usually thought to an ontology which satisfies the adequacy conditions of a philosophical clarification.

19. Let us take another look at the theory. According to it, the statement

fa

predicates f-ness of a by being a concatenating of an 'f' with an 'a'.

20. How does the statement bring *f*-ness to bear on *a*? Because, we are told, '*f*' stands for *f*-ness. This presents us with the following picture. Predicative expressions (schematically '*f*' (e.g. 'red' or 'adjoins') are correlated with singular terms, schematically '*f*-ness'[3] (e.g. 'redness') or 'being adjoined to'.

21. This gives us, schematically, the correlation

predicative expression	*singular term*
f	*f*-ness, being *f*
R	*R*-hood, being *R* to

22. Notice that the theory commits itself to two types of semantical statement with respect to these expressions. On the one hand we are told that

'*f*' *stands for f*-ness,

on the other that

'*f*-ness' is the *name* of *f*-ness.

23. To make things more interesting, the theory adds that in the statement

fa

'*f*' is a *name* of *f*-ness *functioning as* a predicate.

24. How all this is to be understood is crucial to the evaluation of the theory. For while the concept of a name traditionally carries with it the idea that its *nominatum* is an object, it is by no means as clear that the context

--- stands for . . .

requires that . . . be an object.

25. Of course, one is tempted to say that *anything* is an object, a temptation which is reinforced by the fact that 'things are objects' looks like a tautology. But the barest acquaintance with scholastic thought should remind one that the *transcendental* 'anything' should not be confused with the compound expression 'any thing'.[4]

26. Thus the perennial tradition contains the logical space for distinguishing among items which fall under the transcendental 'anything' between those which are objects or things, and those which are not. How such a distinction might be drawn will turn out to be the heart of the matter.

IV

27. Now the context

- - - stands for . . .

belongs to the family of semantical contexts. To say what an expression *stands for* is a way of saying what it *means*. To fix our ideas, therefore, let us put aside the tidy accounts which contemporary semantical theorists, with their set-theoretical sophistication, give of 'semantical relations' and do some first hand reflection on meaning. *Otherwise we may find that we come to the problem of the ontology of predication with dirty hands.*

28. After all, any statement has an equivalent in set theoretical terms. Thus, to take a trivial example,

There are two apples on this table

is necessarily equivalent to

the set of apples on this table has two members

but it is no more synonymous with the latter than

Snow is white

is synonymous with

It is true that snow is white.

29. I refrain from more exotic examples in which sentences are mapped into statements about possible worlds. The reader can find examples in any text-book of Montague grammar.

30. How the variety of ways in which '*p*' can be necessarily equivalent to '*q*' without being synonymous with it is to be botanized, is a problem on which the perennial tradition is still working under watchful eyes. But then it has always been the Quines and the Hobbeses who keep the philosophical enter-prise honest and force comfortable metaphors to assume critizable form. With these sobering thoughts, back to our problem.

31. Consider the meaning statement

'Und' (in German) means *and*.

Ostensibly this has the form 'xRy' and asserts the two items, a German word and the item *and* – whatever that may be – stand in a certain semantical relation, that of meaning. It is, however, received wisdom that this is a misconstrual of the statement, one which gives us a 'false picture'.

32. Now it might be thought that the best way to explain why meaning is *not* a relation is to tell us what it *is*. But this is not what is done. The subject is changed. We are told interesting things about language as a system of tools, as a form of life, as a means of communication, as a medium in which speech acts are performed and a linguistic community is enabled to carry out its communal tasks. Much of what is said in this connection is both true and important, but it does not, at least directly, clarify classical problems of meaning and reference.

33. To do so we must take a longer look at the above statement. To begin with, it is obvious that the word 'and' at the right is not functioning in its normal way. It is not serving as a sentential connective.

34. Now the most familiar way in which a word which is not in the ordinary sense ambiguous can play a radically different role is by being used in material imposition – that is, in effect, by being placed in quotation marks. It might therefore be argued that our meaning statement, grammatically regimented, has the form

'Und' (in German) means 'and'

where the quoted 'and' is the name of the English word 'and'.

35. But since it is obvious that the German word 'und' does not mean the English word 'and', this suggestion soon becomes the idea that the original statement has the more complex force of

'Und' (in G) *means the same as* 'and'.

But while this is of course true, it is by no means synonymous with the original.

36. This becomes clear when we reflect that whereas the original statement *tells us*, and is designed to tell us, what 'und' means, the new formulation requires additional information to do so.

37. Thus if I say that

'Und' (in G) means the same as 'et' (in French)

I *give you* the meaning of 'und' only if, in effect, I add, or it is presupposed, that 'et' in French means *and*, which brings us back full circle.

38. To break out of this circle, the first step is to ask the question, 'How is the German word 'und' functioning in the original statement?' The answer should by now be obvious. Like the English word 'and' in the statement

'and' is a conjunction

it is functioning as a distributive singular term, and is equivalent to

(Standard) 'und's (in G) mean *and*.

And while this does not immediately clarify our problem, it does suggest that if the '*x*' of our putative relational statement, (31), turns out not to be a name, perhaps the same is true of the '*y*'.

39. To make a long story short[5] I propose the following:

a. The 'and' of the original statement be reconstructed as the metalinguistic sortal '·and·' – where the dot-quotes are indeed quotation-marks thus preserving the insight that it is being used in material imposition, but with special criteria.
b. We take into account the flexibility of quotation. Thus while the standard use of quotation marks tends to be tied to the sign designs of the quoted expressions, it is not tied rigidly, for auditory tokens are included in the scope of 'red'. Yet there is an intra-linguistic limitation, for English *redds*, but not German *roats* are included in the scope of 'red'.
c. Thus, in the case of ordinary quotes, the relevant sign designs are taken as *functioning in a given language*. This makes it possible to distinguish between two dimensions of the criteria for being a 'red': the sign design dimension, and the dimension in which it is considered as functioning in a specific manner in the English language.
d. It is a familiar fact that in different functional systems, empirically different objects can play similar roles. Game events can in an important sense be tokens of the same game, even though they are embodied in different materials and motions. This suggests that the sign design *redd* might play a linguistic role in English which is interestingly similar to that played in German by the sign design *roat*.
e. It is but a step from these considerations to suggest that quotation can

play an *inter*-linguistic role. Thus whereas ordinary quotation can transcend the specific sign designs included between them, but not the language to which they belong, we can conceive of a form of quotation – *q*-quotation – in which

$^{q}{\text{red}}^{q}$

applies to tokens in any language which play the role in that language which is played by the sign design *redd* in the language in which quotation marks are used. In the case with which we are concerned this language will be our own background language, English.

f. Notice that nothing has been said so far about the specific natures of these linguistic roles.

40. At the beginning of the previous paragraph I proposed that the 'and' of our original meaning statement be construed as a metalinguistic sortal, and indeed as occurring in a special form of quotation which I represented by the use of dot-quotes, thus:

'und' (in G) means ·and·.

41. But the reconstruction of one of the terms in a problematic context typically has repercussions for the others. Consider, for example the familiar reconstruction of

Some men are mortal

as

There are mortal men.

One might say that to do this is to reconstruct 'some' as 'there are'. But clearly the entire sentence has been involved.

42. In the present case the change of the 'and' of

'und' (in G) means *and*

into ··and·· also requires corresponding changes in the rest of the sentence. What these changes might be is readily determined by reflecting that if ··and·· is a quoted expression, the basic form of sentences involving it is

--- is an ·and·.

43. Since we have already interpreted the "und" as a distributive singular term which transforms into a reference to 'und's, we move smoothly from our original statement to

An 'und' (in G) is an ·and·

or, to flesh it out,

(Standard) 'und's (in G) are ·and·s.

44. *The result is a straightforward explanation of why meaning is not a relation, for, at bottom, it is the copula, and the copula 'is' is not a relation word.*

45. Of course, the recalcitrant can argue that the copula 'is' stands for the relation of exemplification, but (a) it has already been pointed out that the latter is not a relation in the sense in which adjoining is a relation and (b) presumably it is in the latter sense that meaning has been thought to be a relation.

46. Thus, who would have defended the idea that meaning is a relation by pointing to a necessary equivalence between

'und's (in G) are ·and·s

and

'und's (in G) exemplify ·and·-sort?

V

47. Clearly to say that meaning is not a relation (i.e. that 'means' does not stand for a relation) is not to say that meaning does not *involve* relations, even *essentially*.

48. Thus it should be obvious that 'und' (in G) would not mean what it does *if*

p und *q*

in German did not stand in the same consequence relation to

nicht (nicht-*p* oder nicht-*q*)

as does

p and q

to

(not-p or not-q)

does in English.

49. Thus to say that meaning is not a relation is compatible with the idea that for an expression to have a specific meaning, it must stand in specific relations.

50. Thus, 'und' would not mean *and* unless, for example, the above consequence relation (48) obtained.

51. Nor would 'Sokrates' (in G) mean *Socrates* unless in one of its uses, the German word 'Sokrates' stood in certan socio-psychological-historical relations to Plato's snubnosed mentor.

52. Nor, indeed, would 'rot's (in G) be •red•s unless they belonged with 'gelb's, 'blau's, etc. in a family of competing predicates and unless 'das ist rot's were proper perceptual responses to red objects in standard conditions.

53. It should also be clear that in this reconstruction 'sameness of meaning' is simply the *extremum* of similarity of meaning. If to say what an expression means is to classify it, the relevant philosophical point is that classification requires criteria, and that the criteria for classification under a sortal are typically flexible. In one classificatory context a spade may be a spade, in another a shovel.

54. In one context a 'nicht' may be functioning as a classical •not•, in another as an intuitionist •not•. The two functional classifications are *species of negation* as 'biped' and 'quadruped' are species of 'animal with feet'.

55. Thus in a given context

This 'nicht' is a •not•

will be true or false depending upon whether the criteria for being a •not• include or do not include the consequence relations involved in the principle of excluded middle.

VI

56. At this stage the reader, who has been struggling to remain 'in the dialogue' is likely to say: 'All this is interesting, and, perhaps, important, but what does it have to do with predication?' The answer is that what we have been doing is largely to explore features of the diagram and the propositions in terms of which the 'perennial theory' of predication was initially introduced. We have been gaining our bearings.

57. Indeed, a review of the argument to date reminds us that the above discussion of meaning was initiated by an attempt to grasp the implications of the context '--- stands for . . . '

58. Does it, we asked, imply that . . . is an object? Is *standing for* a relation between --- and . . . ?

59. We began by pointing out that an item can be a *something* without being, in any ordinary sense, a *thing* or *object*.

60. But this simply confronts us with the task of drawing an illuminating distinction between the *somethings* which are and the *somethings* which are not *objects*. So far, the only cash we have for this distinction is that objects have *names* whereas the terms which refer to non-objects may be singular terms which are not *names*.

61. And the sole alternative with which we have been presented rests on the fact that the class of singular terms includes not only names but distributive singular terms. How far does this get us?

62. Since the only ostensible name directly involved in the standard theory is, schematically, '*f*-ness', the only startling possibility this alternative opens up would be for there to be reasons to suppose that the latter is a schema, not for names, but for distributive singular terms.

63. This would indeed strike at the very heart of the theory, though the exact import of this fact would remain to be evaluated.

VII

64. I now come to a critical point which will dominate the remainder of the

argument. Although the proposition was not advanced in so many words, the theory is committed to the idea that if

a is red

rather than

a is green

is true, there must be *something* in the extra-linguistic domain which accounts for this fact. This something can not be *a* by itself, for the existence of *a* is necessary to and compatible with the truth of either.

65. What is this *something*? At this point the theory interprets the requirement as the need for a *thing* (*res*), and by virtue of this fact becomes a strong form of *realism*. The 'thing' in question is interpreted as an abstract object, a *property*, the property of being *red*, i.e. *redness*.

66. Of course, just putting another object in the extra-linguistic domain wouldn't satisfy the requirement, for the truth of

b is green

would put greenness there as well.

67. Thus the something turns out to be *the fact that a exemplifies redness*. And, indeed, the theory asks, 'How can there be such a fact, unless there *is* such a thing as redness'.

68. The theory thus presents us with *two* contexts in which the ostensible name 'red-ness' occurs

(a) *a* exemplifies redness
(b) 'redness' stands for redness.

These are both *ostensibly relational contexts* of the form '*xRy*'.

69. Reference to exemplification reminds us of the fact, to which attention was called in (64), that the theory of predication is also a theory of the truth conditions of statements of the form '*fa*'. Does this suggest any additional intuitive principles which might belong with our beginning diagram?

70. One raw candidate might connect 'a exemplifies redness' with

'red' is *true of a*.

But how? The latter clearly has the form

'red' (in E) is true of *a*

and must we not say that it is *because* 'red' (in E) stands for redness and *a* exemplifies redness that 'red' (in E) is true of *a*?

71. But might we not interpret the close connection between

'red' is true of *a*

and

redness is true of *a*

in terms of the idea that 'redness' is the distributive singular term

'the 'red''

so that

redness is true of *a*

would parse as

(Standard) 'red's are true of *a*

and hence, hopefully, as

Sentences which consist of the predicate 'red' concatenated with an individual constant are true in case the constant is '*a*'.

Or, using 'INDCON' as a sortal for individual constants and

'red' INDCON

as a sortal for sentences consisting of a 'red' concatenated with an INDCON,

'red' INDCONS are true '*a*'/INDCON.

72. But – to make a long story short – why not make exactly the same move with the exception that we interpret redness not as

the 'red'

but rather as

the ·red·

for while redness can scarcely by identified with the English predicate 'red' –the property of being red is scarcely parochial to our language community – it is conceptually possible to identify it with a role which is played in English by 'red', in French by 'rouge' and in an interestingly different language by '---', where 'red', 'rouge' and '---' are all (bearing in mind the flexibility involved in likeness of role) ·red·s.

73. If we make this move, however, we confront the fact that

'red' (in E) stands for redness

threatens to become

'red' (in E) stands for the ·red·

and, hence, given our previous strategy with respect to meaning contexts

'red's (in E) are ·red·s.

This would require that

'rot' (in G) means *red*

and

'rot' (in G) stand for redness

have the same depth grammar, namely

rot's (in G) are ·red·s

74. But is this so absurd? Of these two semantical contexts, the former is that which is used to *explain* the meaning of a term. Thus the hearer of

'und' (in G) means *and*

is led to grasp the function of the German word 'und' by rehearsing in imagination that of the English word 'red'. It is a positive virtue of this context, then, that the expression following 'means' is the very word which has in English, the function of the appropriate connective.

75. But meaning also relates to truth. Thus, in addition to the context

'Schnee ist weiss' (in G) means *Show is white*

which is used to explain the meaning of 'Schnee ist weiss' we have

'Schnee ist weiss' (in G) stands for *that snow is white*

which dovetails with the predication of truth

That snow is white is true.

76. Consider also the pair

'dreieckig' (in G) means *triangular*

'dreieckig' (in G) stands for *triangularity*

The latter involves the singular term 'triangularity', by virtue of which it dovetails with

a exemplifies triangularity

and its equivalent

triangularity is true of *a*.

77. Thus it would not be without reason that

'dreieckig's (in G) are ·triangular·s

transforms into both

'dreieckig' (in G) *means triangular*

and

'dreieckig' (in G) *stands for* triangularity

VIII

78. At this point in the argument it might seem as though the most that has been established is that whereas according to the standard theory the *something* in addition to *a* required by the truth of

red *a*

is a *nominatum*, and hence, in a straight forward sense, an *object*, a structurally similar thesis can be constructed in terms of which '*f*-ness' is not a name but a distributive singular term.

79. If we introduce the expression 'distributive object' in such a way that

The K is a distributive object

is material mode of speech for

'The *K*' is a distributive singular term

we might put this by saying that the above suggestion has amounted to the claim that a form of 'moderate realism' can be constructed which is structurally similar to the theory with which we began. The realism would be 'moderate' in that the *res* in question (*f*-ness) would be a distributive object rather than an object proper.

80. But, of course, the suggestion as thus far constructed is much more radical than this, for *f*-ness has been interpreted, not as an *extra-linguistic* distributive object, e.g.

The lion is . . .	Lions are . . .
The triangle is . . .	Triangles are . . .

but rather as a *linguistic* distributive object, e.g.

The ·red· is . . .	·red·s are . . .
The 'und' is . . .	'und's are . . .

81. And this, it might be said, pointing to the family resemblance between thought and language, would take us from moderate realism to conceptualism with all the puzzles *that* entails.

82. To bring matters to a head, if to be a

·red·

is to do, in some language or other, including Mentalese, the job done in our language by 'red', then (a) what is this job? and (b) by virtue of what is a *predicative* job?

83. At this point, the standard theory can be expected to concede that the concepts introduced by its critics are useful and important, but argue that what is called for is a more careful distinction between the *psychological* and the *logical* dimensions of predication. In doing so, it would claim that the existence of the *linguistic distributive object*, *the ·red·*, is compatible with the existence of the *non-linguistic object* redness, and suggest that it is premature to identify the property of being red with the distributive linguistic object, the ·red·.

84. The idea that there is a linguistic 'object' in the neighborhood of the predicate 'red' is suggestive, but surely much honest toil would have to be done to establish that this 'object' is a plausible candidate to be that in the world by virtue of a relation to which *a* is red.

85. This toil begins, appropriately enough, with Bertrand Russell. We must cope with his argument for platonic realism in *The Problems of Philosophy* on which we all cut our teeth.

86. His argument begins by pointing out that to make a basic statement, more is necessary than names of particulars. Thus, it is obvious that a list of names, such as

Cassio, Desdemona

simply raises the question 'Well, what about them?''

87. Russell draws the consequence that to make a statement, a sentence must include an item which is not the name of a particular, thus

Cassio *loves* Desdemona

or, in the non-relational case.

triangular a.

88. It is, of course, obvious that it would be a mistake to equate

item which is *not* the name of a particular

with

item which is the name of a *non*-particular

but since the language includes many items which ostensibly fall in the latter category, and since they pair up nicely in accordance with the pattern:

f	being *f*
triangular	triangularity
next to	next-to-ness

with the non-names we call predicates, the temptation to interpret the distinctive roles in statements of *non-names of particulars* in terms of the distinctive character of being a name of a non-particular (abstract singular term) becomes attractive. It is this line of thought which, if Russell is our guide, underlies the construal of

fa

having the form of

f-ness *a*

in the commentary on the diagram with which we began. In the relational case,

aRb

would be construed as

a R-hood *b*.

89. Interestingly enough, just as our problem – that of predication – was beginning to assume the form

> What do the non-names in a basic statement contribute to the latter's statement making force?

the Russellian line of thought turned these non-names into names. Thus, 'red' became 'redness' and 'loves', 'loving'.[6]

90. Russell's original question

> How does a statement differ from a list of names *of particulars*?

to which the tentative answer was

> by including a non-name

has now become

> How does a statement differ from a list of names *period*?

Thus, how does

aRb

construed as

a R-hood *b*

differ from the list

a, *R*-hood, *b*?

91. The standard theory has made two moves in answer to this question, the most insightful of which was the attempt to collapse the two moves into one.

92. The first, and most obvious, move was to give a primary role to the *non-name*, 'exemplify', thus

a exemplifies redness
a and *b* exemplify loving

93. If asked, 'doesn't the Russellian strategy require construing these as

a exemplification redness
a, *b* exemplification loving

and hence to a regress of exemplifications (in short Bradley's paradox)? The answer tended to be, " 'exemplify' isn't like *that*".

94. Although Wittgenstein was responsible for the insight contained in the second move, it was Russell who put it to use, although in a way which left the situation more confused, if anything, than before.

95. According to Russell, one says that *a* and *b* stand in the loving relation by placing tokens of the names '*a*' and '*b*' and 'loving' in a conventional relation, for example,

loving *a b*.

Thus we express a relationship between objects by constructing a linguistic relationship between tokens of the names of these objects – in this case a concatenation in which the *name* of the universal is to the left of the *names* of the particulars.

96. This suggests that the previous move of introducing a *privileged* non-name, 'exemplifies', to stop Bradley's regress was a clumsy attempt to capture the radical difference between expressing a relation by *using a relation word in addition to the names of the terms*, and expressing it by *relating the names of the terms*.

97. Viewed from this perspective

a, *b* exemplify loving

could be interpreted as a way of relating tokens of '*a*', '*b*' and 'loving', for the tokens of '*a*' and '*b*' to the left, and the tokens of 'loving' to the right stand in the relation of having a case of the sign design *exemplify* between them.

98. Since ostensibly 'exemplify' stands for exemplification as 'red' stands for redness and 'loves' stands for loving, only the account of 'standing for' we have given in (27ff) blocks the resumption of the Bradley regress. For if

'- - -' (in L) stands for exemplification

does not *introduce an object* but simply has the sense of

'- - -'s (in L) are ·exemplifies·s

this would simply tell us that (in L) '- - -'s do the job done in our language by 'exemplify's and this job could perfectly well be that of bringing it about that tokens of '*a*', '*b*' and 'loving' stand in the relation of having a case of a distinctive sign design – in this case *exemplify* – between them.

99. But while this *could* be the job of ·exemplify·s – and the insight that there could be such expressions is an important one – this does not seem to be the case, After all, as we have noticed

a and *b* exemplify loving

tells us

loving is *true of a* and *b*

and hence 'exemplify' seems to play the substantive role of predicating a relation between loving, *a* and *b*, whereas on the above suggestion

a and *b* exemplify loving

would differ from

loving *a, b*,

only in that while both place tokens of '*a*', '*b*' and 'loving' in a conventional triadic relation, the latter does so without the use of an auxiliary symbol.

100. The two statements would have the same syntactical structure, i.e.

token of name of universal triadically related to tokens of names of particulars.

That in the one case the triadic relation makes use of an additional symbol would involve no additional semantical content. The two statements would be synonymous, though they would involve interestingly different ways of relating the name tokens.

101. The thesis that 'exemplify' is a *privileged* non-name would seem to embody two strategies

(a) 'exemplify' does not introduce a new object, for its role is simply that of an auxiliary symbol
(b) 'exemplify' introduces a *peculiar* object (or entity) which does not have a name.

The latter strategy would indeed block the regress generating moves from

a exemplifies *f*-ness

and

a, *b* exemplify loving

to

a exemplification *f*-ness

and

a, *b* exemplification loving.

but it would do so in a patently *ad hoc*-ish manner.

102. Confronting (a) is the fact that 'exemplify' ostensibly stands for the relation *being true of*, and consequently is no mere auxiliary symbol. Confronting (b) is the fact that if 'exemplify' does stand for a relation, then the abstract singular term 'exemplification' would seem to have as much right to be considered its name as 'juxtaposition' to be considered the name of *being juxtaposed*.

IX

103. That exemplification and truth are essential elements of the problem of predication has been clear from the beginning. It would seem obvious, therefore, that we should seek to bring them relevantly into the picture.

104. Once again we are confronted by an orthodox view which can serve as a dialectial foil. In and of itself it brings no solution to the problem of predication. But it does yield a new perspective which will help us evaluate the previous discussion.

105. This orthodox view is the Tarski version of the semantical theory of truth as solidified into what can appropriately be called 'semantic formalism'.

106. It will be remembered that one of the key themes in the early sections of this paper was that meaning is not a relation. The term 'means' was interpreted as a form of the copula 'is' specialized to linguistic contexts. The statement

'und' (in G) means *and*

was reconstructed as

'und's (in G) are ·and·s.

A similar account was given of 'stands for', thus

'dreieckig' (in G) stands for triangularity

became

'dreieckig's (in G) are ·triangular·s.

107. But it turns out that there is another way of assimilating semantical predicates to logical expressions; this time, not the copula 'is' but rather logical connectives (and quantifiers). Consider, for example, 'denotes (in G)' Instead of interpreting

'- - -' (in G) denotes . . .

as

'- - -'s (in G) are ·. . .·s

we construct the following formula

x den (in G) $y \equiv x$ = 'Sokrates' · y = Socrates

or x = 'Griechenland' · y = Greece

or x = 'der Mond' · y = the moon

....

This formula, appropriately extended, is obviously true. If interpreted as a *definition*, however, it defines 'den (in G)' in terms of conjunction, disjunction, identity and a correct list of wedded pairs. (It obviously has little of substance to say about the wedding ceremony). To be married is, effectively, to be listed as married.[7]

108. That

'Sokrates' den-in-G Socrates

turns out to be logically true may startle. But, after all, it does not follow from this that the *non-technical* statement

The German word 'Sokrates' refers to Socrates

is logically true, The latter idea is obviously false, unless to be a German word is simply specified to mean to be an expression belonging to a linguistic system for which the above definition of 'den-in-G' is an axiom – i.e. for which 'refers' is stipulated to have the sense of 'den-in-G'.

109. A parallel point can be made about the concept of satisfaction, where *ostensibly* to say that an object satisfies a predicate, PR, in a language is to say that there is an appropriate matter-of-factual relation between the object and the predicate, as an item in the linguistic institution to which the predicate belongs. Here, once again, definition by listing can give us a stone in place of bread.

110. Thus, after defining 'is a PR-in-G' and 'is an INDCON-in-G' in terms of listings, we now add a further listing

x sat-in-G $y =_{\mathrm{df}} x = a \cdot y =$ 'rot' · red a

or $x = b \cdot y =$ 'blau' · blue b

or ....

These definitions by listing put us in a position to introduce the technical expression 'true-in-G' and define it as follows

PR INDCON is true-in-G $=_{\mathrm{df}}$ INDCON den-in-G $a \cdot a$ sat-in-G 'rot'

or INDCON den-in-G $b \cdot b$ sat-in-G 'blau'

or ....

of which given these stipulations, the T-sentences

'rot *a*' is true-in-G $\equiv$ red *a*

is a consequence.[8]

111. But all this provides little illumination as to how the German sentence 'rot *a*' must be connected with the world in order to be assessed as *true* which, in spite of all the logic that is floating around, is an evaluation.

112. Thus, given listing definitions of 'den-in-E' and 'satisfies-in-E' and the corresponding definition of 'true-in-E,' it becomes a logical truth that

'*fa*' is true-in-E $\equiv$ *fa*.

113. Unless we are going to permit ourselves to be hypnotized by all this rigor, it is essential to remind ourselves that even if

$$p \equiv q$$

is logically true, it need not be the case that '*p*' *explains* '*q*'. (Consider, after all, that

$$2 + 2 = 4 \equiv 3 + 3 = 6.)$$

114. On the other hand, clearly the fact that *a* is red *combined with additional premises* provides an *explanation* of the fact that

'*a* is red' (in E) is true

But what exactly is the something more?

115. If one is hasty and careless in going through the Tarski motions, one can easily be deceived into thinking that the explanation in question is readily forthcoming. After all, using the appropriate definitions of 'den-in-E', 'satisfied-in-E' and 'true-in-E' we understand why, given that *a* is red, it *must be the case* that '*a* is red' is true-in-E.

116. But an explanation of *how we know that*

'*a* is red' is true-in-E

need not be an *explanation of why the English sentence 'a is red' is true.*

117. Clearly the issue hinges on the correct interpretation of the fact that a true statement is necessarily a true statement in a given language. For it would be granted on all sides that to abstract from the fact that a statement belongs to a certain language is to cut off any possibility of determining its truth – let alone its meaning.

118. But to specify the language to which a statement belongs is not the same thing as to give a recursive listing of the expressions of the language, nor a recursive listing of what denotes-in-L what, nor, for that matter, what satisfies-in-L certain predicates in L.

119. Thus it makes perfectly good sense to say that a certain expression belongs to a certain language – or, for that matter, a certain dialect, or even idiolect, even though one is not in a position without investigation, to provide such lists. (Compare attributing a certain law to a certain legal corpus.)

120. Even though one lacks a satisfactory formalist account of what it is for '*a* ist rot' to be a German sentence, one can nevertheless be in a position to explain the truth of the statement, made in German, '*a* ist rot'. I shall be explaining the explanation of truth attributions shortly. But before turning directly to this topic, more groundwork must be laid – and, in particular, more must be said about predication, for already this discussion is threatening to become Hamlet without the Prince of Denmark.

X

121. Thus it is time we reflected further on the fact that whereas Russell began by arguing that *something more* than a list of names is necessary to make an assertion, and interpreted this something more as the presence of a non-name, he wound up by pointing out that we don't need a non-name to say that *C* and *D* stand in the *stand in* relation to loving. All we need to do is to bring about the existence of a linguistic relation between tokens of the names '*C*', '*D*' and '*l*', e.g.

loving *C*, *D*.

Thus a triadic relation between tokens of these names can express a relation between three objects.

122. The principle behind this would be that a statement *can* consist of

names, provided that the sentence is not a *mere* list of names. *Mere lists* have a syntactical structure, which typically involves the use of commas, thus:

Tom, Dick, Australia, McTaggart, ...

which blocks the idea that anything more than a list of names is involved.

123. On the other hand, the grouping

Loving *C*, *D*

is specified by syntactical rules in the PMese dialect of English to be a sentence.

124. Reflection suggests, however, that if

Loving *C*, *D*

makes a statement by placing 'loving', '*C*', and '*D*' in a conventional triadic relation, the same must be true of

C, *D* exemplify loving.

The logical form of two statements by which the same statement is made can scarcely differ so radically that in the one case it involves four extra-linguistic items to which four linguistic elements are semantically tied, whereas in the other case, only three extra-linguistic items (a proper sub-set of the four) are involved, to which three linguistic items, a proper sub-set of the previous four are semantically tied.

125. But to make this move requires that 'exemplify' be a *pure auxiliary symbol* in the sense that its only *semantical* role is to be a means of bringing it about that 'C', 'D' and 'loving' have a token of the sign design 'exemplify' between them. If we use asterisk quotes to form a sortal for the class to which the sign design between them belongs, abstracting from the semantical function of these designs, then we could say that in the above expression 'C', D' and 'loving' are, respectively a **C**, **D** and a *loving* functioning as words having extra linguistic reference, whereas the *exemplifies* is not. The latter simply provides a sample of the sign design to bring it about that the other three items have an *exemplifies* in their midst.

126. On the other hand, we must beware of saying that this *exemplifies* has no *meaning*, no 'semantical status' for it is playing the role of bringing it

about that '*C*', '*D*' and 'loving' stand in a conventional triadic relation by virture of which the statement tells us that *C* and *D* exemplify loving, and hence contributes to the *meaning* – in a pre-analytic sense – of the sentence.

127. Now what this brings out clearly is that *exemplify* is not picked out for *this* role by its meaning in any deeper sense. Any arbitrary sign design would serve the purpose. Thus

C, *D* cadabra loving

would do as well. This strongly suggests that the standard treatment of exemplification represents a mixture of two only tangentially related patterns of thought: (a) a nexus of puzzles pertaining to 'true of' and (b) a nexus of *puzzles* pertaining to predication.

XI

128. But once we appreciate the fact that there are two equivalent ways of expressing a relation between objects:

(a) Using a non-name which stands for the relation
(b) Using a relation between tokens of the name of the objects to express the fact that the objects are so related

it must necessarily occur to us that the same is true at a lower step on the semantical ladder.

129. Thus there could be two languages containing different but equivalent ways of saying that *a* is larger than *b*

English PMese:	Larger than *a*, *b*
Jumblese:	a^b

in the former the non-name 'larger than' is used; in the latter no non-name is used, but that *a* is larger than *b* is expressed by placing a token of '*b*' catacorner to the right of a token of '*a*'.[9]

130. According to our present line of thought, *in each case* a token of '*a*' is placed in a conventional dyadic relation to a token of '*b*'. In the one case, however, the use of an auxiliary symbol would be involved, i.e. *larger than.* In the other case not. Thus the following two statements would have the same syntactical form:

Larger than *a*, *b*

a^b

They would both consist of an '*a*' standing in a conventional dyadic relation to a '*b*'.

131. Well, it might be said, so what? The answer, believe it or not, is so *everything*! In other words here is the key to the problem of predication and, for that matter, the Ariadne's thread to the labyrinth of metaphysics.[10]

132. It has surely become apparent that the standard theory has been pursuing single-mindedly the idea that in

fa

the '*f*' and the '*a*' *stand in distinct and autonomous semantic relations to extra-linguistic reality*. The former stands for *something*, the latter for *something else*. The former something is predicated of the latter something else.

133. Initially the *somethings* in question were both interpreted as *objects*. This interpretation, however, was abandoned when it was seen to lead to puzzles. But the above single-minded idea was maintained.

134. But what, it may be said, of the concession made in (64) above to the standard theory that there must be *something* in extra-linguistic reality which *somehow* is singled out by '*f*' (e.g. 'red') by virtue of a *relation* to which something an object is *f*.

135. Consider, in particular, what becomes of our fledgling theory when it is applied to the predication of qualities as contrasted with relations – in other words when it is applied to 'monadic relations' or, *grammatice*, to adjectival predication?

136. Obviously, the extension of the concept of an auxiliary symbol to cover this case proceeds as follows:

> We assert that fa by giving a token of '*a*' a conventional monadic character involving the auxiliary symboi **f**, e.g. the character of being concatenated to the left with an **f**.

137. From this point of view '*fa*' would be viewed as a token of '*a*' which has the monadic character of being concatenated to the left with an **f**.

138. A person who is naive to the ways of logical form and who can only see

aRb

as attributing the dyadic character of *standing in R* to *a* and *b*, but not as attributing the monadic character of *being R to b* to *a*, will fail to appreciate that the syntactically significant fact about the expression '*fa*' might be the fact that it is an '*a*' which has the monadic character of being concatenated to the left with an **f**, i.e. with a case of the sign design between the asterisks.

139. The heart of the matter is that we can conceive of two syntactical systems which have in a legitimate sense the same syntactical form but where in one case this form involves the use of auxiliary symbols, in the other case not.

140. Thus in these two systems

a^b

and

a adjoins *b*

would have respectively the same syntactical form, each consisting of a token of '*a*' and a token of '*b*' in a conventional dyadic relation.

XII

141. But, it will be expostulated,

> Syntactics, schmintactics! Our problem is a problem in *semantics*, and you have not yet answered the question raised in (134).

142. This challenge is direct and to the point, and answering it will enable us to pull the argument together.

143. First, the blunt answer. Look at the diagram with which we began. Is there anything represented in the diagram which has not played a major role in the argument? Consider this part:

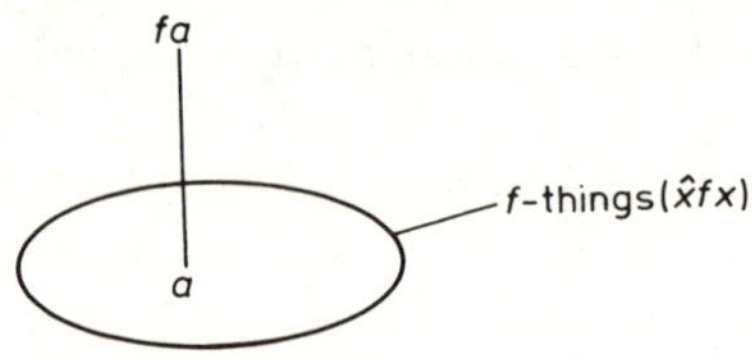

144. I scarcely need to remind you that the term 'thing' is to be watched like a hawk. Thus we all know that the 'problem of the One and Many' is many problems. Thus there is the problem of the unity of one thing having many features and that of many things having one and the same feature.

145. Thus, schematically,

Many triangles – one feature: triangularity$_{21}$

One thing – many features: shape$_{45}$, color$_{25}$.[11]

146. Subjects of predication belong to different kinds and indeed different kinds of kinds, as do the items which are predicated of them.

147. Thus, such a thing-kind predicate as 'horse' as in

Zev is a horse

is interestingly different from such a proper sensible predicate as 'red'. There are complexities here which easily get tangled up with 'the' problem of predication. Thus consider the question 'what is a red thing?' A fire truck is, in one sense, a red thing. But it is also many other things.

148. If we call this *many*, its *features*, we can say that, in some sense, the truck is a *complex object*, of which its features are constituents. Thus, there is a sense in which

red is a constituent of truck 20.

149. Consider the proposition

Truck 20 is red

It has been obvious since Aristotle,[12] that the linguistic expression 'red' can be predicated both of truck 20 and of a certain feature of truck 20. Crudely put, both the truck and its color feature are red – in related, but not identical senses of 'red', much as the All-American boy and Wheaties are both healthy.

150. To use a butcher knife instead of a scalpel let us simply say that the fire truck is a *complexum*, that one of its constituents is a red expanse, and that for the *complexum* to be red is simply for it to *contain* a constituent which is red.

151. Our two senses of 'red' should then be

(a) *Contains in an appropriate manner a constituent (feature) which is red.* Let us speak of this constituent as a case of red or, for brevity, a red. (This amounts to introducing the predicate 'red' as a sortal.) Thus the relevant feature of the truck is an expanse of red, a red.
(b) The sense in which an expanse of red is red. This would be the basic sense, and the other sense definable in terms of it. Thus to be red in sense (a) is to contain a feature which is red (b).

152. This reduces the problem of predicating red in sense (a) of a complexum to that of predicating 'red' in sense (b) of a feature. It is important to realize that traditional discussions of predication have often blurred the distinction between these two modes of predication.

153. Notice that in the case of the truck it is tempting to say that what is predicated is a particular (the red feature). This would be a serious mistake. What is predicated of the truck is the character *being red in sense (a)*. This is, of course, a universal, thus:

Fire truck 21 is red (*a*)

Fire truck 22 is red (*a*)

and so on.

154. To put the matter accurately, we should say that what is predicated of the truck is not a particular, but the character of having a particular which is red (*b*) as a constituent. But it is only too easy to run things together and think of

Truck 21 is red

as predicating a red particular of Truck 21.

155. Let us avoid this mistake and, making a move in the direction of logical atomism let us think of the fire truck as a structure which consists of its features, including the red which is so prominent.

156. And when we speak of red things, let us have in mind red features (reds). In this usage the fire truck will not be a red thing. It will contain a red thing as one of its constituents. In this usage the world continues to contain red things to be sure, but now the 'things' in question are expanses of red, items which are, so to speak, red through and through – unlike the fire truck.

XIII

157. I have distinguished two technical senses in which a *meaning* can be assigned to a singular term which refers to a particular without *meaning* being a matter-of-factual relation, thus

(1) x den-in-L y: where this is defined by a list using identity, conjunction, and disjunction

(2) '- - -' means (in-L) . . .: where this is construed as '- - -'s (in-L) are ˙. . .˙s.

But I have also emphasized that an expression only stands for what it does by virtue of the nexus of relations in which it stands to other expressions and to the extra-linguistic order.

158. Let me now introduce the concept of *linguistic representation* and say, to begin with, that a singular term INDCON_i is the linguistic representative (LR) (in L) of an object, O, if and only if by virtue of its role in L it stands in appropriate psycho-sociological-historical (PSH) relations to O. Thus we have

$$\text{INDCON}_i \text{ is LR (in L) of } O \equiv \text{PSH } (\text{INDCON}_i, O).$$

159. Since the matter-of-factual relations involved in PSH are those to which appeal must be made in justifying ascriptions of meaning, there will also be the equivalences

$$\text{INDCON}_i\text{s (in L) are } \dot{}\ldots\dot{}\text{s} \equiv \text{PSH } [\text{INDCON}_i, O]$$

$$\text{INDCON}_i\text{s (in L) den-in-L } O \equiv \text{PSH } [\text{INDCON}_i, O].$$

160. Thus, even though

'Sokrates' (in G) means *Socrates*

is not a relational statement, it nevertheless implies the truth of a complicated relational statement pertaining to the German expression 'Sokrates' (in a certain usage) and a certain snubnosed Greek philosopher. What this statement is, is, of course, a challenge. It is a statement of a kind which would be expressed by a category in a well organized theory of linguistic representation – if such there were. This category would provide the answer to the question: by virtue of what does a singular term (e.g. 'the moon') pick out a certain object and serve as its linguistic representative? This question is as difficult a question as there is – which is not to say that we don't have vague and open-textured ways of coping with it.

161. We can, in any case, begin our account of the explanation of truth ascriptions by pointing out that 'Sokrates' (in G) means *Socrates* if and only if 'Sokrates' (in G) is the linguistic representative (LR) of Socrates.

162. At the level of predicates a similar distinction is required. We might begin with

'rot' (in G) means *red* ≡ 'rot' (in G) is the LR of *red*.

But if we interpret the left-hand side correctly, as

'rot's (in G) are ·red·s

and ask what is it to be a ·red·? The answer must surely be that it is to be an auxiliary symbol which, appropriately concatenated with an INDCON makes a statement to the effect that the relevant object is red. Thus,

'rot'INDCONs (in G) say of certain objects that they are red

where to be a 'rot'INDCON is to be an INDCON concatenated to the left with a *rot*.

163. Thus, instead of 'rot's being linguistic representatives of something, it is 'rot'INDCONs which are linguistic representatives – of what? *red things*.

164. We come back again to the theme that the basic mistake of the standard theory of predication is that it sees the '*f*' and the '*a*' as having autonomous and parallel semantical relations to extra-linguistic items.

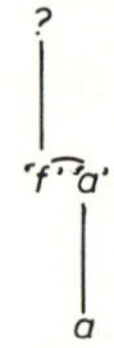

Whereas the truth is that by virtue of being an '*a*' concatenated with an **f**, an '*fa*' is an '*a*' which has a character by virtue of which it is semantically associated with *f*-things. Thus the correct diagram is

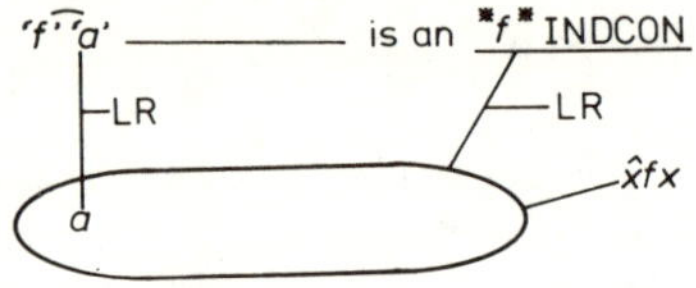

where '*a*' is the LR of *a* and *f* INDCONs are the LRs of *f*-things.

165. In Jumblese,[13] it is not INDCONs which are concatenated with an '*f*' which are LRs of *f*-things, but, for example,

i.e. a wavy '*a*'

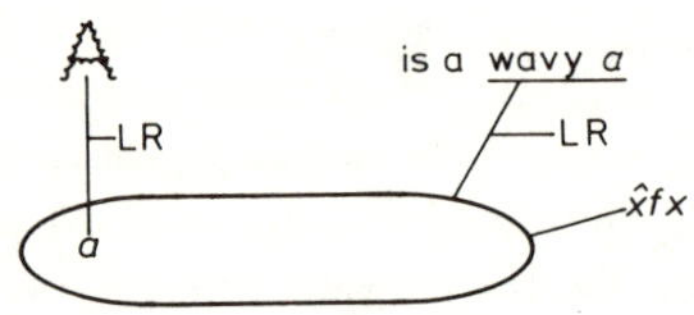

166. Thus when looking for an *explanation* of the truth of 'red *a*' we should concentrate on the fact that

'*a*' is a LR of *a*

and that in the sentence 'red *a*', the '*a*' has a character by virtue of which it belongs to a class of linguistic representatives of red things.

167. Thus, what explains the truth of the English sentence – and, of course, the truth of a sentence is always the truth of a sentence in a language – is not its being, *in the formalist sense*, true-in-E, as defined above, but rather in the fact that

'*a*' is the LR in English of *a*

and that

'*a*'s with a certain character are the LRs in English of red things.

168.

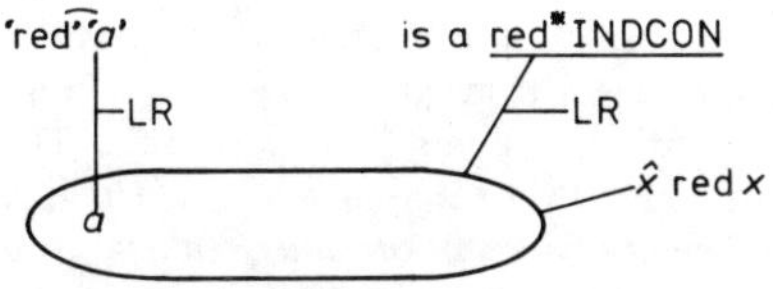

169. Thus the fact that *a* is red explains the truth of 'red *a*' by virtue of the PSH facts which bring it about that

'*a*' is the LR (in E) of *a*

and that

INDCONs (in E) which are concatenated to the left with a *red* (*red*INDCONS) are LRs of red things.

170. To all this we need only add that a theory of truth is a theory of semantic assertability.[14]

'*p*' is true $=_{\text{df}}$ '*p*' is SA.

171. In the case of atomic matter-of-factual statements the core of assertability is *contributing to a map of the world in which we find ourselves*. The concept of linguistic representation is basic to the concept of a linguistic map. But in a deeper sense, the basic concept is that of finding ones way around.[15]

172. To continue would be, of course, to continue the enterprise of philosophy as a whole.

NOTES

[1] For an explanation of this grammatical category, see my 'Abstract Entities', *Review of Metaphysics* **16** (1963). [Reprinted in *Philosophical Perspectives*, Vol. II, Ridgeview. (Reseda, Cal., 1977)].

[2] To emphasize that it is still very much alive.

[3] Where no established expression ending in 'ity', 'hood', '-ness', or '-cy', exists, one can always create an appropriate singular term by prefixing the predicative expression with 'being', thus 'being triangular'.

[4] For an elaboration of this point, see Chapter 1 of my *Naturalism and Ontology* [Reseda, Cal., 1979].

[5] For the longer story, see 'Meaning as Functional Classification', *Synthese* 27 (Reidel, Dordrecht, Holland (1974)).

[6] This development in Russell's thought did not become explicit until his essay 'On Propositions' first published in 1919 and reprinted in *Logic and Knowledge*. For a detailed analysis of this development, see my 'Ontology and the Philosophy of Mind in Russell', in G. Nakhnikian (ed.), *Bertrand Russell's Philosophy*, London, 1974.

[7] I began my exploration of the virtues and dangers of listing definitions in my paper 'Empiricism and Abstract Entities' in the Carnap volume. The point was focussed on semantical issues in 'Truth and "Correspondence" ', *The Journal of Philosophy* 59 (1962) and reprinted in *Science, Perception and Reality*, Routledge and Kegan Paul, London (1963).

[8] An essential feature, of course, of the Tarski strategy is that the definition of 'true-in-L' involves a recursive structure which enables it to apply 'true-in-L' to molecular and quantified sentences-in-L. This feature, however, adds nothing of importance to our understanding of the bearing of the definition of 'true-in-L' on the explanation of matter-of-factual truth ascriptions.

[9] This theme was developed in my 'Naming and Saying', *Philosophy of Science* 29, 1962. (Reprinted as Chapter 7 in *Science, Perception and Reality*.)

[10] The theory of predication to which this is pointing is clearly one that holds that predicates are *merely* auxiliary symbols. To see this, is to *see through* the problem of predication. But we have not yet brought into our account the dimension of truth. This will be the final task – within the limits of this paper.

[11] Triangularity 21 is the twenty-first species of triangularity; shape 45 is the forty-fifth species of shape.

[12] For a discussion of this point see my 'Substance and Form in Aristotle' *Journal of Philosophy* 54 (1957). [Reprinted in *Philosophical Perspectives*, Vol. I Ridgeview (Reseda, Cal., 1977) along with 'Aristotles Metaphysics: An Interpretation', the larger paper from which it was excerpted.]

[13] "Far and few, far and few are the lands where the Jumblies live". Edward Lear.

[14] For an account of truth as semantic assertability see Chapter IV of my *Science and Metaphysics*, London and N.Y., 1968. (Now available in paperback from the Humanities Press.

[15] For an elaboration of this point, see the concluding chapter of *Naturalism and Ontology*; also 'Mental Events', *Philosophical Studies* 39, 1981. The latter paper throws additional light on the problem of predication.

ALAN CODE

APPENDIX ON 'THE THIRD MAN ARGUMENT'

The material in the appendix is taken from unpublished work in progress by H. P. Grice and myself.

I. GENERAL PREMISES UTILIZED IN THE TMA

ALEX-OM: If each member of a plurality of objects is a man, then there is an X such that:
(1) X is a man,
(2) X is separable,
& (3) X is predicable of each member of that plurality.

ALEX-NI: If what is predicable of each member of a plurality is separable, then it is distinct from each member.

II. THE TMA

{assume} (A) Each member of the plurality of sensible men is a man.

{instance of ALEX-OM} (B) If each member of the plurality of sensible men is a men, then there is an X such that:
(1) X is a man,
(2) X is separable,
& (3) X is predicable of each member of that plurality.

{A, B} (C) There is an X such that:
(1) X is a man,
(2) X is separable,
& (3) X is predicable of each member of the plurality of sensible men.

{hypothesis} (D) U | (1) U is a man,
| (2) U is separable,
| & (3) U is predicable of each member of the plurality of sensible men.

James Bogen and James E. McGuire (eds.), How Things Are, 323–326.

Justification	Line		Statement
{import ALEX-NI, & instantiate}	(E)	U	If U is predicable of each member of the plurality of sensible men and U is separable, then U is distinct from each member of the plurality of sensible men.
{D2, D3, E}	(F)		U is distinct from each member of the plurality of sensible men.
{A, D1}	(G)		Each member of the plurality of sensible men is a man, and U is a man.
{SEE III BELOW}	(H)		Each member of the plurality consisting of U together with the sensible men is a man.
{instance of ALEX-OM}	(I)		If each member of the plurality consisting of U together with the sensible men is a man, then there is a Y such that: (1) Y is a man, (2) Y is separable, & (3) Y is predicable of each member of the plurality consisting of U together with the sensible men.
{H, I}	(J)		There is a Y such that: (1) Y is a man, (2) Y is separable, & (3) Y is predicable of each member of the plurality consisting of U together with the sensible men.
{hypothesis}	(K)	V	(1) V is a man, (2) V is separable, & (3) V is predicable of each member of the plurality consisting of U together with the sensible men.
{import ALEX-NI, & instantiate}	(L)		If V is predicable of each member of the plurality consisting of U together with the sensible men, and V is separable, then V is dinstinct from each member of that plurality.

{K2, K3, L}	(M)	U | V |	V is distinct from each member of the plurality consisting of U together with the sensible men.
{K1, M; D1, F}	(N)		(V is a man & V is distinct from each member of the plurality consisting of U together with the sensible men) & (U is a man & U is distinct from each member of the plurality of sensible men).
{N, applying existential generalization twice}	(O)		$(\exists X)\ (\exists Y)$ ((Y is a man and Y is distinct from each member of the plurality consisting of X together with the sensible men) & (X is a man and X is distinct from each member of the plurality of sensible men)).

We may now discharge the hypotheses, (K) and (D), and assert (O) in accordance with the following rule (where U is a parameter that does not occur earlier in the deduction, Ψ contains no occurrences of U, and $\Phi\ (X/Y)$ is just like Φ except that it has an occurrence of X wherever Φ has an occurrence of U):

| $(\exists X)\Phi\ (X/Y)$
| U | Φ
| | •
| | •
| | •
| | Ψ (no formula containing U may be imported across the 'U' barrier
| Ψ

(For this type of rule, see R. H. Thomason, *Symbolic Logic: An Introduction,* Macmillan (1970), p. 192.)

III. COMMENT

Proposition (O) asserts that in addition to the sensible men, there are two distinct items, each of which is a man, one of which is predicable of all sensible men, the other being predicable of this item as well as the sensible men.

The reasoning may be repeated indefinitely to generate an infinite regress without inconsistency.

Semantic meta-principle (A) and a Plurality Principle are both needed to go from (G) to (H). (The meta-principle from Section III of my paper should not be confused with step (A) in Section II of this appendix.

NOTES ON THE CONTRIBUTORS

Marilyn McCord Adams, Professor of Philosophy, UCLA, has published numerous articles on medieval philosophy. Her book, *Ockham*, is forthcoming from the Routledge and Kegan Paul *Arguments of the Philosophers Series*.

Robert Merrihew Adams, Professor of Philosophy, UCLA, writes on the metaphysics of modality, the philosophy of religion, philosophical and theological ethics, and the history of early modern philosophy.

James Bogen, Professor of Philosophy at Pitzer College, Claremont, California, is musical director of the 12th Street Mini-Band (presently, The Real Time Jazz Band) and an honorary member of the Hot Pecans.

Frank A. Lewis, Associate Professor of Philosophy at the University of Arizona, writes on Ancient Philosophy, is presently completing a book on Aristotle's *Metaphysics*, and is the owner of a small wine cellar.

James E. McGuire, of the University of Pittsburgh History and Philosophy of Science Department, is the editor, with Martin Tamny, of a critical edition of Newton's *Certain Philosophical Questions: Newton's Trinity Notebook* (Cambridge, 1983). *Philosophical Themes in Newton's Early Thought* is forthcoming from Reidel. He has published extensively on the history of science and philosophy, and is an authority on diseases of the liver.

D. K. Modrak, Assistant Professor of Philosophy at the University of Rochester, is working on a book on Aristotle's theory of perception and epistemology.

Calvin G. Normore, Associate Professor of Philosophy at the University of Toronto, is currently writing a book on Ockham's logic and is one of the editors of the *Abstractiones* of the Master of the *Abstractiones*. A professed anarchist, he prefers predication to subjection.

Wilfrid Sellars, University Professor of Philosophy at the University of Pittsburgh, is considered by many to be the Wilfrid Sellars of contemporary American Philosophy. Among many other publications, he is the author of *Naturalism and Ontology*, *Science, Perception, and Reality*, and *Foundations for a Metaphysics of Pure Process*, and the Carus Lectures, published in the January 1981 *Monist*. In addition to philosophy, he frequently lectures on baseball statistics.

Robert G. Turnbull, whose family crest is better known than those of

James Bogen and James E. McGuire (eds.), How Things Are, 327–328.

other contributors to this volume, is a Professor of Philosophy at Ohio State University and editor of *Philosophy Research Archives*. He has published papers in Greek, Medieval and Modern Philosophy, Ethics, and Metaphysics. With Peter K. Machamer, he edited *Motion and Time, Space and Matter*, and *Studies in Perception* – both published by the Ohio State University Press. He is also at work on a translation of and commentary on Plato's *Parmenides*.

BIBLIOGRAPHY

Texts marked with an asterisk are cited in this volume by passage or section numbers, rather than page numbers peculiar to any one edition.

Ackrill, J. L., 'Aristotle's Definitions of Psuche', *Proceedings of the Aristotelian Society* **73** (1972–73), 119–33.

Ackrill, J. L., 'Aristotle's Definitions of Psuche', Reprinted in *Articles on Aristotle*, Vol. 4 (edited by J. Barnes, M. Schofield, and R. Sorabji), Duckworth, 1979.

Adams, Marlyn, *Ockham*, Routledge and Kegan Paul (forthcoming).

Adams, Robert Merrihew, 'Leibniz's Theories of Contingency', pp. 1–14, in Mark Kulstad (ed.), *Essays on the Philosophy of Leibniz* **63** (Fall 1977).

Adams, Robert Merrihew, Review of W. L. Rowe, *The Cosmological Argument, Philosophical Review* **87** (1978), 445–50.

Adams, Robert Merrihew, 'Existence, Self-Interest, and the Problem of Evil', *Nous* **13** (1979). Printed with six paragraphs transposed; they begin on p. 60 and should begin on p. 64.

Albert of Saxony. *Perutilis Logica*, Octaviania Scoti, Venice, 1522.

Albritton, Rogers, 'Forms of Particular Substances in Aristotle's Metaphysics', *Journal of Philosophy* **54** (1957), 699–708.

Allen, R. E., 'Participation and Predication in Plato's Middle Dialogues', Reprinted in *Studies in Plato's Metaphysics* (edited by R. E. Allen), New York, 1965.

Allen, R. E. (ed.), *Studies in Plato's Metaphysics*, Routledge and Kegan Paul, 1965, pp. 293–312.

Aristotle, (see Balme, Charlton, Jaeger, Hicks, Kirwan, Ross below for editors and Ranslations of particular works.) *Categories*.*

Aristotle, *De Anima* (*On the Soul*).*

Aristotle, *Metaphysics*.*

Aristotle, *Nichomachaean Ethics*.*

Aristotle, *On Interpretation*.*

Aristotle, *Posterior Analytics*.*

Arnauld, Antoine and Nicole, Pierre, *La Logique, Ou L'Art de Penser* [The Port Royal Logic] (edited by P. Clair and F. Girbal), Critical ed., Presses Universitaries de France, Paris, 1965.*

Balme, D. M., '*Genos* and *Eidos* In Aristotle's Biology', *Classical Quarterly* **12** (1962), 81–98.

Balme, D. M., *Aristotle's de Partibus Animalium I and de Generatione Animalium I*, Oxford: 1972.

Bonitz, H., *Index Aristotelicus*, Berlin, 1890.

Brandom, Robert B., 'Leibniz and Degrees of Perception', *Journal of the History of Philosophy* **19** (1981), 447–70.

Broad, C. D., *Leibniz: An Introduction* (edited by C. Lewy), Cambridge, 1975.

James Bogen and James E. McGuire (eds.), How Things Are, 329–336.

Brody, Baruch, 'Leibniz's Metaphysical Logic', pp. 43–55, in Mark Kulstad (ed.), *Essays on the Philosophy of Leibniz* **63** (Fall 1977).

Burnyeat, Myles *et al.*, *Notes on Book Zeta of Aristotle's Metaphysics*, Published as Monograph 1, Study Aids, Sub-faculty of Philosophy, Oxford, 1979.

Buridan, Johannes, *Questiones Super Octo Libros, Physicorum Aristotelis*, Pierre Dru, 1509.

Buridan, Johannes, *Sophismata* (edited by T. K. Scott), Fromann-Holzboog, Stuttgart, 1977.

Burleigh, Walter, *De Puritate Artis Logicae, Tractatus Longior* (edited by Philotheus Boehner), St Bonaventure, New York, 1955.

Carnap, Rudolf, 'Empiricism, Semantics, and Ontology', Supplement A in Rudolf Carnap, *Meaning and Necessity*, Chicago, 1956.

Charlton, W., *Aristotle's Physics I, II*, Oxford, 1970.

Cherniss, Harold, *Aristotle's Criticism of Plato and the Academy*, Vol. 1, Johns Hopkins, 1944.

Cherniss, Harold, 'The Relation of the Timaeus to Plato's Later Dialogues', *The American Journal of Philosophy* 78 (1957), 225–66. Reprinted in R. E. Allen (ed.), *Studies in Plato's Metaphysics*, Routledge and Kegan Paul, London, 1965.

Code, Alan, 'The Persistence of Aristotelian Matter', *Philosophical Studies* **29** (1976).

Code, Alan, 'No Universal Is a Substance: An Interpretation of Metaphysics Z13, 1038b8–15', *Paideia* (1978), pp. 65–74.

Code, Alan, 'Aristotle, Essence and Accident', in *Philosophical Grounds of Rationality: Intentions, Categories, Ends* (edited by R. Grandy and R. Warner), Oxford, 1983.

Code Alan, 'Aristotle on the Sameness of Each Thing with Its Essence', Unpublished.

Cohen, S. Marc, 'Essentialism in Aristotle', *Review of Metaphysics* **21** (1978), 387–405.

Cohen, S. Marc, 'Individual Essence in Aristotle's Metaphysics', *Paideia, Aristotle Issue, Buffalo* (1978), pp. 75–8.

Cohen, S. Marc, 'The Logic of the Third Man', *Philosophical Review* **80** (1971), 448–57.

Couturat, Louis, *La Logique de Leibniz*, Alcan, Paris, 1901.

Couturat, Louis, 'On Leibniz's Metaphysics', pp. 19–45 (translated by R. A. Ryan), in *Leibniz: A Collection of Critical Essays* (edited by Harry G. Frankfurt), Doubleday, Garden City, 1972.

Crusius, Christian August, *Entwurf der Notwendigen VernunftWahreiten*, 2nd ed, Leipzig, 1753.

Dante, *Divine Comedy*.*

Denniston, J. D., *The Greek Particles*, 2nd ed, Oxford.

Descartes, René, *Oeuvres de Descartes*, Vols. 1–12 (edited by Adam and Tannery), Paris, 1897–1910; Index, Paris, 1931.

Descartes, René, *Philosophical Works*, Vols. 1 and 2 (edited and translated by Haldane and Ross, Dover 1955.

Priscoll, John A., '*Eide* in Aristotle's Earlier and later Theories of Substance', *Studies in philosophy and Aristotle* (edited by D. J. O'Meara), Catholic University press, 1981.

Dummett, Michael, *Frege: Philosophy of Language*, 2nd ed., Harvard University Press, Cambridge, Massachusetts, 1981.

Earman, John, 'Perceptions and Relations in the *Monadology*', *Studia Leibnitiana, Band* 9 (1977), 212–30.

Ferrer, Vincent, *Tractatus de Suppositionibus* (edited by J. A. Trentman), Fromann-Holzboog, Stuttgart, 1977.

Fine, Gail, 'Aristotle and the More Accurate Arguments', *Language and Logos: Studies in Ancient Greek Philosphy Presented to G. E. L. Owen* (edited by M. Schofield and M. C. Nussbaum), Cambridge, 1982.

Fine, Gail, 'Owen, Aristotle, and the Third Man', *Phronesis* (Forthcoming).

Fine, Kit, 'Aristotle on Substance' (unpublished).

Frankfurt, Harry G. (ed.), *Leibniz: A Collection of Critical Essays*, Garden City, 1972.

Frege, Gottlob, Geach, Peter, and Black Max (eds.), *Translations from the Philosophical Writings*, 2nd ed. Blackwell, 1966.

Furth, Montgomery, 'Transtemporal Stability in Aristotelian Substances', *Journal of Philosophy* 75 (1978), 624–46.

Gál, Gideon, 'Adam of Wodeham's Question on the "Complexe Singificabile" as the Immediate Object of Scientific Knowledge', *Franciscan Studies* **15** (1977).

Geach, Peter T., 'The Third Man Again', Reprinted in R. E. Allen (ed.), *Studies in Plato's Metaphysics*, New York, 1965.

Gueroult, Martial, *Leibniz: Dynamique et Metaphysique*, Aubier, Paris, 1967.

Harter, Edward D., 'Aristotle on Primary *Ousia*', *Archiv fur Geschichte der Philosophie* 57, pp. 1–20.

Hartman, Edwin, *Substance, Body, and Soul: Aristotelian Investigations*, Princeton, 1977.

Hartman, Edwin, 'Aristotle on the Identity of Substance and Essence', *Philosophical Review* **65** (1976), 545–61.

Hayduck, Michael, *Alexandri Aphrodisensis in Aristotelis Metaphysica Commentaria*, Vol. 1, *Commentaria in Aristotelem Graeca*, Berlin, 1891.

Hicks, R. D., *Aristotle, de Anima*, Cambridge, 1907.

Hooker, Michael (ed.), *Leibniz: Critical and Interpretive Essays*, University of Minnesota Press, 1982.

Ishiguro, Hide, *Leibniz's Philosophy of Logic and Language*, Duckworth, London, 1972.

Ishiguro, Hide, 'Leibniz's Theory of the Ideality of Relations', in Harry G. Frankfurt (ed.), *Leibniz: A Collection of Critical Essays*, Doubleday, Garden City, 1972.

Jaeger, Werner, *Aristotle* (translated by R. Robinson), 2nd ed., Oxford, 1948 (First German ed., Berlin, 1923).

Jaeger, Werner, *Aristotle, Fundamentals of the History of His Development*, Oxford, 1948 (First German edition, *Aristoteles, Grundlegung Einer Geschichte Seiner Entwicklung*, Berlin, 1923).

Jaeger, Werner, *Aristotelis Metaphysica*, Oxford Classical Texts, Oxford, 1957.

Janke, Wolfgang, 'Das Ontologische Argument in der Früzeit Das Leibnizschen Denkens', *Kantstudien* **54** (1963), 259–87.

Kauppi, Raili, *Über die Leibnizsche Logik – Fasc. 12 of Acta Philosophica Fennica*, Helsinki, 1960.

Kirwan, Christopher, *Aristotle's Metaphysics: Books of Gamma, Delta, and Eta*, Oxford, 1971.

Kulstad, Mark (ed.), *Essays on the Philosophy of Leibniz*, Vol. 63 (Fall 1977).

Kung, Joan, 'Can Substance Be Predicated of Matter?' *Archiv für Geschichte der Philsophie* **60** (1978), 140–59.

Kung, Joan, 'Aristotle on Thises, Suches, and the Third Man Argument', *Phronesis* **26** (1981), 207–47.

Lear, Jonathan, *Aristotle and Logical Theory*, Cambridge, 1980.

Leibniz, G. W., (A) *Samtliche Schriften und Briefe*, Academy ed. Darmstadt and Berlin, 1923.

Leibniz, G. W., (C) *Opuscles et Fragments Inedits de Leibniz*. Edited by Louis Couturat, Paris, 1903.

Leibniz, G. W., (Conf) *Confession Philosophy* (edited and translated into German with a Commentary by Otto Samme), Frankfurt, 1967.

Leibniz, G. W., (DM) *Discourse on Metaphysics**, in G. W. Leibniz, *Basic Writings* (translated by Montgomery), La Salle, 1962.

Leibniz, G. W., (G) *Die Philosphischen Schriften von G. W. Leibniz* (edited by C. I. Gerhardt), 7 vols, Berlin, 1875–90.

Leibniz, G. W., (GR) *Textes Inedits* (edited by Gaston Grua), Paris, 1948.

Leibniz, G. W., (L) *Philosophical Papers and Letters* (translated and edited by LeRoy E. Loemker), 2nd edition, Dordrecht and Boston, 1969.

Leibniz, G. W., (L–A) *The Leibniz–Arnauld Correspondence* (edited and translated by H. T. Mason), Manchester, 1967.

Leibniz, G. W., (L–C) *The Leibniz–Clarke Correspondence* (edited by H. G. Alexander), using Clarke's translation of Leibniz's letters, Manchester, 1956.

Leibniz, G. W., (LW) *Lettres de Leibniz a Arnauld* (edited by Genevieve Lewis), Paris, 1952.

Leibniz, G. W., (Mon) *Monadology*, in G. W. Leibniz, *Basic Writings* (translated by Montgomery), La Salle, 1962.*

Leibniz, G. W., (NE) *New Essays Concerning Human Understanding** (translated by Langley), LaSalle, 1949.

Leibniz, G. W., (P) *Logical Papers* (edited and translated by G. H. R. Parkinson), Oxford, 1966.

Leibniz, G. W., (RML) *Malebranche et Leibniz: Relations Personnelles* (Presented by André Robinet), Paris, 1955.

Leibniz, G. W., (T) *Theodicy**.

Lennox, James, 'Teleology, Chance, and Aristotle's Theory of Spontaneous Generation'. *Journal of the History of Philosophy* (July 1982).

LeRoy, Georges, *Leibniz: Discours de Metaphysique et Correspondence Avec Arnauld: Introduction, Texte et Commentaire*, Vrin, Paris, 1957.

Lesher, James H., 'Aristotle on Form, Substance, and Universals: A Dilemma'. *Phronesis* **16** (1971), 169–78.

Lewis, David, 'Counterpart Theory and Quantified Model Logic', *Journal of Philosophy* **65** (1968), 113–26.

Lewis, Frank A., 'What Is Aristotle's Theory of Essence?', *Canadian Journal of Philosophy*, Supplementary vol. 9 (edited by Francis Jeffry Pelletier), 1983.

Lewis, Frank A., 'Accidental Sameness in Aristotle', *Philosophical Studies* (forthcoming).

Lewis, Frank A., 'Substance and Predication: Some Themes from Aristotle's Metaphysics Z12–16' (unpublished).

Loeb, Louis E., *From Descartes to Hume*, Cornell University Press, 1981.

Long, Peter, 'Formal Relations', *Philosophical Quarterly* **32** (April 1982).

Loux, Michael J., 'Form, Species, and Predication in Metaphysics Z, H, and Theta', *Mind* **88** (1979), 1–23.

Lukasiewicz, Jan, 'The Principle of Individuation', *Proceedings of the Aristotelian Society*, Supplementary vol. 27 (1953), pp. 69–82.

Maier, Anneliese, *Zwischen Philosophie und Mechanik*, Rome, 1958.

Malcolm, John, 'On the Generation and Corruption of the Categories', *Review of Metaphysics* **33** (1981), 662–81.

Mansion, Suzanne, 'The Ontological Composition of Sensible Substances in Aristotle (Metaphysics VII, 7–9), in *Articles on Aristotle* (edited by Jonathan Barnes, Malcolm Schoefield, and Richard Sorabji), Duckworth, London, 1979.

Mates, Benson, 'Individuals and Modality in the Philosophy of Leibniz', *Studia Leibnitiana* **4** (1972), 81–118.

Mates, Benson, 'Leibniz on Possible Worlds', in Harry G. Frankfurt (ed.), *Leibniz: A Collection of Critical Essays*, Doubleday, Garden City, 1972.

McGuire, James E., 'Labyrinthus Continui: Leibniz on Substance, Activity, and Matter', in *Motion and Time, Space and Matter* (edited by Peter Machamer and Robert G. Turnbull), Columbus, 1976.

Modrak, D. K., 'Forms, Types, and Tokens in Aristotle's Metaphysics', *Journal of the History of Philosophy* **17** (1979), 371–81.

Mondadori, Fabrizio, 'Leibniz and the Doctrine of Inter-World Identity', *Studia Leibnitiana* **7** (1975), 21–57.

Mondadori, Fabrizio, 'The Leibnizian "Circle" ', pp. 69–96, in *Essays on the Philosophy of Leibniz* (edited by Mark Kulstad), **63** (Fall 1977).

Moravcsik, J. M. E. (ed.), *Aristotle: A Collection of Critical Essays*, Doubleday, New York, 1967.

Moravcsik, J. M. E., 'Hercleitian Explanation', *Monist* (forthcoming).

Mourelatos, Alexander P. D., 'Aristotle's "Powers" and Modern Empiricism", *Ratio* **9** (1967), 97–104.

Nehamas, A., 'Self-Predication and Plato's Theory of Forms', *American Philosophical Quarterly* **16** (1979), 93–103.

Normore, Calvin G., 'The Logic of Time and Modality in the Later Middle Ages', Ph.D. dissertation, University of Toronto, 1976.

Ockham, William of, *Summa Logicae* (edited by P. Boehner, S. Brown, and G. Gal), Franciscan Institute, St. Bonaventure, New York, 1974.

Owen, G. E. L., ' A Proof in the Peri Ideon', *Journal of Hellenic Studies* 77 (1975), 103–11 (part 1). Reprinted in R. E. Allen (ed.), *Studies in Plato's Meaphysics,* Routledge and Kegan Paul, 1965, pp. 293–312.

Owen, G. E. L., 'Dialectic and Eristic in the Treatment of the Forms', in *Aristotle on Dialectic: The Topics* (edited by G. E. L. Owen), Oxford, 1968.

Owen, G. E. L., 'Logic and Metaphysics in Some Earlier Works of Aristotle', pp. 13–32. Reprinted in *Articles on Aristotle*, Vol. 3 (edited by J. Barnes, M. Schofield, and R. Sorabji).

Owen, G. E. L., 'Particular and General', *Proceedings of the Aristotelian Society* 79 (1978–79), 1–21.

Owen, G. E. L., 'The Platonism of Aristotle'. Reprinted in *Articles on Aristotle*, Vol. 1 (edited by J. Barnes, M. Schofield, and R. Sorabji), Duckworth, London, 1975.

Parkinson, G. H. R., *Logic and Reality in Leibniz's Metaphysics*, Clarendon Press, Oxford, 1965.

Paul of Venice, *Logica Magna, Part II, De Significato Propositionis*, Del Punta and Adams, Oxford, 1977.

Plato, *The Parmenides*.*

Plato, *Phaedo*.*

Plato, *Phaedrus*.*

Plato, *Philebus*.*

Plato, *Republic*.*

Plato, *The Sophist*.*

Plato, *Statesman*.*

Plato, *Timaeus*.*

Plotinus, *Enneads*.*

Porphyry, 'On the Life of Plotinus and the Order of His Books', pp. 3–85 (translated by A. H. Armstrong), in *Plotinus*, Vol. 1, Loeb Classical Library, Cambridge, 1966.

Prior A. N., *Past, Present and Future*, Oxford, 1967.

Rescher, Nicholas, *The Philosophy of Leibniz*, Prentice-Hall, Englewood Cliffs, 1967.

Risse, Wilhelm, *Die Logik Der Neuzeit*, 2 Vols, Stuttgart-Bad Cannstatt, 1967, 1970.

Robinet, Andre, *Systeme et Existence Dans L'Ouvre de Malebranche*, Vrin, Paris, 1965.

Robinson, H. M., 'Prime Matter in Aristotle', *Phronesis* 19, 168–88.

Rorty, Richard, 'Genus as Matter: A Reading of Metaphysics Z-H', pp. 393–420, in *Exegeis and Argument* (edited by E. N. Lee, A. P. D. Mourelatos, and R. M Rorty), The Netherlands, 1973.

Ross, W. D., *Aristotles Metaphysics*, Text and Commentary in 2 Vols, Oxford, 1924.

Ross, W. D., *The Works of Aristotle Translated into English*, 12 Vols; *Select Fragments*, Vol. 12, Oxford, 1952.

Ross, W. D., *Aristotelis Fragmenta Selecta*, Oxford Classical Texts, Oxford, 1955.

Ross, W. D., *Aristotelis Topica Et Sophistici Elenchi*, Oxford Classical Texts, Clarendon Press, Oxford, 1958.

Rowe, William, *The Cosmological Argument*, Princeton, 1975.

Russell, Bertrand, *A Critical Exposition of the Philosophy of Leibniz*, 2nd ed., George Allen and Unwin, London, 1937.

Russell, Bertrand, 'On Propositions', pp. 283–320, in Bertrand Russell, *Logic and Knowledge* (edited by R. C. Marsh), London and New York, 1956.

Russell, Bertrand, 'The Philosophy of Logical Atomism', pp. 177–281, in Russell Bertrand, *Logic and Knowledge* (edited by R. C. Marsh), London and New York, 1956.

Schofield, M. and Nussbaum, M. C. (eds.), *Language and Logos: Studies in Ancient Greek Philosophy Presented to G. E. L. Owen*, Cambridge, 1982.

Schofield, M., 'Metaphysics Z 3: Some Suggestions', *Phronesis*, Vol. 17, pp. 97–101.

Sellars, Wilfrid, 'Abstract Entities', *Review of Metaphysics* 16 (1963). Reprinted in Wilfrid Sellars, *Philosophical Perspectives*, Vol. 2, Ridgeview, Reseda, California, 1977.

Sellars, Wilfrid, 'Naming and Saying', in Wilfrid Sellars, *Science, Perception, and Reality*, Routledge and Kegan Paul, 1963.

Sellars, Wilfrid, *Science, Perception, and Reality*, Routledge and Kegan Paul, London, 1963.

Sellars, Wilfrid, 'Truth and Correspondence', in Wilfrid Sellars, *Science, Perception, and Reality*, Routledge and Kegan Paul, London, 1963.

Sellars, Wilfrid, 'Meditations Leibniziennes', *American Philosophical Quarterly* **2** (April 1965), 1–14.

Sellars, Wilfrid, 'Vlastos and the "Third Man".' *Philosophical Review* **64** (1955), 405–37. Reprinted in R. E. Allen, *Studies in Plato's Metaphysics*, New York, 1965.

Sellars, Wilfrid, 'Aristotle's Metaphysics: An Interpretation', in Wilfrid Sellars, *Philosophical Perspectives*, Springfield, Illinois, 1967.

Sellars, Wilfrid, *Science and Metaphysics*, London and New York, 1968.

Sellars, Wilfrid, 'Meaning as Functional Classification', *Synthese* **27** (1974).

Sellars, Wilfrid, 'On the Introduction of Abstract Entities', in *Forms of Representation*, (edited by Freed, Marras, and Maynard), Amsterdam, 1975.

Sellars, Wilfrid, 'Ontology and the Philosophy of Mind in Russell', in *Bertrand Russell's Philosophy* (edited by G. Naknikian), 1975.

Sellars, Wilfrid, 'Substance and Form in Aristotle', *Journal of Philosophy* **54** (1975). Reprinted in Wilfrid Sellars, *Philosophical Perspectives*, Vol. 1, Ridgeview, Reseda, California, 1977.

Sellars, Wilfrid, *Philosophical Perspectives*, Vols. 1 and 2, Ridgeview, Reseda, California, 1977.

Sellars, Wilfrid, *Naturalism and Ontology*, Ridgeview, Reseda, California, 1979.

Sleigh, Robert, 'Truth and Sufficient Reason in the Philosophy of Leibniz', in *Liebniz: Critical and Interpretive Essays* (edited by Michael Hooker), University of Minnesota Press, Minnesota, 1982.

Spade, Paul, 'Ockham's Distinction Between Absolute and Connotative Terms', *Vivarium* **13** (1975), 55–76.

Spade, Paul, 'Some Epistemological Implications of the Burleigh–Ockham Dispute', *Franciscan Studies* **35** (1975), 212–22.

Spinoza, Benedict, *The Chief Works*, Vols. 1 and 2 (translated by R. H. M. Eswes), New York, 1951.

Sykes, R. D., 'Form in Aristotle: Universal or Particular?' *Philosophy* **50** (1975), 311–31.

Thomas of Erfurt, *Grammatica Speculative* (edited and translated with a Commentary by G. L. Bursill-Hall), Longdon, London, 1972.

Trentman, John, 'Ockham on Mental', *Mind* 79 (1970), 586–90.

Turnbull, R. G., 'Aristotle's Debt to the "Natural Philosophy" of the Phaedo', *Philosophical Quarterly* 8 (1958), 131–43.

Turnbull, R. G., 'Physics I: Sense Universals, Principles, Multiplicity, and Motion', pp. 28–56, in *Motion and Time, Space and Matter* (edited by Peter Machamer and R. G. Turnbull), Ohio State University Press, Ohio, 1976.

Turnbull, R. G., 'Knowledge and the forms in the Later Platonic Dialogues', *Proceedings and Addresses of the American Philosophical Association* **51** (1978), 735–58.

Turnbull, R. G., 'Review of Gregory Vlastos, *Plato's Universe, Journal of Philosophy* **16** (1978), 97–103.

Turnbull, R. G., 'The Role of the "Special Sensibles" in the Perception Theories of Plato and Aristotle', pp. 3–26, in *Studies in Perception* (edited by Peter Machamer and R. G. Turnbull), Ohio State University Press, Ohio.

Tweedale, Martin, *Abailard on Universals*, North-Holland, Amsterdam, 1976.

Van Benthem, J. F. A. K., *The Logic of Time*, Xeroxed lecture notes, Filosofisch Institute, Rijksuniversiteit, Groningen, The Netherlands, 1980.

Van Fraasen, G. B., 'Meaning Relations Among Predicates', *Noûs* **1** (1967), 161–79.

Vlastos, Gregory, *Platonic Studies*, Princeton, 1973.

Vlastos, Gregory, 'Plato's "Third Man" Argument (Parm. 132a1–b2): Text and Logic', *Philosophical Quarterly* **19** (1969), 289–301. Reprinted in Gregory Vlastos, *Platonic Studies*, Princeton: 1973, pp. 342–60 with appendices, pp. 361–5.

Weidemann, Hermann, 'In Defence of Aristotle's Theory Predication', *Phronesis* **25** (1980), 76–87.

Weisheipl, J. A., 'The Concept of Matter in the Fourteenth Century', in *The Concept of Matter in Greek and Medieval Philosophy* (edited by McMullin), Notre Dame, Indiana, 1963.

White, Nicholas, 'A Note on Ekthesis', *Phronesis* **16** (1971), 164–68.

White, Nicholas, 'Origins of Aristotle's Essentialism', *Review of Metaphysics* **26** (1972), 56–85.

William of Sherwood, *Introduction to Logic* (translated by Norman Kretzmann), University of Minnesota, 1966.

Wilson, Margaret, 'Possible Gods', *Review of Metaphysics* **32** (June 1979), 717–33.

Winterbourne, A. T., 'On the Metaphysics of Leibnizian Space and Time', *Studies in the History and Philosophy of Science* **13** (1982), 208.

Woods, M. J., 'Substance and Essence in Aristotle', *Proceedings of the Aristotelian Society* **75** (1974–5), 167–80.

Woods, M., 'Problems in Metaphysics Z, Chapter 13', pp. 215–38, in *Aristotle: A Collection of Critical Essays* (edited by M. M. E. Moravcsik), Doubleday, New York, 1967.

INDEX OF LABELED EXPRESSIONS

Labeled expressions are listed under the names of the authors who use them. Please note that in some cases, the same label is used for different expressions by different authors.

Adams, M.
- (T1), (T2), p. 175
- (T3)–(T6), p. 178
- (T7), p. 179
- (T8)–(T10), p. 186

Adams, R. M.
- (A), (B), (C), p. 259

Bogen
- (C1) (cf. Turnbull, 'Zeno's Stricture'), p. 6

Code
- (A), p. 107
- ALEX-NI, p. 105
- ALEX-OM (1)(2)(3), p. 105
- (B), p. 107
- (C), p. 108
- DEF 1, p. 103
- DEF 2, p. 103
- NEG-NI, p. 102
- (NI) (Non-Identity), p. 101
- (SP) (Self Predication), p. 101

Lewis
- (A1) ('Form and Predication'), p. 63; ('Plato's Third Man'), p. 161
- (A2)–(A4), p. 72
- (AL), p. 156
- (A)–'large', p. 144
- ALEX-PL (1)(2)(3), p. 138
- (B)–'large', p. 144
- (C)–'large', p. 145
- (D1) ('Form and Predication'), p. 72; ('Plato's Third Man'), p. 160
- (D2) ('Form and Predication'), p. 72; ('Plato's Third Man'), p. 161
- (D3) (D4), p. 72
- (D5)–(*D10), p. 73
- (*D11), p. 77
- HOMOGENEITY (1)(2)(3), p. 139
- (P1), p. 80
- (P2), p. 80
- SEPARATION, p. 141
- (T1)–(T3), p. 161
- (T4), p. 162

Modrak
- (S), p. 91
- (S*), p. 92
- (U), p. 91
- (U*), p. 92

Turnbull
- Zeno's Stricture, p. 24

NAME INDEX

Entries marked with a ° superscript occur in the text as examples.

Abelard, Peter 201
Ackrill, John 62, 79, 96, 98, 99, 123, 128
Adam 237°, 240°, 241°, 281°
Adam, Charles 233
Adams, Robert M. 12, 14, 58, 221, 225, 233, 237, 253, 281, 327
Adams, Marilyn McCord 10, 11, 12, 184, 185, 188, 195, 196, 202, 283, 327
Albert of Saxony 201
Albritton, Rogers 66, 79
Alexander of Aphrodisias 40, 104, 124, 125, 133, 136ff, 142, 152, 154, 164, 167, 168, 171, 172, 173
Alexander, the Great 227°
Allen, Reginald 169
Alice 3°, 4°
Aristotle 1, 3, 5, 6, 8, 9, 10, 12, 14, 17, 18, 21–58 *passim,* 59–83 *passim,* 85–99 *passim,* 101–131 *passim,* 133–174 *passim,* 193ff, 197ff, 202, 316
Armstrong, A. H. 55, 56, 57
Arnauld, Antoine 12°, 13°, 205, 210, 211, 214, 224°, 230, 233, 235, 237°, 240, 241, 242, 243, 245, 246°, 249°, 253, 254, 256, 258°, 259, 265°, 268, 269, 270, 271, 275, 277°, 281, 282, 283
Augustine 21, 57

Balme, D. M. 77
Barnes, J. 123
Baumgarten, Alexander 282
Bayle, Pierre 279
Berkeley 15, 21
Black, Max 78
Blumenfeld, David 283
Boehner, Philotheus 187
Bogen, Aubrey 1°
Bogen, Carlota 1°
Bogen, James 83, 98, 99, 174, 327
Bonitz, H. 77, 78, 167, 174
Bradley, F. H. 197, 305, 306
Brandom, R. 233
Broad, C. D. 208, 209, 232, 283
Brody, Baruch 253
Burge, Tyler 83, 283
Buridan 3, 10, 11, 12, 180, 181, 189–203
Burleigh, Walter 187
Burnyeat, Miles 129
Bursill-Hill, G. L. 188

Caesar 236°, 237°, 238°, 239°, 264°, 267°
Callias 35°, 54°, 60°, 62°, 93°, 115°, 116°, 117°, 128°, 135°
Carnap, Rudolf 2, 18
Cassio 303°
Charlton, W. 78, 99
Cherniss, Harold 166, 168
Christ 187°, 195°, 238°
Clarke, Samuel 207, 227, 253, 273, 283
Code, Alan 5, 7, 10, 78, 79, 80, 97, 99, 123, 133–174 *passim,* 327
Cohen, S. Marc 70, 81–82, 98, 170, 173
Couturat, L. 232, 253, 258, 281
Crusius, Christian 282

Dante 56
Darius 227°
David 206°, 207°
Del Punta 184, 185, 188
Democrites 180
Denniston, J. D. 129
Des Bosses 207, 217, 221, 229, 233
Descartes, Rene 14, 15, 21, 56, 210, 250
Desdemona 303°
De Volder 219, 221, 227, 263, 282
Dick 311°
Dionysodorus 23, 50
Driscoll, John A. 83, 99, 123
Dummett, Michael 253–4

Earman, John 232, 283
Edwards, Cindy Sweetroll 125
Eleatic Stranger 42, 52, 53, 55
Euclid 53
Eudemus 166
Euthydemus 23, 50

Ferrer, Vincent 201
Fine, Gail 126, 171, 172
Fine, Kit 78, 82, 83
Frankfurt, Harry 232
Frede, Michael 130
Frege, Gottlob 60, 62, 78, 253–254
Furth, Montgomery 60, 81, 82, 124

Gál, Gideon 188
Geach, Peter 78, 168, 170
Gerhardt, C. I. 232, 282
Grandy, R. 123
Gregory of Rimini 184–5
Grice, H. P. 123, 129

Hackforth, R. 50
Hamlet 311°
Harry, Deborah 7°
Harter, Edward D. 65, 79
Hartman, Edwin 65
Haslanger, Sally 126
Hayduck, M. 124
Helen 26°, 28°
Henry 55
Herzog, Whitey 1°
Hitler, Adolph 275°, 276°, 277°
Hobbes, Thomas 291
Hugolino 188
Hume, David 15, 262

Ishiguro, H. 232, 233

Jaeger, Werner 82
Jowett, Benjamin 52

Kant, Immanuel 261
Kauppi, Raili 252, 256, 281
Kirwan, Christopher 127, 128, 174
Kretzmann, Norman 187
Kung, Joan 79, 153, 164, 172

Lamy, Francois 271
Lear, Edward 322
Lear, Jonathan 130
Lee, E. N. 51
Leibniz 12, 13, 14, 15, 16, 17, 18, 48, 58, 205–233 *passim*, 235–283 *passim*
Lennox, James 19
Leo, 287°
LeRoy 283
Lesher, James H. 68, 79
Lewis, David 236, 283
Lewis, Frank 5, 7, 9, 10, 79, 80, 81, 82, 85–99 *passim*, 110, 123, 125, 131, 166, 172, 173, 174, 327
Liddell 54
Lloyd, A. C. 83
Locke, John 262
Loeb, Louis 262, 282
Loemker, L. E. 232
Long, Peter 232
Loux, Michael 70, 71, 74, 75, 97, 98
Lucretius 53
Łukasiewicz, Jan 60
Lycophron 61

Machamer, Peter 54, 58
Maier, Anneliese 188
Malcolm, John 97
Malebranche 268, 269, 272
Mansion, S. 60
Marsh, R. D. 52
Mates, Benson 232, 257–8, 281
McKenna 55, 57
McTaggert 311°
Melissus 180
Mirecourt, John 179ff
Modrak, D. K. 5, 9, 66, 79, 81, 83, 94, 97, 99, 327
Monk, Thelonius Sphere 1°
McGuire, James E. 12, 14, 15–16, 283, 327
Mondadori, Fabrizio 258, 281
Montague, Richard 291
Moore, G. E. 39
Moravcsik, Julius 18
Mourelatos, A. D. P. 51, 79

Nakhnikian, George 322

Nehamas, Alexander 50, 129
Nestook 191°
Newton, Isaac 56, 262
Nicole, Pierre 242, 243, 254, 281, 282
Normore, Calvin 10, 11, 12, 99, 327
Nussbaum, Martha 126

Ockham, William of 3, 10, 11, 12, 175–188 *passim*, 189ff, 195ff, 199, 201
Olivi 181
O'Meara, Dominic 83, 99, 123
Owen, G. E. L. 97, 101–103, 106, 110, 114, 119, 123, 124, 125, 126, 127, 130, 133–174 *passim*

Palamides of Elea 50
Parkinson, G. H. R. 232
Parmenides 22, 23, 25, 49, 50
Paul of Venice 183, 184, 188
Paulys 55
Peter 238°, 247°
Peter of Ailly 186
Phaedo 25°, 26°, 31°, 52°
Philip 227°
Plato 2, 4, 6, 7, 8, 11, 12, 13, 14, 18, 21–58 *passim*, 61, 108, 126, 133–174 *passim*, 175°, 179°, 296
Plotinus 6, 14, 18, 21–58 *passim*
Porphyry 40, 55, 57
Price 58
Prometheus, A Certain 27
Protarchus 27
Ptolemy, Claudius 46

Quine, Willard Van Orman 183, 291

Reier, David 128
Rescher, N. 232
Risse, Willhelm, 282
Robinet, Andre 283
Robinson, H. M. 99
Rorty, Richard, 51
Ross, W. D. 54, 60, 61, 62, 69, 70, 74, 77, 82, 97, 129, 130
Rowe, William 253
Russell, Bertrand 31, 39, 50, 52, 58, 258, 303, 304, 305, 311, 322

Sandro, Nicholas J. 125
Sayre, Kenneth 52
Schofield, M. 99, 123, 126
Schwyzer 55
Scott 54
Sellars, Wilfrid 10, 11, 12, 14, 19, 65, 124, 199, 208, 210, 211, 232, 233, 285–322, 327
Sextus 260°
Silverman, Alan 129
Simmias 52°
Sleigh, Robert 252
Spade, Paul 187, 188, 201
Spellman, Lynne 83
Spinoza 43, 49, 56, 263
Socrates 1°, 3°, 5°, 6°, 8°, 9°, 10°, 12°, 14°, 17°, 21, 25, 35°, 50, 51, 52°, 53°, 60°, 62°, 64°, 67°, 69°, 73°, 75°, 82°, 85°, 87°, 88°, 89°, 103°, 104°, 112°, 113°, 114°, 115°, 116°, 117°, 118°, 124°, 126°, 127°, 128°, 134°, 135°, 140°, 144°, 145°, 147°, 148°, 153°, 154°, 166°, 170°, 171°, 175°, 177°, 178°, 179°, 187°, 192°, 194°, 195°, 197°, 225°, 287°, 296°, 308°, 309°, 319°
Solomon 206°, 207°
Sophia 229
Sorabji, Richard 123
Sykes, R. D. 79

Tannery, Paul 233
Tarski, Alfred 308, 310, 322
Theaetetus 31°, 52°, 55
Thomas of Erfurt 181, 188
Tom 311°
Trentman, John 201
Turnbull, Robert G. 6, 18, 52, 54, 58, 126, 323–4
Tweedale, Martin 201

Wagner, Michael 57
Warner, R. 123
Wedderkopf, Magnus 253
Weisheipl, J. A. 201
White, Nicholas 171
William of Sherwood 187

Wilson, Margaret 236, 240, 241, 281
Winterbourne, A. T. 233
Wodeham, Adam 185, 188
Woods, M. 66, 79, 114ff, 120, 123, 126, 127, 128, 129

Xantippe 17°

Zeno 6, 16, 17, 21ff
Zev 316°

SUBJECT INDEX

Note: In addition to this index, the reader is referred to the introductory essay which includes partial summaries of the papers in this volume.

accident, *see* predication, accidental
anything 290ff
aporia and philosophical method 21
a priori/a posteriori 282
artifact terms 87

being
 in Plato 27, 29; in Plotinus 41 ff; Plotinian exfoliation of 42; in Aristotle 70 ff
 see is/exists, Is/Has
Bradley's regress 197ff, 305

cannonical notations 3, 6, 285–321 *passim; see* Predication, cannonical forms in Aristotle
change 195ff
compounding, Aristotelian 61ff, 95–96

definitions in Aristotle 38
division, method of in *Philebus* 27ff

essence 9
 and universals in Aristotle 65–76; and the thesis of Z/vi 110ff, 114ff, 119ff, 127, 154ff, 157ff; Aristotle's phrase for 111
 see predication, essential
eucharist, *see* transubstantiation
exfoliation 45, 221

forces in Leibniz 262ff, 266ff
form/matter 9
 in Plotinus 44, 45, 46ff
 in Aristotle 59ff, 65ff; predicated of matter in Aristotle 68–71; matter as subject in Aristotle 79, 90–96, form and species 87ff
 see predication, of forms

forms, Platonic
 Aristotle's opposition to 8, 38ff, 102ff, 108ff; and Zeno's stricture 6–7, 16; as reified natures 28
 see third man argument

God 13, 216
 'possible Gods' 237ff; and trans-world identity 239; His knowledge 251, 279; and laws of nature 265ff; actions reserved to 272ff

identity
 trans-world in Leibniz 235ff, and counterpart theory 236, thought to involve a contradiction 240, and the conceptual containment theory of truth 255ff, 261ff, theological motives for denying 240, of God 239
 transtemporal in Leibniz 262ff
Indcon 299
is/exists 27
Is/Has 25ff, 103

Jumblese 313ff

language of thought 189
LR 309

matter *see* form/matter
meaning 291ff
 not a relation 295; sameness of 296, 318ff
miracle
 in Leibniz 264ff, 268ff
 of Cana 272
modes (hows) 181–186, 189–203

monads 15ff, 223
  causality and their predicates 220; causal connections between 208-9, 213, and representation 209ff, 212ff, 218, 279; their formal being 213; pre-established harmony 215, 279ff; their activity 211ff; ideal influences on one another 279

necessity and contingency in Leibniz 12ff; *see* possible worlds, truth, principle of the best, principle of sufficient reason
nominalism 10ff, papers of M. Adams, C. Normore, W. Sellars *passim*
  and qualities 11; its parsimony 11ff, 179ff, 199; *see* modes

ontology 189
  and syntax 2ff; and semantics 4ff

participation 6ff, 31
  problems in *Parmenides* 7ff, 25ff
particulars, *see* universals/particulars
phenomenalism, Leibnizian 15ff, 210ff, 214ff
  and extention 227ff, 231
physical objects
  their status 14ff, 44; in Leibniz 208, 210, 227; *see* phenomenalism
possible worlds 237ff
  and God's goodness 238; laws of 260ff
PR 309
predication
  accidenta 6, 72ff, 103, 194ff
  cannonical forms in Aristotle 36
  causal basis in Leibniz 220
  and completeness of concepts in Leibniz 256ff
  and division in *Philebus* 27ff
  essential 9, 72ff, 103
  of forms 26ff, 28
  Is/Has *see* Is/Has
  of kinds 68ff, 86ff
  levels of 16ff, 226ff
  linguistic/metaphysical 1ff, 5
    in Aristotle 4-5, 64, 89ff, 103; of essence 111ff; and Zeno's stricture 6; 'Socrates is a man' 82, 89ff
  and the problem of the one and the many 287ff, 316
  PMese 285ff
  and *Port Royal Logic* 254-5
  of qualities 11ff, 176ff, 194ff, 298, 316ff
  of relations 207ff
  said of/in *see* said of/in
  self predication 101, 106, 152ff, 167, 317ff
  'standard theory of' 285-289, 305
  subject of in Aristotle 63
principle of the best 217, 218, 238
principle of sufficient reason 215ff
  and the denial of trans-world identity 241; and Leibniz' conceptual containment theory of truth 253
PSH 318

qualities, *see* nominalism, predication
quotation, Sellarsian
  dot quotes 293ff; Q quotes 294; ordinary quotes 293-4; * quotes for sign design 306

realism, Russell's argument for 303-304 304
relations 305ff
  in Leibniz 206ff, 209ff, 213ff; and perception 274-280; relations/ties (nexus) 286ff

said of/in 34, 62
sameness
  in Aristotle 113; and Ockham 175; of meaning in Sellars 296
schmyntactics 315
scientific explanation and predication 7, 14ff

semantics
and ontology 4ff; Tarskian 308ff; Sellarsian 285ff
signification 189ff
soul, in Plotinus 47
substance Aristotelian 62, 158ff
differences between early and later conceptions of 59, 62; as composite 60 ff; primary/secondary 62; as an explanatory factor 69; in 14th century 194
supposition 190ff
syntax and ontology 2ff

terms 190
absolute/connotative 176; syncategorematic 180ff
'Third Large' Argument 144ff
Third Man Argument (TMA) 101–131 *passim:* 133–174 *passim*
Alexander's commentary on 104ff, 136ff; Owen's account of, 101ff, 133ff, 155ff
TMA, *see* Third Man Argument
transubstantiation 179ff, 196ff
truth
Leibniz' conceptual containment theory 12ff, 209, 217, 220, 235, 241–248, 250–252, 255ff; and falsity in Plato 33; Ockham and categorical propositions 175ff, 177ff; existence and the truth of categorical propositions 248ff; Frege on the correspondence theory 253ff; 'true in *G*' 309; some of Sellars' theory of 320ff
type/token 288ff

universals/particulars 23–24, 31ff, 39, 65ff
Aristotle's definition 66, 93ff; in Aristotle 102ff; in Russell 203ff; universals and essence in Aristotle 9, 65–76

Zeno's Stricture 21ff
responses of Plato, Aristotle, and Plotinus 29; and Plato 24ff; and Aristotle 34ff
Zeta vi, thesis, *see* essence